Heritage, Affect and Emotion

Heritage and its economies are driven by affective politics and consolidated through emotions such as pride, awe, joy and pain. In the humanities and social sciences, there is a widespread acknowledgement of the limits not only of language and subjectivity, but also of visuality and representation. Social scientists, particularly within cultural geography and cultural studies, have recently attempted to define and understand that which is more-than-representational, through the development of theories of affect, assemblage, post-humanism and actor–network theory, to name a few. While there have been some recent attempts to draw these lines of thinking more forcefully into the field of heritage studies, this book focuses for the first time on relating heritage with the politics of affect. The volume argues that our engagements with heritage are almost entirely figured through the politics of affective registers such as pain, loss, joy, nostalgia, pleasure, belonging or anger. It brings together a number of contributions that collectively – and with critical acuity – question how researchers working in the field of heritage might begin to discover and describe affective experiences, especially those that are shaped and expressed in moments and spaces that can be, at times, intensely personal, intimately shared and ultimately social. It explores current theoretical advances that enable heritage to be affected, released from conventional understandings of both heritage-as-objects and objects-as-representations by opening it up to a range of new meanings, emergent and formed in moments of encounter. While representational understandings of heritage are by no means made redundant through this agenda, they are destabilized and can thus be judged anew in light of these developments. Each chapter offers a novel and provocative contribution, provided by an interdisciplinary team of researchers who are thinking theoretically about affect through landscapes, practices of commemoration, visitor experience, site interpretation and other heritage work.

Divya P. Tolia-Kelly is Reader in the Department of Geography at Durham University, UK.

Emma Waterton is Associate Professor in the Geographies of Heritage at Western Sydney University, Australia.

Steve Watson is Professor of Cultural Heritage at York St John University, UK.

Critical Studies in Heritage, Emotion and Affect
Series editors:
Divya Tolia-Kelly (*Durham University*) and Emma Waterton (*Western Sydney University*)

In Memory of Professor Steve Watson (1958–2016)

This book series, edited by Divya P. Tolia-Kelly and Emma Waterton, is dedicated to Professor Steve Watson. Steve was a pioneer in heritage studies and was inspirational in both our personal academic trajectories. We, as three editors of the series, started this journey together, but alas we lost his magnificent scholarship and valued counsel too soon.

The series brings together a variety of new approaches to heritage as a significant affective cultural experience. Collectively, the volumes in the series provide orientation and a voice for scholars who are making distinctive progress in a field that draws from a range of disciplines, including geography, history, cultural studies, archaeology, heritage studies, public history, tourism studies, sociology and anthropology – as evidenced in the disciplinary origins of contributors to current heritage debates. The series publishes a mix of speculative and research-informed monographs and edited collections that will shape the agenda for heritage research and debate. The series engages with the concept and practice of Heritage as co-constituted through emotion and affect. The series privileges the cultural politics of emotion and affect as key categories of heritage experience. These are the registers through which the authors in the series engage with theory, methods and innovations in scholarship in the sphere of heritage studies.

Published

Heritage, Affect and Emotion
Politics, practices and infrastructures
Edited by Divya P. Tolia-Kelly, Emma Waterton and Steve Watson

Heritage, Affect and Emotion

Politics, practices and infrastructures

Edited by Divya P. Tolia-Kelly, Emma Waterton and Steve Watson

LONDON AND NEW YORK

First published 2017
by Routledge

2 Park Square, Milton Park, Abingdon, Oxfordshire OX14 4RN
711 Third Avenue, New York, NY 10017

Routledge is an imprint of the Taylor & Francis Group, an informa business

First issued in paperback 2018

Copyright © 2017 selection and editorial material, Divya P. Tolia-Kelly, Emma Waterton and Steve Watson; individual chapters, the contributors

The right of Divya P. Tolia-Kelly, Emma Waterton and Steve Watson to be identified as the authors of the editorial material, and of the authors for their individual chapters, has been asserted in accordance with sections 77 and 78 of the Copyright, Designs and Patents Act 1988.

All rights reserved. No part of this book may be reprinted or reproduced or utilised in any form or by any electronic, mechanical, or other means, now known or hereafter invented, including photocopying and recording, or in any information storage or retrieval system, without permission in writing from the publishers.

Notice:
Product or corporate names may be trademarks or registered trademarks, and are used only for identification and explanation without intent to infringe.

British Library Cataloguing-in-Publication Data
A catalogue record for this book is available from the British Library

Library of Congress Cataloging-in-Publication Data
Names: Tolia-Kelly, Divya Praful. | Waterton, Emma. | Watson, Steve.
Title: Heritage, affect and emotion: politics, practices and infrastructures / edited by Divya P. Tolia-Kelly, Emma Waterton and Steve Watson.
Description: Milton Park, Abingdon, Oxon; New York, NY: Routledge, 2016. | Includes bibliographical references and index.
Identifiers: LCCN 2016002731| ISBN 9781472454874 (hardback) | ISBN 9781315586656 (e-book)
Subjects: LCSH: Cultural property–Psychological aspects. | Historic preservation–Psychological aspects.|Affect(Psychology)–Socialaspects. | Emotions–Social aspects. | Cultural property–Protection–Research. | Historic preservation–Research. | Cultural property–Protection–Social aspects. | Cultural property–Protection–Political aspects. | Historic preservation–Social aspects. | Historic preservation–Political aspects.
Classification: LCC CC135 .H44 2016 | DDC 152.4–dc23
LC record available at http://lccn.loc.gov/2016002731

ISBN: 978-1-4724-5487-4 (hbk)
ISBN: 978-1-138-54734-6 (pbk)

Typeset in Times New Roman
by Sunrise Setting Ltd, Brixham, UK

Steve Watson, 14 June 1958–22 January 2016

Steve died on January 22, 2016, aged 57 and has left a large crater in our lives. His influence was international, and yet his spirit was immediate and personal. This spirit is etched in our everyday practice, both at home and at work. Steve's modesty steered us, and continues to pedagogically shape our practice. His intellect served to enrich our own, as he touched us with care and warmth. He practiced his vocation with elegance and humanity, and a collegiality that is rare. We hope that we can do justice to his memory by continuing in the spirit of his engagement with the world, without any hint of superiority, disrespect or derision. Steve was a true scholar and friend; one we doubt we will see the like of again soon. This volume was conceived and shaped with his lead and completed with his companionship: we dedicate it to him.

Photo courtesy of the Watson family

Contents

List of figures ix
Notes on contributors xi
Acknowledgements xvii

Introduction: heritage, affect and emotion 1
DIVYA P. TOLIA-KELLY, EMMA WATERTON AND STEVE WATSON

1 **Making polysense of the world: affect, memory, heritage** 12
JOY SATHER-WAGSTAFF

PART I
Memories 31

2 **Race and affect at the museum: the museum as a *theatre of pain*** 33
DIVYA P. TOLIA-KELLY

3 **Affecting the body: cultures of militarism at the Australian War Memorial** 47
JASON DITTMER AND EMMA WATERTON

4 **Affect and the politics of testimony in Holocaust museums** 75
STEVEN COOKE AND DONNA-LEE FRIEZE

5 **Museum canopies and affective cosmopolitanism: cultivating cross-cultural landscapes for ethical embodied responses** 93
PHILIPP SCHORCH, EMMA WATERTON AND STEVE WATSON

6 **Constructing affective narratives in transatlantic slavery museums in the UK** 114
LEANNE MUNROE

viii Contents

PART II
Places 133

7   Overlooking affect? Vertigo as geo-sensitive industrial
    heritage at Malakoff Diggins, California 135
    GARETH HOSKINS

8   The castle imagined: emotion and affect in the
    experience of ruins 154
    DUNCAN LIGHT AND STEVE WATSON

9   From Menie to Montego Bay: documenting, representing
    and mobilising emotion in coastal heritage landscapes 179
    SUSAN P. MAINS

10 Touching time: photography, affect and the digital archive 201
    LÁSZLÓ MUNTEÁN

11 Commemoration, heritage, and affective ecology: the case
    of Utøya 219
    BRITTA TIMM KNUDSEN AND JAN IFVERSEN

12 Social housing as built heritage: the presence and absence
    of affective heritage 237
    SOPHIE YARKER

PART III
Practices 255

13 'Please Mr President, we know you are busy, but can
    you get our bridge sorted?' 257
    KEITH EMERICK

14 Dark seas and glass walls – feeling injustice at the museum
    Practitioner perspectives: Rosanna Raymond 276
    AN INTERVIEW WITH ROSANNA RAYMOND BY
    DIVYA P. TOLIA-KELLY

    Index 293

Figures

3.1	"Tail End Charlie"	48
3.2	The *Lone Pine* diorama	59
3.3	The *First World War* galleries, prior to their redevelopment	60
3.4	The Kapyong faces: in photographs and the *Kapyong* diorama	61
3.5	The ghostly figures of the *Menin Gate at Midnight*	63
3.6	"Anzac Hall," *Striking By Night* sound and light show	64
3.7	"Dust Off" in the Vietnam War gallery	66
3.8	The Bridge, HMAS Brisbane	68
3.9	The Discovery Zone helicopter	69
4.1	*History You Can't Erase* poster	82
5.1	*Signs of a Nation/Ngā Tohu Kotahitanga* exhibition within Te Papa	99
5.2	*Leaving Home* long-term exhibition, Immigration Museum, Melbourne	100
5.3	*Te Hau ki Turanga* in the *Mana Whenua* exhibition	102
5.4	*Leaving Dublin: Photographs by David Monahan* touring exhibition 2012–13, Immigration Museum, Melbourne	105
7.1	Panoramic strip of view from Chute Hill Campground Overlook	138
7.2	Chute Hill campground overlook sign	142
7.3	Marker posts on the Diggins Loop Trail	146
8.1	Landscape with castles: Dunstanburgh, Northumberland	156
8.2	The castle (re-)assembled, Bamburgh, UK: Norman with later restoration	162
8.3	The castle experienced: the remains of a staircase at Bolton Castle, UK	167
8.4	Dizzy heights at Bolton Castle, UK	169
8.5	Dangerous places, Monolithos, Rhodes	172
8.6	The castle imagined: a moment of reverie at Warkworth, UK	174

9.1 A still from *Jamaica for Sale* showing a wall around the construction of an all-inclusive hotel that has cut off beach access to local residents 191
10.1 *Fortepan*/Ferencvárosi Helytörténeti Gyűjtemény 209
10.2 Zoltán Kerényi, Window to the Past, ablakamultra.hu 214
11.1 Heart of Candy in the Café Building 228
11.2 Grassroots memorial with view of Utøya 231
13.1 Creets Bridge following its reconstruction, seen from the village 260

Contributors

Steven Cooke is a cultural and historical geographer who has published widely on issues relating to the memorial landscapes of war and genocide, museums and national identity, and maritime heritage and urban redevelopment. He is Course Director for the Cultural Heritage and Museum Studies postgraduate programs at Deakin University, Australia, where he is a Senior Lecturer in Cultural Heritage. He is the author of *The Sweetland Project: Remembering Gallipoli in the Shire of Nunawading* (2015) and the co-author of *'The interior of our memories': A History of Melbourne's Jewish Holocaust Centre* (2015), with Donna-Lee Frieze.

Jason Dittmer is Professor of Political Geography at University College London. He is the author of *Popular Culture, Geopolitics, and Identity* (2010) and *Captain America and the Nationalist Superhero: Metaphors, Narratives, and Geopolitics* (2013). He is also the (co-)editor of several volumes, including the *Ashgate Research Companion to Media Geography* (2014) and *Geopolitics: An Introductory Reader* (2014). His current research is on assemblage theory and diplomacy.

Keith Emerick is a practising cultural heritage manager based in the north of England, currently working for Historic England (the national conservation and heritage agency) as an Inspector of Ancient Monuments. The post involves working and advising on ruins, archaeological sites and battlefields with their respective stakeholders. The date range covers the Palaeolithic to the Cold War, so Keith has an interest in anything and everything. Keith also has several years' prior experience as an archaeologist, principally with the York Archaeological Trust. Keith has special responsibility for Fountains Abbey and Studley Royal, World Heritage Site and advises on Saltaire, World Heritage Site, West Yorkshire. He is a Member of the Steering Group that produced the Fountains Abbey and Studley Royal World Heritage Site Management Plan (2000) and its first review, published in 2009. Keith's principal research and work interests are: community heritage; Conservation and Management Plans; the use of World Heritage Management Plans, guidelines, charters and conventions; social value; intangible heritage; conservation and heritage management in post-war reconstruction; adaptive re-use; public history;

physical and intellectual access; CHM theory; origins and development of conservation practice (with particular interest in conservation practice in colonial and post-colonial contexts); and origins of conservation practice in the USA. In addition to working in England, Keith has worked in Cyprus, Croatia, Ireland, the United States (chiefly South Carolina), Romania and the Occupied West Bank. He has a doctorate in Conservation Philosophy from the University of York, where he is a Research Associate, and from 2014 was a Research Associate at Savannah Technical College, Georgia, USA.

Donna-Lee Frieze is a genocide studies scholar specializing in memory and aftermath studies, a Research Fellow at Deakin University, Australia, and a Visiting Scholar at the Centre for the Study of Genocide, Conflict Resolution and Human Rights at Rutgers University, New Jersey, USA. She taught a graduate unit on genocide for more than ten years and has published widely on the Armenian genocide, the Holocaust and the Bosnian genocides in relation to testimony, film and philosophy. She has been an academic advisor for several NY films and genocide exhibitions and is the Exhibition Curator Scholar for 'Raphael Lemkin: The Quest to end Genocide' with the Center for Jewish History for the Google Cultural Institute. She was the 2013–14 Prins Senior Scholar at the Centre for Jewish History in NYC. She is the editor and transcriber of Raphael Lemkin's autobiography, *Totally Unofficial* (2013) and co-author with Steven Cooke of *'The interior of our memories': A History of Melbourne's Jewish Holocaust Centre* (2015). Donna is a board member of the Institute for the Study of Genocide, New York University and former First Vice President for the International Association of Genocide Scholars.

Gareth Hoskins is a geographer based at Aberystwyth University, UK. He has published numerous academic articles, book chapters and magazine pieces on heritage operations in the United States, the United Kingdom and South Africa. He was principal investigator on a study of mining-related heritage sites funded by the AHRC and has recently completed another AHRC-funded project comparing the logics and protocols of value as conceived by UK and US state preservation agencies.

Jan Ifversen holds a PhD in cultural studies from Aarhus University, Denmark. He is currently vice dean of knowledge exchange at the faculty of Arts, Aarhus University. From 2004 to 2011, he was head of the Department of History and Area Studies at the same university. He is the author of two books in Danish, a history of current European history (*Hjem til Europa*, 1992) and a book on power, democracy and discourse (*Om magt, demokrati og diskurs*, two volumes, 1997). He has published many articles on conceptual history, European history, European identity politics and history politics. He is one of the founding members of the History of Concepts Group, an international research group on conceptual history. From 2000 to 2009 he served as its executive secretary.

Duncan Light is Senior Lecturer in the School of Tourism, Bournemouth University, UK. He has research interests in the practices and performances of heritage tourists and the role of imagination in the tourism experience. He is also interested in relationships between tourism and national identities, and has explored this issue with particular reference to 'Dracula tourism' in Romania (a country he has visited regularly for the past 20 years). He is the author of *The Dracula Dilemma: Tourism, Identity and the State in Romania* (2012).

Susan P. Mains is a Lecturer in Human Geography at the University of Dundee, UK. Her work explores transnational identities and media representations of mobility, borders and security in the context of heritage tourism, Caribbean migration and creative industries in Jamaica and Scotland. She is co-editor of *Mediated Geographies and Geographies of Media* (2015) and has published in a range of international journals including *Social and Cultural Geography*, *GeoJournal*, *Journal of Geography in Higher Education*, *The Singapore Journal of Tropical Geography*, *Small Axe: A Caribbean Journal of Criticism* and *Caribbean Geography*. Her previous work has been funded through the University of the West Indies, the Association of American Geographers and the American Geographical Society. Currently she is the coordinator for a Carnegie Trust funded project, 'Ties to the Tay', which utilises participatory filmmaking as part of an interdisciplinary project between a group of academics at the University of Dundee, the Nethergate Writers, artists and community organisations. She is also a partner in a Royal Society of Edinburgh-funded collaborative project exploring representations of place and landscape connections between the Scottish Highlands and the Caribbean.

Leanne Munroe is a PhD student at the University of Cambridge, UK, in the Division of Archaeology. Her research examines the construction and presentation of narratives of transatlantic slavery in UK museums and heritage sites, with a view to understanding the complex relationship between narratives, agents and social arenas involved in this process. Leanne also obtained her BA in History and MPhil in Archaeological Heritage at the University of Cambridge. Before starting her PhD, she was the manager of an eighteenth-century museum which examined the lives of abolitionists John Newton and William Cowper. She was also the research co-ordinator for a HLF project which worked with East African communities in London to design an exhibition on the slave trade in Zanzibar.

László Munteán is an Assistant Professor of Cultural Studies and American Studies at Radboud University Nijmegen, the Netherlands. His research focuses on intersections of cultural memory, visual culture and the built environment. Drawing on diverse theoretical apparatuses, his publications have focused on the memorialization of 9/11 in literature and the visual arts, American cities and architecture, rephotography and the memorialization of the allied bombing of Budapest in the Second World War. In a broader sense, his scholarly work revolves around the juncture of literature, visual culture, heritage studies and cultural memory in American and Eastern European contexts. Currently, he is working on a book project based on his dissertation on trauma and taboo in the context of 9/11.

xiv *Contributors*

Rosanna Raymond's activities over the past twenty years have made her a notable producer of, and commentator on, contemporary Pacific Island culture in Aotearoa New Zealand, the UK and the USA. She specializes in working within museums and higher education institutions as an artist, performer, curator, guest speaker, poet and workshop leader. Raymond was recently appointed as Honorary Research Associate at the Department of Anthropology and Institute of Archaeology at University College London. Raymond has achieved international renown for her performances, installations, body adornment and spoken word. She is a published writer and poet; her works are held by museums and private collectors across the globe. Since returning to New Zealand to live she has held her first solo exhibition and has been invited to take part in the 8th Asia Pacific Triennial of Contemporary Art opening at the Queensland Art Gallery of Modern Art (QAGOMA), Brisbane, Australia.

Joy Sather-Wagstaff is a Visiting Research Scholar with the Collaborative for Cultural Heritage Management and Policy (CHAMP) at the University of Illinois at Urbana-Champaign, USA. As an interdisciplinary sociocultural anthropologist, her research areas include critical heritage studies, memorial landscapes and museums, popular and individual memory, genocide, slavery and disaster, museum evaluation, consumption and commodification, and tourism studies. She is the author of *Heritage that Hurts: Tourists in the Memoryscapes of September 11* (2011).

Philipp Schorch is Marie Curie Research Fellow (European Commission) at the Institute of Social and Cultural Anthropology at Ludwig-Maximilians-University Munich, Germany, and Honorary Research Fellow at the Alfred Deakin Research Institute at Deakin University, Australia. He is currently conducting a multi-sited, collaborative ethnographic investigation of contemporary Indigenous curatorial practices in three Pacific museums (Bishop Museum, Hawai'i; Museum of New Zealand Te Papa Tongarewa; and Museo Rapa Nui, Easter Island). Philipp recently co-convened (with A/Prof. Conal McCarthy and Prof. Eveline Dürr) the symposium *Curatopia: Histories, Theories, Practices – Museums and the Future of Curatorship*, and is co-editing (with Prof. Eveline Dürr) the forthcoming volume *Transpacific Americas: Encounters and Engagements between the Americas and the South Pacific*.

Britta Timm Knudsen is an Associate Professor in the Department of Aesthetics and Communication, ARTS at Aarhus University, Denmark. Her recent previous books include *Affective Methodologies* (co-ed, 2015); *Global Media, Biopolitics, and Affect* (co-authored with Carsten Stage, 2015), *Enterprising Initiatives in the Experience Economy* (co-ed, 2014) and *Re-investing Authenticity, Tourism, Place and Emotions* (co-ed, 2009). She has published extensively on difficult heritage and tourism, experience and event culture. She teaches on MAs in Experience Economy, Event Culture and Sustainable Heritage Management. She has been special issue editor of 'Re-enacting the past' in the *Journal of Heritage Studies* (2014) and 'Postmodern Crowds' in *Distinktion* (2013). She is currently taking part in two projects connected to the ECOC (European Capital of Culture) 2017 on new ways of motivating and evaluating affective mobilization and engagement in sustainability issues through living experiments (*Greening the City* 2014–17 and *Reclaiming Waste* 2015–16).

Divya P. Tolia-Kelly is a Reader in the Geography Department at Durham University, UK. Divya has focussed on identities, visual culture, heritage and landscape throughout her career. She has convened research collaborations with artists Melanie Carvalho, Graham Lowe, Rosanna Raymond and Kahutoi Te Kanawa and contributed to scholarship on visual culture, material culture, race and affect, visual methodologies and the production of images that continue in circulation in national and international art collections. These canvases feature migrant sensibilities and their experience of 'nurturing ecologies' in the landscape. These projects and linked exhibitions include 'Describe a Landscape' (1997–2003); 'Nurturing Ecologies' (2003–9); SPILL (2006–7); 'Tales of the Frontier' (2007–11); and 'Archaeologies of Race at the Museum' (2007–14). Divya's current research is focussed on 'race', memory and affect/emotion at the museum. She has particular interest in engaging critical race theory and postcolonial theory at the museum cabinet. Her key question of concern is: how do we exhibit national cultures in a post-Imperial world that is free seeing the world through an Imperial taxonomy? This current research is leading to a monograph entitled *An Archaeology of Race at the Museum* with Routledge. This monograph builds on the AHRC-funded exhibition 'An Archaeology of "Race"' and her monograph *Landscape, Race and Memory: Material Ecologies of Citizenship*.

Emma Waterton is an Associate Professor based at the Western Sydney University in the Institute for Culture and Society and the School of Social Sciences and Psychology. Her research explores the interface between heritage, identity, memory and affect. Her most recent project, 'Photos of the Past', is a three-year examination of all four concepts at a range of Australian heritage tourism sites, including Uluru Kata-Tjuta National Park, Sovereign Hill, the Blue Mountains National Park and Kakadu National Park. She is author of *Politics, Policy and the Discourses of Heritage in Britain* (2010), and co-author of *Heritage, Communities and Archaeology* (with Laurajane Smith; 2009) and *The Semiotics of Heritage Tourism* (with Steve Watson; 2014).

Steve Watson is Professor of Cultural Heritage in the Business School at York St John University, UK, where he teaches a range of subjects in the cultural sphere. His research is concerned primarily with the representation and experience of heritage, especially through tourism, and he is active in the development of theory that explores the relationship between representational practices and the performative encounters and engagement of tourists with heritage places. He has explored these issues in Greece, Spain and the United Kingdom and he has a particular interest in Spanish travel writing. His most recent book is *The Semiotics of Heritage Tourism* (with Emma Waterton, 2014). He recently established HAVRC, the Heritage, Arts and Visitor Research Collaborative, as a way of engaging with communities of interest in the heritage sector.

Sophie Yarker is a Teaching Fellow in Human Geography at Newcastle University, UK, in the Centre for Urban and Regional Development Studies (CURDS). She completed her PhD with CURDS, exploring the nature of local belonging and attachment in contemporary urban neighbourhoods.

Acknowledgements

This volume has been a long time in the making; as such, we find ourselves with an unsurprisingly long list of people and organizations we need to thank for providing us with the conditions and support that made it possible. The majority of the contributions in the collection emerged from sessions organized at three different conferences held in 2013: the Association of American Geographers Annual Meeting, the Fourth International and Interdisciplinary Conference on Emotional Geographies and the RGS-IBG Annual Conference. Our double session in the latter was generously supported by the Social and Cultural Geography Research Group (SCGRG), and our own participation across these events was enabled by Durham University, Western Sydney University and York St John University. In addition, the Australian Research Council's DECRA Scheme (project number DE120101072) provided financial support for Emma's attendance at all three events, and the Durham University Geography Research Fund likewise provided assistance for Divya.

We would also like to extend a special thanks to those who have given us permission to reprint their photographs and images. In particular, we would like to acknowledge the Jewish Holocaust Centre, Museum Victoria, and the Museum of New Zealand Te Papa Tongarewa. We are also grateful to be able to include a still from the film *Jamaica for Sale*, written and produced by Esther Figueroa and Diana McCaulay. We would also like to thank photographers Benjamin Healley, Jon Augier and Wendy Lee, and Ingvild-Anita Velde from the Norwegian Broadcasting Cooperation. For the images found in Chapter 10 we would like to thank Fortepan (www.fortepan.hu), and for the photo montage we thank Zoltán Kerényi (Window to the Past, www.ablakamultra.hu).

Our thinking on the relationship between heritage, affect and emotion has been enormously improved by numerous conversations on the topic with colleagues. We would thus like to take this opportunity to also thank and acknowledge those who have been so helpful in this endeavour: Denis Byrne, Annie Clarke, Sarah De Nardi, Hayley Saul, Philipp Schorch and Andrea Witcomb. In particular, we are grateful to both Sarah and Hayley for providing us with helpful comments on the volume's introduction at very short notice! Finally, we would like to offer special thanks to Valerie Rose for her enthusiastic response to our initial proposal and for seeing the volume almost to fruition – and we would also like to thank Faye Leerink for picking up so seamlessly where Val left off. Both Ashgate and Routledge have been immensely supportive throughout the editorial process and we thank them for sharing so freely their professional knowledge, expertise and experience. Omissions, errors and shortcomings, however, we claim entirely for ourselves.

Introduction

Heritage, affect and emotion

Divya P. Tolia-Kelly, Emma Waterton and Steve Watson

There has been a shift in the heritage landscape in recent years. It has been a palpable, visceral shift that challenges the format, engagements and paradigms through which we articulate heritage at sites, in scholarship and in practice. Fuelling this shift is a groundswell of research that attends to the value, power and politics of affect and emotion, and shapes heritage landscapes as experienced, as curated and as foundational to our relationship with the past. These sensibilities, evoked and experienced, also co-constitute meaning in memory, identity and heritage, past and present (Crang and Tolia-Kelly, 2010). This edited collection was developed to capture this shift, and it does so by interrogating the very underpinnings of heritage and the moments, engagements and economies that shape and enable its presence in the twenty-first century. Earlier explorations of visual representation in heritage research (see Waterton and Watson, 2010), along with a more recent call to expand the palate of heritage theory (Waterton and Watson, 2013), had already signalled the turn to a consideration of more-than-textual embodied approaches to heritage research. With this volume, we seek to put a little more might behind that turn and propel heritage studies away from simpler 'two-dimensional' textual readings and narrative accounts towards engaging with experience, the sensory realm and the affective materialities and atmospheres of heritage landscapes.

As editors, our aim with this volume is to curate an intellectual space that can provision three key agendas: (1) bringing to bear a fuller understanding of the embodied aspects of heritage experiences; (2) interpreting the spaces of heritage so as to highlight the affective relationships that we have with our pasts; and (3) acknowledging the 'rolling maelstroms of affect' (Thrift, 2004) that shape and are located in, articulated by and palpably accumulated at heritage sites. Our purpose, therefore, is to articulate a realm of experience, thinking and being; one that has formerly been considered inarticulatable. That is not to say that this line of thinking has been without powerful antecedents: indeed, Stuart Hall, in 1999, argued that '[heritage] is one of the ways in which a nation slowly constructs for itself a sort of collective social memory' (see Hall, 2005 p. 25). But to this discursive consideration many of the scholars in this volume bring forth the challenging idea of heritage as a materialized social memory, thought through the nodal points of the body and its being and doing in a world that is both felt and expressive. Yet, while these lines of thinking clearly continue to probe at Hall's original question of 'whose heritage', they also bring to the debate questions such as: What counts as heritage? How is heritage encountered? How might it be engaged with? And why is it valued?

Collectively, these sorts of questions drive home the point that in scholarship that tackles heritage, affect and emotion, emerging theories are not situated as fashionable 'add-ons' to the project of heritage, but instead are positioned as a way of engaging with materials of social memory that are variously occluded, marginalized or, indeed, core to the projects of conservation, preservation and self-determination for all societies. Rodney Harrison (2010, 2013), for example, has challenged many taken-for-granted perspectives about heritage, advancing Smith's (2006) 'authorized heritage discourse'. We build on this account a politics that challenges traditional ways of understanding how we 'do heritage', one that is propelled, moved and mobilized by a range of feelings, affordances and capacities that have worked outside the mainstream and conventional renderings of the heritage debate. To do so, we have taken a lead largely from those working in contexts of resistance between the imperial and the colonized, and subsequent critiques of the histories, practices and repercussions of cultural display. A key text in this regard has been McCarthy's (2007) *Exhibiting Māori: a history of colonial cultures of display*, in which the question of Māori heritage is positioned as revolving around securing a space for 'curiosities' within an ethnological frame of understanding. This positioning of 'heritage' as a display of an 'Other' to the progressive settler colony was constructed, McCarthy argues, through a European lens that violated aesthetic, spiritual and historical relationships and affective economies and, indeed, closed down an account of the atmospheres of everyday living that came before. The book opens with the following provocative statement:

> Exhibitions today seem natural to us only because they are dressed in a familiar style. Historical exhibitions, on the other hand, expose the continuing artifice of display. They are instructive because they show us how differently our ancestors saw the world and how contested the process of exhibiting was.
> (McCarthy, 2007, p. 1)

Despite a shift in curating Māori culture, from primitive ethnographic artefact to the possibilities of being judged as art and elite culture, McCarthy reminds us that the articulation remains a deadened account that omits the plurality of narratives, values and time-space necessary to fully appreciate Māori culture. There are thus powerful lessons to be learnt from McCarthy's observations, and in this volume we seek to think through how 'Otherness' plays out via matrices of power *as well as* those feelings that shape encounters and move us beyond looking *onto* 'cultures of display'. Sherman (2008, p. 4) argues succinctly how 'the use of alterity as a structuring concept entails taking the relationship between Self and Other as an irreducible component of cognition, desire, power, and ethics'; museums are constituted by, and themselves constitute, frameworks that use alterity as an organizing intellectual logic. This will be an argument already familiar to some readers, particularly those who have engaged with the gap between 'Self' and 'Other' identified in Tony Bennett's (2004) account of museums in Britain, the United States and Australia, in which the two could not be bridged (at the colonial museum), as there is an active de-historicization of 'Others' and a lack of belief that the 'Other' is situated in geological time and not modernity. In the Australian context, this de-historicization has situated heritage as reading Aboriginal culture as a prehistoric living presence in which the 'Other' is not imagined as visiting the site. That 'Other' Aboriginal body is singular, without historical sensibilities; to be looked upon and categorized. There is no seeing-with or *being-with* the 'Other' as a possibility; indeed, the Other is not felt, known or understood beyond consideration as material artefact.

In this volume, however, alterity is heterogeneous and situated at the site of bodies, not artefact or aesthetics. Moreover, the politics of difference in the heritage sphere is presented here as felt, embodied, intense and dynamically co-constituting the practices of meaning-making and world-sense-making. What is articulated in this volume, then, is the process of recognition, understanding and experiencing *self* at heritage sites. This *self* in our analysis is presented as singular and collective, and figured through multiple space-times rather than the fixed, dioramic representations of cultures outside of European modernity and thus European space-time. In the accounts that populate this volume, *feeling the past* through embodied presencing of geological/environmental space-time is core to understanding identity, difference and alterity at heritage sites. In this regard, memory is posited as an *affective* tool for the co-constitution of embodied, political narratives. While memory is at stake in contemporary heritage studies, practices and management, as Fairclough *et al.* (2008) remind us, the *object* of memory is as transformative as is the experience of encountering it. We thus point to an agentic relationship between object and visitor, which through affective energies shapes the envisioning of environment, meanings and futures. The politics of affective memory in this volume attends to the power of memory not to *translate* cultural objects, but to acknowledge their power to articulate pasts, identities, events and create atmospheres of experience and creative heritage. This moves the debate beyond an authorized heritage/alternative heritage binary; affective memory, when forged at heritage sites, shatters singular readings and narratives. Benton (2010) expands this further by situating heritage as dynamic and being as much about the recent modern past as it is about a 'national' culture or 'ethnic' culture. Memory, as advanced by Benton, becomes a means of doing heritage inclusively, through understanding intangible components of the value of heritage including beliefs, feelings and practices. We, in the investigation of practices at heritage sites, extend the focus on remembering and feeling to consider practitioner articulations of affective heritage relationships, focused on inclusion and the plurality of narratives, meanings and memories that are co-constituted through experiencing heritage sites and places.

The problematizing of material culture has shifted the ways in which the 'object' has been examined as symbolic or functional (Meskell, 2005), and anthropological theories on material cultures (Miller, 1998) and the 'social life of things' (Gell, 1986) have shifted the ground in terms of how we think through artefacts and materialities at heritage sites. However, by *Thinking Through Things*, Henare *et al.* (2007) have aimed to take seriously Tim Ingold's account of how 'culture is conceived to hover over the material world but not to permeate it' (cited in Henare *et al.*, 2007, p. 3). Here, interpretation is not the sole role of the anthropologist; rather, it is to consider the agency of things in articulating meaning, thus raising the possibility that they themselves create life-worlds. Heritage, then, becomes less about 'ways of seeing' centred upon anthropocentric values, and more about giving power to the thing itself and making space for resonances not before encountered. In opening up this new terrain, the collection engages critically with the body of work that has been developing in the realms of cultural geography and elsewhere, where affect and emotion are thought, unproblematically, through a 'universal' frame (Tolia-Kelly, 2006). Thus, questions of history are at the heart of research, as are the politics of race, racism, equality, social justice and 'other' ways of experiencing the heritage landscape. In terms of the volume's contribution to critical heritage studies, there is no assertion of a post-stucturalist lens peering into the literature on heritage, but instead a weaving together of the critical views of philosophers that challenge dominant (including archival and textual) narratives and discourse, such as can be found in the work of Foucault, Derrida, Sedgwick, Deleuze and Guattari, with that of scholars of heritage, geography and culture such as Raphael Samuel, Tony Bennett, Nigel Thrift and Stuart Hall.

In this new heritage landscape, there are also different heritage economies at play, where 'feeling' and 'being' are important trajectories of engagement at heritage sites and museums. A key argument in this edited volume, therefore, is that heritage and its economies are driven by affective politics and consolidated through sensibilities such as pride, awe, joy, pain, fear. Most of all, however, the research collected here exemplifies the ways in which theories of emotion and affect need to engage with the historical, and be situated within matrices of power, so that affective logics of history and heritage are sensitive to differently positioned narratives, memories, emotions and, indeed, material cultures. In tune with the need for the humanities and social sciences to acknowledge the limits of language and subjectivity, affective registers, atmospheres, emotional contours and embodied memories are about pluralizing the terms of engagement, through format, media and embodied accounts. At the same time, ever renewed efforts to 'return' to the powerful possibilities of discourse, visuality and performativity remain active within the academic literature (Crouch, 2015; Haldrup and Bærenholdt, 2015; Harrison, 2013; Waterton and Watson, 2013, 2014; Wu and Hou, 2015). In other words, 'representation' has not yet run its course but has been re-energized by the very countenance of its limitations. One of the ways in which it has achieved this reanimation is through recent attempts to define and understand the more-than-representational realm. This project continues to unfold within the wider social sciences, particularly within cultural geography and cultural studies, where such attempts to advance contemporary theory have seen the addition of theories of affect, assemblage, post-humanism, new materialism and actor network theory. While there have been some recent efforts to draw these lines of thinking more forcefully into the field of heritage studies (see Waterton, 2014, 2015; Waterton and Watson, 2013, 2014, 2015; Witcomb, 2012), our interest is focused and purposeful in attending specifically to the theoretical potentialities of affect and emotion in the experience of heritage. Below, we lay out how this has been achieved.

The collection's structure

Captured across the following contributions, readers will find a shared exploration of current theoretical advances that aim to enable heritage to be affected and released from conventional understandings of both 'heritage-as-objects' and 'objects-as-representations', thereby opening it up to a range of new meanings, emergent and formed in moments of encounter. While we have acknowledged that representational understandings of heritage are by no means made redundant through this agenda, they are destabilized and can thus be judged anew in light of these developments. For the various contributions collected together here, the notion of affecting heritage will play out in myriad ways: as writers that employ rhetoric, metaphor, aesthetics, form, narrative, description; as borders and genres that are crossed and melded; as surfaces and depths which are thought and felt in the act of writing and analysis; as a subject that is liberated, interrogated, positioned, disturbed; and in the ways that heritage refuses to be conventionalized, to be coherent, rational or ordered, and emerges as disparate, contradictory and multivalent. In order to gain some conceptual traction over the contributions that make up this volume, we have parcelled them into three groups, or Parts, arranged around the themes of memory, place and practice. In addition to providing a sense of order for the volume itself, these Parts serve to contextualize the provocative contributions each author lends to our re-theorization of heritage. Indeed, they illustrate the ways in which a number of both well-established and emerging researchers and practitioners are thinking theoretically about affect, whether that is via landscapes, practices of commemoration, visitor experience, site interpretation or other heritage work.

Part I: memory

An impulse to engage research and *think* heritage more democratically has resulted in a focus on memory. Before being canonized, authorized or, indeed, made material in the public domain, memory is at heart inclusive, accessible and a way of 'doing' heritage from below. Memory within this volume offers a route to counter authorized accounts at particular sites and practices, but also affords recognition of the striated nature of stories that *belong* to places. Memory expresses and articulates a plurality of attachments located in people's hearts and minds, so that memories elicit affects at heritage sites, and vice-versa. Memory, identity and affect are thus co-constituted within the experiential landscapes of monuments, places and spaces where we preserve an account of the past for ourselves and others, and for future generations. As Joy Sather-Wagstaff (Chapter 1, this volume) articulates, a new theorization of heritage through affective registers

> is critical to unfolding the latent individual and social meanings of experiences as a form of embodied knowledge and cogent responses, even when not fully articulated, as well as the potential politics of affect through ambiguous and constantly shifting articulations and social effects.

Through thinking memory and affect simultaneously, as Sather-Wagstaff goes on to articulate, we can fully recognize the power, place and dynamic nature of heritage sites as they signify a plurality of valuable heritage narratives that contribute to history and visions of futures. Likewise, Jason Dittmer and Emma Waterton (Chapter 3) engage with memory through their attempts to demonstrate how there is an affective homology that is material and visceral in the museum encounter. In their example of the Australian War Memorial, the memory of past bodies at war is simultaneously to be 'seen', felt and embedded through 'a multiplicity of lines of flight unspooled in and through us'. This is how memory is treated beyond a sanctioned ocular association, but through a visceral shudder, felt at the heritage space itself. Memory is four-dimensional in their account; the Memorial is worked through and disturbed, resulting in a plural account where past encounters are enlivened and brought to bear on future understandings of nation and identity, *as well as* the possibilities for sovereignty over these memories themselves. Memory, memorial and the visitor event thus can be seen as affective registers through which particular narratives gain momentum and meaning and hold resonance.

Steven Cooke and Donna-Lee Frieze (Chapter 4) take affective memories and futures further by considering the question: 'What happens to Holocaust memory when there are no survivors left?' In their chapter, the transformational potential of school children meeting Holocaust survivors is considered where forces of negation are tempered by the politics of witness and testimony. They discuss a methodology for keeping the affective alive and in contemporary debate through a 'pedagogy of feeling'. This is understood as learning through experience, which is transformative in attuning visitors to the project of learning to remember the Holocaust. Philipp Schorch, Emma Waterton and Steve Watson (Chapter 5) extend this account of the transformative potential of the heritage space through to thinking of it as a *canopy* under which cosmopolitan sensibilities can emerge, be articulated and, indeed, be understood. The transformative logics in their chapter extend an account of social mixing to appreciating the development of multicultural feeling, expression and tolerance. In investigating the ways in which the space of the museum choreographs conversations across cultural differences and boundaries, the authors simultaneously elucidate the ways that bodies, memories, enactments and expressive cultures take place at heritage sites and thus enfranchise progressive sensibilities through the affordance of in-process and emergent affective cosmopolitanisms.

Our Part on 'memory' closes with a critical inspection of affective memory via an exploration of slave-memory at the museum (Chapter 6). Here, Leanne Munroe argues that affect and emotion suffuse processes of narrative construction at the museum, and the 'narrative' *per se* is powerful, affective and transformative. As such, Munroe argues that we can no longer consider affect at the museum simply in 'call' and 'response' mode, but rather as a relationship that is co-constituted at the site *through* the encounter, charged with emotion and inspiring connectivities in a dialogic way. Munroe argues for the recalibration of the relationship between representational and more-than-representational theories to enhance explorations of heritage narratives that are always contested, perceptual and at risk of exclusion.

Part II: places

In the second Part of the volume, 'places' are considered as affective engagements. Iconic places such as castles, ruins and sites that somehow elide modernity are thought through as sites of feeling and affect. By situating the heritage encounter as affectively charged, our usual narratives about heritage sites are troubled, enriched and made more inclusive. In this Part, readers will find post-human engagements with heritage places where they become ecologies with agencies, intensities and capacities, and through their agency actively co-create the landscape (see De Nardi, 2014). This sort of theoretical engagement commences with the work of Gareth Hoskins (Chapter 7), who takes affective memory into dialogue with posthuman sensibilities. In his account, the agency of geology is privileged, challenging (human) ocular-centric accounts of the heritage landscape of Malakoff Diggins, California. Hoskins' heritage of place is understood by illustrating the embodied affective relationships that take place at a gold mine. Vertigo experienced at this site becomes a way of reflecting how we extend into the world and how the world extends into us in a post-humanist framework. Through the experience of vertigo, we no longer simply look outward onto a scene, but are physically troubled internally by the space itself. This examination extends heritage landscapes beyond traditional 'ways of seeing' the picturesque towards an understanding of heritage as an encounter with active agentic life-worlds.

Duncan Light and Steve Watson's account (Chapter 8) achieves what affective theory does not always quite lend itself to, and that is theorizing the power of affect *in situ* within a particular temporal and spatial framework. Their account considers the coordinates of affective registers and their power to co-constitute experience and, indeed, the narratives that are also bounded within a space. Castles are thus received and achieved in an assemblage of knowledge and experience. The abstract nature of theory is synthesized to communicate the nature of castles, their layered experience and how the space articulates symphonically, unpicking the competing understandings that are presenced at the encounter. The castle as material heritage is thus expressed in a coalescence of haptic moments, smells, shadow, light and atmosphere together with feelings of achievement, risk and foreboding.

From castles to coasts, we encounter benign representations and others that are not so. In her chapter, Susan Mains (Chapter 9) focuses on heritage and contestation;

8 *Divya P. Tolia-Kelly et al.*

she explores this via the importance of affective geographies in opposed representations of coastal landscapes, and how these mobilize conflicting notions of heritage, development and sovereignty. The *place* of heritage in this chapter is co-narrated through the registers of loss, anger, fear and discontent; significations that are often considered oppositional to the place of leisure that is so often associated with coastal heritage. The currency of affect in these economies of heritage is articulated through two international examples – Jamaica's North Coast and north-east Scotland – of coastal sites of discordancy and struggle between locals and others. The politics of feeling drive the economies of heritage, and in Mains' account they are laid bare to expose hierarchies of power, control and enfranchisement that trigger conflict at heritage sites. Coastal sites are situated here as liminal spaces as well as held within circuits of mobility, transfer and exchange, and as such lay redundant notions of 'universal' heritage experiences. Through illustrating affective registers at play in place, Mains illustrates the politics of value and logics of heritage that counter naturalized representations of coastal heritage sites as benign and framed through parochially situated nostalgia. Dominant discourses are thus troubled through contestation and counter-discourse.

A completely different texture of place is presented in László Munteán's (Chapter 10) account of digital heritage in the twenty-first century. Encounters with heritage through the digital archive are more and more becoming a dominant interface between 'visitors' and heritage collections. This chapter illustrates how digital documentary heritage produces new auratic environments for visitors' own affective engagements with heritage. The archive of photographs from Hungary entitled *Forteplan* is examined here as a means of democratizing heritage through making it accessible, but also through recording a variety of sensibilities that are made possible through the preservation of the *Forteplan* archive. The photographs offer us a collection of amateur pictures which would not normally be of the sort encountered at a heritage site, and as such they democratize textures, iconographies and indeed the orientation from which heritage is done, felt and experienced. Commemoration occurs in the visual engagement with the archive in this account; conversely, in Britta Timm Knudsen and Jan Ifversen's (Chapter 11) account of commemorations of the Oslo and Utøya massacre in Norway in 2011, millions of people engaged with remembering the dead and the event post-massacre. A grass-roots memorialization of the event is discussed in their chapter alongside a notion of the homogeneity of national trauma troubled by the authors. Death, massacre and loss become the collective registers of commemorating a public event connected by the politics of nation and affect. However, in their research the traumatic effect of the memorial itself is engaged with as a site of re-traumatization. The chapter problematizes the ethics of making visitors 'feel' trauma in the context of a 'national' loss, and here the question of commemoration radically alters the affective ecologies of locale, place and nation.

The north-east of England is the site for Sophie Yarker's account of the affective heritage landscape (Chapter 12). In her chapter, Yarker argues for the place of heritage in the urban environment, suggesting that it ought to be more inclusive of working-class values. The dominant heritage discourse that Yarker encounters in place is only sympathetic to a particular 'set' of affective values, whereby some architecture is more valuable than others. Yarker's research attends to the affective heritage of the built environment as a way to understand the dissonances between an authorized heritage discourse (Smith, 2006) and the values of those residents who live among it in the everyday. Here, there is a call for an attunement to affective charges that celebrate textures of the urban as heritage rather than simply architecture, especially when valued by poorer majority communities. Thinking heritage *in situ* through the affective registers of counter-cultures, majority cultures and indeed occluded voices is privileged in these accounts.

Part III: practices

Praxis is exemplified in the final Part of the collection through the work of Keith Emerick and Rosanna Raymond, both of whom are practitioners within the heritage realm. The critical tension in both of these accounts resides in practice rather than in theory, and is exemplified through their articulations of how power, narration, display and ontologies themselves need to shift if we are to get beyond 'Self' and 'Other' in heritage displays and discourse. Giving power back to communities enables a reflection on the spectrum of affective relations that are part of democratic heritage practice. In his contribution, Emerick (Chapter 13) exposes the tensions between formal heritage frameworks and everyday values of heritage sites. He articulates how conflicts can arise between the public and the heritage professional, seen here from the perspective of the practitioner. He shows how heritage is more than the site, but equally about narration, co-production and everyday feelings about a heritage site. Emerick demonstrates how 'narratives' are contested, and often occlude the non-textual encounters that make sites meaningful and valued. The storying of place and 'doing' of heritage is pitched here as crucial in heritage practice rather than in the fabric of the site itself. Raymond (Chapter 14) is also a published poet, writer and founding member of the *SaVAge K'lub*, with art works held in museum and private collections around the world. Through her performances and art, we can learn what a practitioner-led heritage space could look like. How is an inclusive approach to Maori, Polynesian and Oceanic heritage possible? And what are the missing accounts in our current exhibitions housed in national museums? These are two questions central to Raymond's art practice. Pain, loss, defilement and guilt are part of the process of not doing heritage respectfully; bringing heritage back to life, to its rightful place, offers an alternative heritage landscape.

Conclusion

When we started this project it was unclear where it would lead us. Three conference sessions later, in Los Angeles, Groningen and London, we knew of at least a few more people who were interested in heritage, affect and emotion, and their contributions have formed the core of this collection. At the time of writing, we are aware that there is as yet no other comprehensive text available that deals specifically with these sorts of emergent theorizations in the context of heritage, and so it is to this gap that we turn with this volume. But in truth, that gap also provides for us, as editors, something of a personal opportunity, for our own engagements with heritage are almost entirely figured through affective registers and their expressive corollaries. What we were less certain about when we embarked on this project was how these engagements could be best understood and, more importantly, what new kinds of thinking were needed to capture what is immediate, embodied and performative about them. We are pleased, therefore, to bring together more certainty with this collection of contributions, all of which hone in – with great critical acuity – on questions of how researchers working in the field of heritage might begin to discover and describe affective experiences, especially those that are shaped and expressed in moments and spaces that can be, at times, intensely personal, intimately shared and ultimately social. The critical heritage debate provided momentum and motivation; the emerging theoretical canon around affect and emotion provided the framework for our thinking and research; and a concern with encounter and engagement gave us a point of departure, a place from which to begin this and to which we find ourselves returning in our conversations and debates in the heritage field. More than anything else, then, we are pleased to be part of the beginnings of a debate that has seemed a long time coming, and if this book achieves that goal then we will be more than satisfied with it. Clearly, there is much more to say, but the thinking and ideas presented here will, we hope, help to inform and enliven those conversations.

References

Bennett, T. (2004) *Pasts beyond memory: evolution, museums, colonialism*. London and New York: Routledge.

Benton, T. (ed) (2010) *Understanding heritage and memory*. Manchester and New York: Manchester University Press.

Crang, M. and Tolia-Kelly, D.P. (2010) Nation, race and affect: senses and sensibilities at national heritage sites. *Environment and Planning A*, 42(10): 2315–31.

Crouch, D. (2015) Affect, heritage, feeling, in E. Waterton and S. Watson (eds) *The Palgrave handbook of contemporary heritage research* (pp. 177–90). Basingstoke: Palgrave Macmillan.

De Nardi, S. (2014) Senses of place, senses of the past: making experiential maps as part of community heritage fieldwork. *Journal of Community Archaeology and Heritage*, 1(1): 5–23.

Fairclough, G., Harrison, R., Jameson, J.H. and Schofield, J. (eds) (2008) *The heritage reader*. London and New York: Routledge.

Gell, A. (1986) Newcomers to the world of goods: consumption among the Muria Gonds, in A. Appadurai (ed) *The social life of things: commodities in cultural perspective* (pp. 110–38). Cambridge: Cambridge University Press.

Haldrup, M. and Bærenholdt, J.O. (2015) Heritage as performance, in E. Waterton and S. Watson (eds) *The Palgrave handbook of contemporary heritage research* (pp. 52–68). Basingstoke: Palgrave Macmillan.

Hall, S. (2005) Whose heritage? Un-settling 'The Heritage', re-imagining the post-nation, in J. Littler and R. Naidoo (eds) *The politics of heritage: the legacies of 'race'* (pp. 23–35). London: Routledge.
Harrison, R. (ed) (2010) *Understanding the politics of heritage*. Manchester and New York: Manchester University Press.
Harrison, R. (2013) *Heritage: critical approaches*. London: Routledge.
Henare, A., Holbraad, M. and Wastell, S. (2007) Introduction: thinking through things, in A. Henare, M. Holbraad and S. Wastell (eds) *Thinking through things: theorising artefacts ethnographically* (pp. 1–31). London: Routledge.
McCarthy, C. (2007) *Exhibiting Maori: a history of colonial cultures of display*. New York: Berg.
Meskell, L. (ed) (2005) *Archaeologies of materiality*. Oxford: Blackwell.
Miller, D. (1998) Why some things matter, in D. Miller (ed) *Material cultures: why some things matter* (pp. 3–21). London: Routledge.
Sherman, D.J. (2008) Introduction, in D.J. Sherman (ed) *Museums and difference* (pp. 1–21). Bloomington and Indianapolis: Indiana University Press.
Smith, L. (2006) *Uses of heritage*. London: Routledge.
Thrift, N. (2004) Intensities of feeling: towards a spatial politics of affect. *Geografiska Annaler. Series B. Human Geography*, 86 B(1): 57–78.
Tolia-Kelly, D.P. (2006) Affect – an ethnocentric encounter? Exploring the 'universalist' imperative of emotional/affectual geographies. *Area*, 38: 213–17.
Waterton, E. (2014) More-than-representational heritage? The past and the politics of affect. *Geography Compass*, 8(11): 823–33.
Waterton, E. (2015) Visuality and its affects: some new directions for Australian heritage tourism. *History Compass*, 13(2): 51–63.
Waterton, E. and Watson, S. (eds) (2010) *Culture, heritage and representation: perspectives on visuality and the past*. Farnham: Ashgate.
Waterton, E. and Watson, S. (2013) Framing theory: towards a critical imagination in heritage studies. *International Journal of Heritage Studies*, 19(6): 546–61.
Waterton, E. and Watson, S. (2014) *The semiotics of heritage tourism*. Bristol: Channel View Publications.
Waterton, E. and S. Watson (2015) A war long forgotten: feeling the past in an English country village. *Angelaki*, 20(3): 89–103.
Witcomb, A. (2012) On memory, affect and atonement: the Long Tan memorial cross(es). *Historic Environment*, 24(3): 35–42.
Wu, Z. and Hou, S. (2015) Heritage and discourse, in E. Waterton and S. Watson (eds) *The Palgrave handbook of contemporary heritage research* (pp. 37–51). Basingstoke: Palgrave Macmillan.

1 Making polysense of the world
Affect, memory, heritage

Joy Sather-Wagstaff

Attention to the lived and narrated experiences of heritage consumers, constructors, and producers is one means to critically re-think the powerful discursive authority of heritage practice. One thematic aspect to such an approach is a critique of academic devaluations of experiential, senses-inclusive meaning-making in the world over presumably purely cognitive forms of knowledge construction. Across disciplines, much of this critique has centered on what has become a sort of standard shorthand—a convenient gloss, if you will—for the experiential in all of its rich dimensions: embodiment. Any truly rigorous focus on embodiment requires unpacking the performative human body, even if only partially, in order to avoid reifying a universal or essentialist definition of embodiment. In this chapter, I propose a careful re/theorizing of the role of the sensory and performative body in knowledge construction and (re)production. In this process, I address the potential of sensoria-attentive research focused on narrated experiences, using an expanded conceptualization of discourse analysis that focuses on the senses for a critical heritage theory and practice. The contextual focus here is on bodies of work that critically interrogate the relationships between memory, memorywork, affect, emotion, subjectivities, and the material world in relation to ways of knowing that occur in the dialogical and slippery interface between the bodily and the cognitive.

Critical to understanding such relationships is a very brief discussion of *affect* as the potential to elicit intense embodied, physiological responses with very powerful effects. I do so in the context of evaluating and weaving together multiple threads of current and past work in cultural geography, anthropology, tourism studies, and other humanistic fields that attend to the performative, agentic, and sensing body, affect (even if only implicitly), emotion, and meaning-making in the world, work exemplified by scholars such as Denis Byrne, Mike Crang, Britta Timm Knudsen, Sharon Macdonald, Nadia Seremetakis, David Stoller, Divya Toila-Kelly, Emma Waterton, and Steve Watson, among others. I find this particular set of approaches to be exceptionally fruitful given that attention to affect is becoming a means to thinking through a number of challenges posed in heritage research. They also add a crucial element that allows us to go beyond the theory to application and further (re)theorizing: Ethnographic and participant-based methodologies that generate, in all their human messiness, knowledge about what people are actually experiencing. This is critical to unfolding the latent individual

Making polysense of the world 13

and social meanings of experiences as a form of embodied knowledge and cogent responses, even when not fully articulated, as well as the potential politics of affect through ambiguous and constantly shifting articulations and social effects.

Affective experiences translate into multiple effects, one being knowledge (and by extension memory and in some cases, heritage) and the other an excess residual that may never be fully categorized cognitively. As will be discussed, affect has been addressed in relationship to emotion and memory, particularly in terms of theorizing geographies of either emotion, feeling, or memory (see Crang and Tolia-Kelly, 2010; Crang and Travlou, 2001; Davidson *et al*., 2005; Picard and Robinson, 2012; Thien, 2005; Thrift, 2008). Yet the senses, as an element of affect-in-action, have rarely been explicitly discussed as part of the constellation of the interdependent sociocultural *and* biological phenomena that engender emotion and memory. It is the incorporation of the broadest array of the sociocultural and biological aspects of the sensory that I argue for here, particularly in terms of attention to the politics of affect and the senses and emotion as entangled in the processual construction of memory and knowledge, both of which are crucial elements to the dynamic construction and performance of heritage.

A strong case for the significance of the sensory and sensorial for theory and practice in critical heritage studies requires situating such in the context of current approaches to the body, senses, embodied performativity, and affect. Such approaches include those from anthropology, cultural history, and cultural geography, as well as studies that traverse traditional disciplinary boundaries including heritage, tourism, performance, museum, and memory studies. I then concentrate on that which is called "negative" or "difficult" heritage—the tangible and intangible results of war, genocide, terrorism, poverty, crime, and other human-made atrocities as represented in memorial museums and landscapes as heritage sites. I do so due to a public and scholarly propensity to sometimes criticize the consumption of such heritage as morbid, undesirable, inauthentic, or destructive rather than as part of the productive and dynamic social construction of heritage and knowledge, even if it is heritage built upon the darkest aspects of the human experience. A focus on difficult heritage also allows for a critical disruption in how we think about dark heritage sites, moving away from assuming that they have limited or predictable affective capacities and toward a perspective that considers the highly variable individual and social effects of visiting these sites over time and space. By utilizing an expanded conceptualization of critical discourse analysis that also considers forms of communication and meaning that are beyond the spoken, we may move toward an emergent understanding of the highly polysensorial, affective experiences these sites generate.

I aim to engender a multimodal, multidimensional approach to difficult heritage (or any sense of heritage) as processually formed, performed, and reformed through dialogic processes of affective, sensory stimuli and responses, cognitive sense-making, narratives, and reverie as memory performance, as well as other forms of bodily performance and interpretation over time and space. Such an understanding further complicates not only notions that heritage, knowledge, and memory are somehow contained in the built environment, objects, landscapes, or people, but also approaches to affective experiences as "beyond discourses." The first step toward this understanding is to interrogate why attention to the body, embodied knowledge, and the senses is germane to any human experience-based discipline, and particularly so for those theorizing heritage from a critical standpoint. Over the past few decades there has been a highly significant turn in social sciences disciplines to the body as a vehicle for the construction of knowledge and the display and performance of a wide range of sociocultural phenomena. This turn is grounded in a deeper historical interest in the body and the bodily, but takes a much more expansive and reflexive approach to the bodily as a part of making sense of our worlds. Briefly unpacking this turn and its underpinnings shall initiate discussion of the value of attention to an expanded theorization of human sensorium as polysensory and intrinsic to analyzing the role of embodied experiences for knowledge and heritage construction.

The senses unbound: toward a polysensory model

Given that interest in the body and the senses also has a history as deep as that of humanity itself, a truly comprehensive overview is impossible to present here. Such an overview may be found collectively in the rich ethnographic and historical works of Classen (1993, 2012), Classen *et al.* (1994), Howes (2005, 2008), Kwint (1999), Landsberg (2004), Lock (1993), and Stoller (1989, 1997). However, a very brief outline of some of the key contours of the study of the bodily in cultural contexts over part of the past century as they influence work in the present is necessary to contextualize the proposition for a polysensory approach to heritage. A key starting point comes from anthropology, where cultural attention to the body has a deep, shifting, and sometimes contentious history. As Howes notes, Douglas's *Purity and Danger* (1966) evidenced the establishment of an anthropology of the body, albeit one that was "curiously desensualized" (2008, p. 443), focusing on the body and bodily strictly in terms of symbolic meanings rather than through the senses as bodily experienced. Part of this desensualization stems from historically deeper notions regarding social hierarchies of knowledge and power where conceptualizations of doing and being in the world derived in part from highly racialized cultural evolutionist paradigms of the late nineteenth century (see Howes, 2008; Lock, 1993). In such a paradigm, emotions (and most things bodily or sensuous) were marked as "primitive" while cognitive "knowledge proper" and "rationality" were associated with "civilization," a position that still underpins various popular and scholarly conceptions of what constitutes human knowledge. In the nineteenth century, the individual senses were even organized into a racialized hierarchy, with "civilized" Europeans associated with vision and "primitive" Africans with the body, specifically touch (Howes, 2008, p. 444).

Avoiding analysis of the sensual in favor of a focus on the symbolic (rather than sensory) dimensions of the bodily was thus a means for anthropologists to distance themselves from such theories without a significant critique of such. This also accounts for the fact that, as Lock notes, the "body's explicit appearance has been sporadic throughout the history of the discipline [of anthropology]" (1993, p. 134). However, in the 1980s and 90s, Seremetakis (1994) and Stoller (1989, 1997) established a foundation for a highly sensuous anthropological scholarship that embraces the cultural and corporeal lushness of the senses. Their work engendered critical attention to the everyday importance of performed and interpreted sensory worlds, the processes through which this occurs, and the broader tangible and intangible social effects of the sensory in the world. These works also stimulated interest in how the critical interrogation of local epistemologies of the senses reveals different interpretive classifications as well as types and hierarchies of the senses, allowing for productive critiques of what Tolia-Kelly identifies as a "Westnocentric" sensory repertoire (2006, p. 214). How many senses one is believed to have and how they are interpreted (that is, identified and translated into emotion or other types of embodied performative responses) should indeed be understood to depend upon the cosmology of the cultural systems to which one belongs, despite the biological universality of the sensory system as a foundational aspect of human physiology. From the late 1970s through the 1990s, corporeal and linguistic turns in several disciplines generated attention to the body and embodied performance as a form of meaning-making, influencing work on topics including material culture past and present, human geographies, media, heritage, learning, and expressive culture such as dance and music.

In heritage and museum studies in particular, a turn toward critically thinking about the embodied performance occurred at least partially in response to perspectives in the literature that considered visitors to facilities to be "passive, uncritical consumers of 'heritage'" (Bagnall, 2003, p. 87). Attention to embodiment and performance currently also centers frequently on how engaging with heritage is very often a physical effort, requiring both "an effort of organization and a full-body presence" (Macdonald, 2013, p. 234). In contrast to works that focus solely on the representational analysis of heritage on display, much of the key literature addresses heritage experiences as a two-way street—visitors perform physical consumption, cognitive production, and emotional labor both on-site and beyond in response to the performative representations of museums and heritage sites themselves (see Bagnall, 2003; Bærenholdt *et al.*, 2004; Crouch, 2010, 2012; Dicks, 2003; Jackson and Kidd, 2011; Smith, 2006; Staiff *et al.*, 2013). Despite a significant focus on the performative and embodied in heritage studies and beyond, the body does continue to be somewhat desensualized, with senses frequently mentioned but not always the central factor in analyses. Several exceptions exist, including, most notably, powerful work on the senses, heritage, subjectivity, and emotion by Crang and Tolia-Kelly (2010), Bryne (2009, 2013), Knudsen and Waade (2010), and Macdonald (2013). Macdonald (2013, p. 79), for example, accords critical attention to "feeling the past" through "embodiment and emplacement, materiality and affect," whereby

> ... specific and embodied constellations of affect accompany some forms of past presencing ... [some of which] are harder to characterize—the mix of melancholy and pleasure in touching, and being touched by, the indentations

on an old chest of drawers or spade handle left by years or use; or the sense of being pulled into wistful recollection by the scent of hyacinths or the notes of a street piano.

Macdonald notes that a key issue with the analysis of embodiment and materiality, particularly in getting at the effects of affect and feeling, is that in the current "affective turn," the body becomes the privileged vehicle for creating "authentic" knowledge and this "has the effect of separating 'the felt' from the linguistically expressed," thus dismissing discourse (2013, p. 81). Yet, as Waterton and Watson ask, "*what* happens to our bodies" that changes us through processes of experiences "within spaces of heritage, whether they are *physical, discursive or affective*" (2013, p. 551–2, emphases added)? I, like Macdonald (2013), argue that we must bring discourse and the sensing body together into both analyses and "theories *in, of* and *for* heritage" (Waterton and Watson, 2013, p. 547, emphasis in original), considering multiple performative, discursive forms beyond that of just oral narratives as linguistic expressions. Imagining ways of doing this aims to meet, in part, Waterson and Watson's challenge to heritage studies to take an "approach that pays due respect to—and draws from—a number of disciplinary sources of theory" without generating complete stand-alone theories and which attends to the "range and purpose of various theoretical interventions in order to apply them usefully in appropriate contexts" (2013, p. 547). However, this first requires more specific discussion on what constitutes the sensorium and its relationship to affect and meaning-making in the world.

That the lived sensorium is sometimes overlooked as specific or primary information for the analysis of heritage experiences may also reflect a reluctance to engage in approaches that could be construed or interpreted as biologically deterministic. However, if we take embodied performance to mean physical (re)actions in space and time, humans are always engaged in embodied activity at biological levels and, during our waking hours, involved in such through a complex interface of the sociocultural and the biological. As an example from the human sensorium, we might examine this through the multiple senses and associated systems supporting one of our most basic biological needs, food: the preparation and presentation of food engenders sensory acts of "making it right" in terms of physical technique in space[1] as well as aromas and visual appeal which trigger various autonomic, predigestive responses. One prepares food and eats for biological sustenance but also as a way of facilitating, making sense of, and reinforcing social relations. Acts of food preparation and consumption are wrapped up in a biosocial constellation that may involve all five of the basic physical senses, from taste and smell to touch, vision, and hearing. And through the sensorium, foodways, as synesthetic and kinesthetic heritage, also provoke various affective modes of sociocultural identification (including memory and remembering) with communities of belonging in the present as connected through individual and collective knowledge and memories of temporally and/or spatially distant communities of the past.

Given this complex nature of the sensorium in practice, I utilize the term "polysensory" rather than "multisensory." While it is normative to employ prefixes for words from the same language of origin, the employment of a non-isomorphic prefix does occur as a means to expand or disrupt normative, long-standing definitions for a term. An example of this is the term "polyvocal" (a Greek-derived prefix modifying a Latin-derived root word) as used in semiotics to understand (if not encourage) ambiguity and slippage in meaning-making through the use of multiple, sometimes overlapping or intertextual narratives. This contrasts with a standard, normative understanding of "multivocal" as simply describing multiple narratives or voices. As such, a polysensory theory and approach requires that we incorporate a more fluid and dynamic consideration of the slippage between and complexity within bodily stimuli and responses (intersensorality) along with the diverse interpretive schemas for such.[2] It also allows for thinking about the elusive registers of the affective and a constellation of meaning-making processes rather than clearly delineated, discrete clusters of fully identifiable, universal end-point or "final" knowledge. A polysensory approach also encompasses imagined (versus actually experienced) sensory stimuli or responses as well as acknowledges the power that acts of oral or written narration have for invoking sensory responses; the performance of language itself joins this expanded repertoire of sensory stimuli. The term "polysensory" itself also captures the essence of the sensorium as a highly complex, culturally mediated, and thus varied, biologically grounded (but not biologically determined) processual meaning-making phenomenon.[3]

The sensorium incorporates a much wider array of stimuli and responses than the five most recognized senses of hearing, smell, taste, touch, and vision. It includes the kinesthetic, from proprioception—an awareness of the orientation or positioning of the body and body parts in space, including that of one's posture, muscle activity, limb positioning, or facial expressions—to other forms of body movement through or in space, such as walking or gesture. Equilibrium, pain, temperature awareness, various responses of the skin (such as blushing), and an array of internally based stimuli and responses, some of which are categorized biologically as the sympathetic parts of the autonomic nervous system related to fight-or-flight responses, are also aspects of this dynamic polysensory system. While these are all biologically based stimuli and responses, the sensorium as a whole system requires cultural mediation of such when experienced. Consider the affective, sensory responses that may generate or be generated by both passion and disgust or love and hate under differing contexts of experience. Responses may include a combination of increased circulatory activity, flushing, perspiration, agitation, rapid breathing, and goosebumps. The individual and social effects of the experience are context-bound as interpretation of these bodily responses is based on an individual's biography of previous experiences and culture-specific knowledge. The sensing person interprets these autonomous responses, translating them into emotion, knowledge, and subjectivity within the limits of their specific, culturally determined categories; what fails to be explained or categorized is one facet of what Massumi (2002) calls the excesses of affect.

Making polysense of the world: affect, emotion, memory, heritage

There exist several definitions for, theorizations of, and approaches to affect, all of which share some commonalities as well as differences.[4] Here I focus on Massumi's (2002) approach to affect, but in particular only as it can be linked to understanding the experiential aspects of human being-ness in the world through numerous modes of interpersonal interaction and interaction with the tangible and intangible culture of our worlds. Massumi (2002) proposes that we take very seriously the importance of sensation in our encounters in and dynamic sense-making of the world, but only if we concede that our explanations will always be incomplete and shifting due to the slippery natures of both affect and the sensory. Affect here is our ever-changing capacity or potential for a range of sensory responses through interactions with others as well as landscapes, the built environment or other material culture and the intangibles of memorywork. These sensory responses may become feelings and emotions and perhaps produce tangible effects or actions in the world. Affect is that which is experienced through the body and is non-conscious (as a capacity for being in the world) yet is not ever fully separable, in a Deleuzian sense rather than a Cartesian sense, from the cognitive in how we make sense of the changing and multiple worlds in which we live over space and time. Sensory experiences may become feeling through an individual's assessment and cultural categorization of a set of sensations; feeling may then be displayed as emotion; and emotion may further be propelled into forms of knowledge and social actions that change over time with every new experience—that which fails interpretation and categorization is the excess of affect. It is this grounding of feelings and emotions in sensorial, embodied experiences that allows for a theorization of the affective and polysensory as necessary to the construction of memory through physical, emotional, and narrative responses and then, by extension, heritage.

A polysensual approach thus centers on the dynamic relationship between the senses, feeling, emotion, cognition, and memory as continually in process. It also requires attention to culturally specific epistemological models for the senses and sensory experiences as well as differences between bodies in terms of affective capacities, both being phenomena not interrogated by Massumi. Indeed, the latter phenomenon is at the center of constructive critiques of affect theories such as those of Massumi (2002) and Thrift (2008). These critiques consider such theorizations to ignore how "affective registers have to be understood within the context of power geometries" of our lived worlds which results in the "political fact of different bodies having different affective capacities" (Tolia-Kelly, 2006, p. 213). As Tolia-Kelly argues, "racialized, gendered and sexualized markedness," as well as bodies marked by historical violence, "magnetize various capacities for being affected" whereby the sensory and the emotional is differentially experienced due to social position disparities in the world that are shaped by power, politics, and history (2006, p. 215). Likewise, critiques of Thrift's (2008) "transhuman" framework that replaces emotion and feeling with affect center on a feminist reading of the transhuman experience as fundamentally masculinist, associating emotion with the feminine "touchy-feely," thus implying that a masculine rationality underpins affect (Thien, 2005, pp. 450–1). In contrast, a polysensory approach to the politics of affect, memory, and heritage as proposed here considers dimensions of social difference and affective capacities and does not attempt to transcend or replace emotion as an intrinsic part of the effects of affective experiences.

Consideration of the complexities of affect and the sensory is crucial to my ethnographic work addressing the politics of power, human activity, interpretation, emotion, and memory-making related to memorial sites as part of the ongoing social construction of such places through visit experiences and post-visit memorywork. Commemorative museums, monumental architecture, and landscapes represent the dark side of human heritage—the negative, brutal, and violent side of humanity. This has been called by many names, including "difficult heritage" (Logan and Reeves, 2009; Macdonald, 2009), "dissonant heritage" (Tunbridge and Ashworth, 1996), and a "heritage that hurts" (Sather-Wagstaff, 2011; Schofield et al., 2002; Uzzell and Ballantyne, 2008). Such dark heritage is grounded in "difficult knowledge" (Lehrer et al., 2011) produced by heritage institutions and socially constructed through embodied encounters with landscapes and artifacts, producing dynamic memories and histories (both individual and collective, vernacular and official) whose meanings are often highly contested and which may change over time and in space. A key element to the affective power of such institutions is their potential to intentionally evoke a range of powerful emotions and memories, most notably through affective, polysensory modes of encounter with difficult artifacts; material culture in a diverse array of forms on display that, in lived experience, do elicit sensory engagement beyond just that of the visual. For institutions to take such a bodily, material-based "approach to memory speaks to memory as both meta-sensory capacity and as a sense organ in-its-self"; thus memory "as a distinct meta-sense transports, bridges and crosses all the other senses" (Seremetakis, 1994, p. 9).

Likewise, in challenging notions that "authentic" historical consciousness (or any knowledge) is purely cognitive, Landsberg also proposes that we seriously consider the basic senses—vision, hearing, touch, movement, smell, taste—and the various emotional states of being produced through such as critical components in the ways that we negotiate and construct our knowledge of the past. By embedding the cognitive in a dialogic interface with affect, sensuousness, and the mimetic (Landsberg, 2004), sensory, experiential modes of meaning and memory-making have numerous effects, including the construction and reconstruction of individual and collective cultural and historical consciousness of the past, one aspect to heritage. Landsberg (2004, p. 2) thus proposes a powerful and very useful notion of "prosthetic memory," a form of sensorially based memory that "emerges at the interface between a person and a historical narrative about the past" and specifically, I add, their encounters with "difficult objects" in memorial landscape or at museums. Prosthetic memory is that which is "experienced with one's own body by means of a wide range of cultural technologies and as such becomes part of one's personal archive of experience, informing not only one's subjectivity but one's relationship to the present and future" (Landsberg, 2004, p. 26).

Prosthetic memory is not a first-hand witnessing to or participation in an event and its aftermath as it unfolds immediately in time and space, but a form of second-hand witnessing to institutionally mediated representations of the past. In the case of my research, such second-hand witnessing may include visual and textual media forms (television, print media, photographs, or film/video), hearing oral narratives of events or other audioscapes, or, as I focus on here, encounters with various types of material culture in museums and landscapes. Witnessing in and of itself, be it first- or second-hand, implicitly requires bodily, sensory engagement in some form or forms (even imagined), from seeing, touching, and hearing to movement through space, tasting, and smelling. For example, one may look at a photograph and imagine one's self (or someone beloved) in the event visually documented, potentially evoking an empathic, deeply—even bodily—felt emotional and affective response on some register. This may generate what Hirsch (1999) identifies as heteropathic memory, which, like prosthetic memory, is a means complicating the tidy, authorized, and official narratives about difficult visual and material artifacts that are a part of creating knowledge through affective, polysensory experiences. Through such various experiences, one composes multiple strata of memory, from the first-hand experience of visiting a site and prosthetic, heteropathic memories of past events as represented at the site to future layers of memory through recollection and narration in changing subjective contexts.

The polysensuality of difficult heritage

Works on difficult heritage that center on individuals' experiences at sites, specifically their encounters with difficult objects, range from those that only implicitly or minimally acknowledge the sensory to those that focus tightly on embodied, sensory experiences and responses. Representing the former, Ballantyne's (2003) work at the District Six Museum includes only one mention of sensory, emotional responses, something unexpected given the physically brutal history of Apartheid. This regarded a memory cloth upon which visitors leave written comments; Ballantyne noted that it is common that viewers are "moved to tears as they confront words [on the cloth] that hint at the real impact and hurt of the [forced] removals" of black District Six residents (2003, pp. 281–2). In contrast, Knudsen (2011) examines three different intertextual second-hand, sensorially based witnessing modes (testimonial, landscape, and kinetic immersion) experienced by visitors at three different difficult sites. At Auschwitz-Birkenau, the experience of going up into a guard tower produced responses "expressed with some degree of horror by the tourists," evoked by the polysensory, heteropathic experience of having touched the same handrail once touched by Nazi guards in the past (Knudsen, 2011, p. 65).

For nearly a decade and a half I have engaged in research with visitors who have been to similar types of sites, specifically those memorializing victims of violence, terrorism, and genocide, including (but not limited to) the Oklahoma City National Memorial, various September 11 exhibits, and the United States Holocaust Memorial Museum. These and other institutions that deal with violent and tragic episodes in the human experience such as slavery, genocide, or war have been called "museum(s) of conscience" (Kirshenblatt-Gimblett, 2000; Ballantyne, 2003, p. 291) or, more commonly by such museums themselves, institutions of conscience. These institutions are tasked with the presentation of what Lehrer, Milton, and Patterson (2011) define as "difficult knowledge": knowledge that is traumatic, unsettling, challenging, uncomfortable to apprehend, and a means to generate social conscience and action in the present and future. Through my work at these sites I have focused on understanding post-visit knowledge production and the reverberating sociocultural effects of encounters with the material aftermath of tragedy as a means to critically think through the value of these sites as institutions for social conscience and potential social change. All of my research sites of interest were constructed in the last quarter of the twentieth century or later and, in contrast to monument and museum constructions of the more distant past, they favor design and content that allows for "bodily visitor experiences that are sensory and emotional rather than visual and impassive; interpretive strategies that utilize private, subjective testimony over official historical narratives" (Williams, 2007, p. 3).

However, memorial landscapes are indeed still designed with intentional, specific symbolic elements and museum exhibits are forms of authoritative curatorial discourses, yet these are, in the end, subject to visitor interpretation through acts of embodied encounter. And such interpretations, as well as the sensory experience and the capacity for affective encounters, are indeed mediated, even constrained, by individuals' subjectivities (Lorimer, 2008). Yet how can we understand individuals' slippery senses and their affective capacities in practice? While both the senses and affect may be argued to be pre-linguistic or extra-discursive, individuals' narrations of experiences across time and space can begin to reveal the power and politics (both ethical and social) of polysensory experiences far beyond the original experiences at memorial heritage sites. This requires seeking out traces of the affective through a combination of careful visual observation and application of expanded discourse analysis methods.

In terms of analytical methods, discursive analysis of both site contents (landscapes, artifacts and media) and visitor narratives has become central in my approach to the polysensuality of difficult heritage and its social effects. In practice this requires, at the most basic, analytical attention to how individuals attempt to narrate the sensorium of their experiences. This includes how they felt emotionally and physically during encounters, what sense they made of their experience (or not), what memories they carry forward, what stimuli precipitate the explicit recall of those memories, and how they feel during the act of narration. This approach necessitates implementing a form of critical discourse analysis that attends to the minutiae of the narrative as well as the non-spoken performances and responses that accompany them, all in the context of the power and agency of objects, their display, and the institutional discourse of exhibits. Such an analysis is grounded in spoken discourse, attending to how visitors describe the sensorial aspects to their encounters and emotional responses, both then and in the present, to the experience as well as their descriptions of self/subject positions. Added to this are analyses of kinesthetic and paralinguistic aspects accompanying spoken narratives such as hand gestures, posture and other bodily positioning, facial expressions, and narrative silences. This extends even further to what might otherwise be categorized as extra-linguistic but yet is quite communicative—autonomic nervous system responses such as flushing or shaking.

As acts of discourse, the performance of both spoken narratives and silent reverie precipitates sensory responses. Narration and reverie can both produce autonomic bodily responses such as perspiration, flushing, rapid respiration, nausea, gooseflesh, or chills. These characteristics and more may be found in nearly every narrative regarding intensely polysensory experiences and one's power of observation is critical to noting as much as possible in the performance of narration including and beyond that of the spoken word. Post-visit conversations with visitors have thus been a key part of my ongoing work and the narratives produced in these conversations have often been full of rich and imaginative descriptions of their sensory experiences at the sites—descriptions that likely capture only a minute part of their entire registers of feeling. Because artifacts can and do engender particularly powerful, affective experiences, I focus here on material culture on display. I find that many visitors to institutions for conscience are perhaps not best characterized, as Sturken argues, by an "innocent pose and distanced position" (2007, p. 10), but rather as agents engaging in a complicated polysensory dialog with the difficult objects they encounter as well as the landscapes and built environments that contain such. Returned visitors with whom I spoke frequently employed direct descriptions of sensations in their recollection narratives—"My hands went cold and numb" or "I felt nauseous, dizzy." They also employ sensory metaphors such as "I was frozen with anger" and "felt shot through the heart" as ways of making sense of otherwise preconscious sensory experiences of feeling literal or figurative paralysis, disequilibrium, sadness, and pain. Perhaps one of the most intriguing aspects to these narratives have been recollections that spoke also to imagined sensorial stimuli that led to very real physical and emotional responses as a part of the affective on-site experience.

The United States Holocaust Memorial Museum's exhibits are, for many visitors, extremely powerful on a polysensual level. In my work, many of the key memory narratives from past visitors to the Museum concerned the railcar located in the permanent exhibit, which is positioned so that one may physically walk through the artifact. They spared no words in expressing to me how physically encountering the railcar made them feel. Some thought they felt it moving as though they were on their way to the camps, evoking responses of anxiety, fear, or panic. Others recalled imagining the car full of people so realistically that they believed they heard breathing, and some were utterly convinced that they could hear the voices of those who had been transported to their deaths. Our conversations took place sometimes many years after their visits, yet the acts of recalling and narrating their experiences evidenced the power that these acts have as precipitates for a range of bodily responses. When we spoke, many were rendered wordless every now and then, pausing, sighing. Some were shaking or physically agitated, fists clenched, brows furrowed, flushing, on the verge of tears or crying. Those who often appeared the most deeply affected were visitors with some direct link to or a deep historical knowledge of the Holocaust, and thus one's historical and personal biography within diverse webs of power does indeed shape capacities for affect as well as interpretation of affective experiences.

Difficult objects are also a crucial part of the displays at the Oklahoma City National Memorial to the 1995 bombing of the Murrah Federal Building. In one section representing the aftermath of the bombing, victims are represented by watches, keys, and damaged coffee mugs, all carefully grouped by type in individual display cases. These are mundane, everyday and familiar objects, items that any one of us might carry with us, wear, or use as a part of our daily routines. They are not unlike the everyday material culture—shoes or other clothing, suitcases, books, identification cards, family photographs, toys, wallets, purses, name tags, and coins—displayed in other memorial exhibits that, as powerful artifacts, are intended to represent the victims of unspeakable violence. They do so in powerfully sensory, affective, and subjective ways potentially unforeseen by the institutions where they are exhibited. For example, I once spoke at length with an elderly man who had visited this museum while spending a holiday with family in Oklahoma. When I asked what impacted him the most in the museum, he recalled the display case of keys. He began to tell me how his attention was caught by a very large round key ring nestled among the numerous smaller key chains and many loose keys in the case. He said that seeing this particular object precipitated memories of his days as a school janitor: the weight and noise of a quite similar (but even larger) key ring hung on his belt loop day in and day out, banging on his hip and jingling with every move as he worked.

Both during his encounter with the display and again during our conversation, he wondered about what these keys indexed: Had the person to whom the keys on display belonged to felt those same sensations? Did they survive? What doors did the keys lock and unlock? He told me that he had felt as though the weight was hanging from his belt at the very moment he was there in the exhibit; while he spoke with me over a year later, he kept patting his right hip as if he was searching for his old key ring. The effects of his polysensual experience were highly heteropathic as he used his own biography of sensory experience to continually index what someone else—the owner of the key ring—might have felt. He told me that nearly every time he remembers his encounter with the key display, sometimes triggered by the sound of jangling keys, he feels "something," an empathic yet, to him, unidentifiable "something." This is, perhaps, what Massumi understands to be the excess of affect (and Thrift the "non-representational"): a bleeding and porous territory that is yet, if ever, to be transformed into individual, identifiable, and discrete cognitive knowledge that can be narrated into complete representation.

Railcars, shoes, keys, cups, and suitcases are objects presumed, as Kidd proposes, to be a "taken-for-granted means of 'accessing' the past," yet in practice their utility requires a "recognition of the collision between their role as perceived objective manifestations of knowledge and as 'things' whose use value and/or tangibility is being demonstrated: they are 'alive'" (2011, p. 29). They are, as Byrne so poetically describes, the material culture of love, "the kind of objects that affect sticks to" such as a "particular shirt worn by the lover ... a sofa, a car, the key to an apartment and the door handle of the apartment" (2013, p. 599). These objects "extend into us as we extend into them" and the agency they have "in relation to us has been acquired through a specific history" (Byrne, 2013, pp. 600–1), including both the histories of the places in which they originate or come to be displayed and those that inform the subjectivities of those who engage with the objects. They become difficult, "sticky" heritage objects due not solely to their status as carefully collected and curated artifacts of violence and tragedy but through the polysensory, affective encounters of visitors with the agency of their display. Such objects are indeed interpreted in the highly subjective contexts of individual visitors' polysensory experiences in the museum. They also serve, paradoxically, to represent a new canon of authoritative objects for contemporary museum exhibits, their affective presence expected, normative, and intentionally meant for highly individualized interpretation.

Conclusion: implications of the polysensory for critical heritage studies

Polysensory experiences are far more than the physical consumption, if you will, of heritage presumably pre-made and pre-interpreted; they are part and parcel of constructing and performing heritage over time and space. To thus engage in a polysensory approach with attention to the politics of affect in the construction of heritage, we must always ground such in more than simple observation or textual analysis—it requires critical interrogations into and sensitive analyses of individuals' actual polysensory, affective experiences in-the-world, even if we can only ever make partial sense of such. Deep and constantly shifting objects of inquiry such as the relationships between affect, the senses, memory, and effects also require a geographical and temporal re-siting of what constitutes our fields of study, not so much in terms of disciplinary fields but for our fieldwork. Narratives—conversation and storytelling—that may take place many months and years post-encounter are valuable for teasing out the unfolding meanings of polysensory processes of knowledge-making and the potentially transformative, always-in-process politics of affect. In a way, attention to the polysensory is also an interesting means to understand how persistently, particularly in the Western world, we humans try to literally "mind" the body, valiantly attempting to force a sense-making out of bodily responses that may resist neat cultural categorizations or have no categorization schema. Yet as humans we fail at doing this in so many ways. This failure also represents the excess of affect that extends over time, perhaps eventually becoming knowledge or persisting in resisting any sense-making. Nevertheless, the senses do ground affective knowledge-making and have real social effects, making the politics of the polysensory quite powerful for opening up possibilities for new political imaginaries and emergent realities.

In the specific case of difficult heritage, it is not simply about "never forgetting" atrocities and traumas of the past, but the polysensorial recollection (and re-remembering over time) of the intensity of one's experiences and transforming that potential into social effects, that may change the world. However, there is a caveat to this, as Thrift (2008) points out: Methods and techniques for the intentional manipulation of affect implemented in multiple venues (including media) may be informed by intentions to change the world but that change may not necessarily be for the better, particularly in terms of social justice and other highly political or politicized consequences. This is a valuable observation, informing us not only of the promise of the power of polysensual experiences to evoke actions in the world but also of the perils of such. Such an observation leads us back into thinking about heritage from a critical perspective, particularly difficult heritage, as it relates to museum practice and visitors' "second-hand witnessing" in museums. Considering the different affective capacities of individuals complicates already existing questions regarding what material, visual, or auditory media to exhibit—what may be potentially too traumatic? What might evoke affective responses that result in effects opposite to those intended by curators or site designers?

Ongoing discussions on the collections acquisition process for the future Smithsonian National Museum of African American History and Culture, scheduled to open in 2016, provide a concluding example here, one that points to the importance of discursive attention to polysensory experiences in-the-world. In 2009, National Public Radio began running a series of discussions with museum director Lonnie Bunch on the acquisition of items for the museum's permanent collection. One story featured several "positive heritage" objects, including jazzman Louis Armstrong's trumpet, along with a highly sobering item: a set of slave shackles from the eighteenth century. While these shackles are clearly instruments of denigration and forced bondage, Bunch noted that he takes every possible opportunity to touch them, as this embodied, sensory experience is, for him, "the closest I come to understanding my slave ancestors" (Stamberg, 2009). He accesses this past through an affective, sensory, emotional, and imaginative encounter, evoking the "immanence of ancestors" (Byrne, 2009, p. 249) and their violent pasts as they resonate in the present. For Bunch, as well as future museum visitors, these shackles not only represent one of the darkest eras in human history but are indeed a potentially powerful artifact for constructing a sensory, prosthetic memory through the museum encounter and, hopefully, engendering or reinforcing anti-racist subjectivities. Yet this is not to say that all encounters will be equivalent or have this particular effect, given that personal biographies and social subjectivities always bear on the potential intensity of the sensory experience and its reverberating effects.

In November 2010, the program's discussion centered on two other distinctly different objects that have similar affective capacities. The first was a powerful letter written by abolitionist John Brown to his wife while staying at the home of Frederick Douglass, an antislavery proponent and social reformer. The other, quite difficult, artifact was a Ku Klux Klan (KKK) banner from the 1920s, representing a specific Klan unit in Indiana and evidencing the broad national scope of the KKK in the United States during the early part of the twentieth century, well post-emancipation. During the show, Bunch was asked by the host, Guy Raz, "could you celebrate African-American history and culture at the museum without showing things like this?" (National Public Radio, 2010). Bunch replied:

> I don't think so. It's a real issue for us [the museum]. Do you put out for display things that are hateful, that are painful, that in some ways there are still people who see this as a positive? But I think that part of a museum's job is to tell the unvarnished truth. And you can't understand black resiliency or achievement or concern about race without understanding how powerful the Klan was and how important it's been as a force for almost a hundred years.
> (National Public Radio, 2010)

We have yet to know the polysensory, affective power this banner and other difficult objects may actually have for visitor understanding of an African-American heritage that incorporates the deep trauma of persecution and violence with histories of resiliency and resistance and unique, extraordinary social, cultural, and political achievements. This will require a commitment to engaging with the visitors to the museum after the banner unfurls, should it be displayed, both on-site and well after their return to numerous communities of belonging.

This kind of engagement with heritage site visitors requires careful attention to the polysensual dimensions of the discourses they craft and perform about their encounters, memories, and memorywork. It is indeed a potentially fruitful means for tracing affect as it shapes the slippery contours of the highly complex social and political machinery of memory and heritage-making. In general, this approach may well be one productive "route from the non-representational to the representational and back again," one that looks beyond heritage objects and their institutional authority to incorporate the performative, affective relationships these objects have with lived experience and thus individual subjectivities through acts of encounter (Waterton and Watson, 2013, pp. 557–8). For critical heritage studies, such a practice allows for a richer yet more complicated understanding of difficult heritage, the immanent power of difficult objects, the polysensory construction of a heritage that hurts, and, as Landsberg (2004) proposes, the potential for producing both empathy and social transformation. The politics of affect, the senses, and emotion are truly entangled in the construction of memory and knowledge, both of which are truly "critical in the politics of identification" and thus all "contemporary global politics of migration, race, and heritage" (Tolia-Kelly, 2004, p. 323).

Notes

1 Knowledge about the preparation of many heritage foods is often passed on through physical demonstration. In thinking about "making it right," learned bodily techniques are key, and these may include how to pat *masa* into the perfect thickness and shape for tamales or twirl thinly rolled *lefse* dough onto a flat wooden stick in preparation for laying on the griddle. I take this from my own experience of making heritage food which, in my case, involves knishes and pierogies. My grandmother had a very specific way of rolling, cutting, and filling the dough, followed by a pinching process that magically closed one side of the knish or sealed the half-moon pierogie. Despite the number of times we made these together, I never mastered her magnificent pinching technique and thus depend upon other means (even mechanically with a press). Every time I make these foods I remember her graceful ease when doing so and my repeated failures to replicate her movements.
2 Another example of different prefixes to infer different meanings comes from textile production in the form of resist dying. Batik, a form of resist dying, is considered a polycolor technique given that the wax used in the design to block dye cracks during the processing allows for an element of color seepage and layering. This is in contrast to multicolored or particolored techniques which produce clear delineations of color.
3 McCormack (2007) is one of the few scholars who, by focusing on the relations between the molecular and affective, is critically and productively engaging in provocative thinking about the relationships between the physical body, emotion, and affect in a manner that is not biologically reductive.
4 Thrift (2008) provides the most comprehensive and foundational overview of theories of and approaches to affect, particularly in Chapter 8.

Bibliography

Bærenholdt, J.O., Haldrup, M., Larsen, J. and Urry, J. 2004. *Performing tourist places*. Aldershot: Ashgate.
Bagnall, G. 2003. Performance and performativity at heritage sites. *Museum and Society*, 1(2): 87–103.
Ballantyne, R. 2003. Interpreting Apartheid: visitors' perceptions of the District Six Museum. *Curator*, 46(3): 279–92.
Byrne, D. 2009. A critique of unfeeling heritage, in L. Smith and N. Akagawa (eds) *Intangible heritage* (pp. 229–52). London and New York: Routledge.
Byrne, D. 2013. Love and loss in the 1960s. *International Journal of Heritage Studies*, 19(6): 596–609.
Classen, C. 1993. *Worlds of sense: exploring the senses in history and across cultures*. London and New York: Routledge.
Classen, C. 2012. *The deepest sense: a cultural history of touch*. Champaign, IL: University of Illinois Press.
Classen, C., Howes, D. and Synnott, A. 1994. *Aroma: the cultural history of smell*. London and New York: Routledge.
Crang, M. and Tolia-Kelly, D.P. 2010. Nation, race and affect: senses and sensibilities at national heritage sites. *Environment and Planning A*, 42(10): 2315–31.
Crang, M. and Travlou, P.S. 2001. The city and topologies of memory. *Environment and Planning D: Society and Space*, 19: 161–77.

Crouch, D. 2010. The perpetual performance and emergence of heritage, in S. Watson and E. Waterton (eds) *Culture, heritage and representation: perspectives on visuality and the past* (pp. 57–71). Farnham: Ashgate.

Crouch, D. 2012. Meaning, encounter and performativity: threads and moments of space-times in doing tourism, in L. Smith, E. Waterton and S. Watson (eds) *The cultural moment in tourism* (pp. 19–37). London and New York: Routledge.

Davidson, J, Bondi, L. and Smith, M. 2005. *Emotional geographies*. Burlington: Ashgate.

Dicks, B. 2003. *Culture on display: the production of contemporary visitability*. Maidenhead: Open University Press.

Douglas, M. 1966. *Purity and danger*. New York: Praeger.

Hirsch, M. 1999. Projected memory: Holocaust photographs in personal and public fantasy, in M. Bal, J. Crewe and L. Spitzer (eds) *Acts of memory: cultural recall in the present* (pp. 3–23). Hanover and London: University Press of New England.

Howes, D. (ed), 2005. *Empire of the senses: the sensual culture reader*. Oxford and New York: Berg.

Howes, D. 2008. Can these dry bones live? An anthropological approach to the history of the senses. *The Journal of American History*, 95(2): 442–51.

Jackson, A. and Kidd, J. (eds), 2011. *Performing heritage: research, practice and innovation in museum theatre and live interpretation*. Manchester: Manchester University Press.

Kidd, J. 2011. Performing the knowing archive: heritage performance and authenticity. *International Journal of Heritage Studies*, 17(1): 22–35.

Kirshenblatt-Gimblett, B. 2000. Keynote address: the museum as catalyst. *Museums 2000: Confirmation or Challenge*. Vadstena, Sweden, Sept. 29. www.nyu.edu/classes/bkg/web/ (accessed January 2013).

Knudsen, B.T. 2011. Thanatourism: witnessing difficult pasts. *Tourist Studies*, 11(1): 55–72.

Knudsen, B.T. and Waade, A.M. (eds), 2010. *Re-investing authenticity: tourism, place and emotions*. Bristol: Channel View.

Kwint, M. 1999. Introduction: the physical past, in M. Kwint, C. Brewer and J. Aynsley (eds) *Material memories: design and evocation* (pp. 1–16). Oxford and New York: Berg.

Landsberg, A. 2004. *Prosthetic memory: the transformation of American remembrance in the age of mass culture*. New York: Columbia University Press.

Lehrer, E., Milton, C.E. and Patterson, M.E. (eds), 2011. *Curating difficult knowledge: violent pasts in public places*. New York: Palgrave Macmillan.

Lock, M. 1993. Cultivating the body: anthropology and epistemologies of bodily practice and knowledge. *Annual Review of Anthropology*, 22: 133–55.

Logan, W. and Reeves, K. (eds), 2009. *Places of pain and shame: dealing with 'difficult heritage'*. London and New York: Routledge.

Lorimer, H. 2008. Cultural geography: non-representational conditions and concerns. *Progress in Human Geography*, 32: 551–9.

Macdonald, S. 2009. *Difficult heritage: negotiating the Nazi past in Nuremberg and beyond*. London and New York: Routledge.

Macdonald, S. 2013. *Memorylands: heritage and identity in Europe today*. London and New York: Routledge.

Massumi, B. 2002. *Parables for the virtual: movement, affect, sensation*. Durham, NC and London: Duke University Press.

McCormack, D.P. 2007. Molecular affects in human geographies. *Environment and Planning A*, 39: 359–77.
National Public Radio, 2010. Museum's latest find: a love letter from John Brown, *All Things Considered*, Nov. 7. www.npr.org/2010/11/07/131145982/museums-latest-find-love-letter-from-john-brown (accessed June 2012).
Picard, D. and Robinson, M. (eds), 2012. *Emotions in motion: tourism, affect and transformation*. Farnham: Ashgate.
Sather-Wagstaff, J. 2011. *Heritage that hurts: tourists in the memoryscapes of September 11*. Walnut Creek: Left Coast Press.
Schofield, J., Johnson, W.G. and Beck, C.M. (eds), 2002. *Matériel culture: the archaeology of twentieth-century conflict*. London: Routledge.
Seremetakis, C.N. 1994. The memory of the senses, part I: marks of the transitory, in C.N. Serematakis (ed) *The senses still: perception and memory as material culture in modernity* (pp. 1–18). Chicago: University of Chicago Press.
Smith, L. 2006. *Uses of heritage*. London and New York: Routledge.
Staiff, R., Bushell R. and Watson, S. (eds), 2013 *Heritage and tourism: place, encounter, engagement*. London: Routledge.
Stamberg, S. 2009. Assembling artifacts of African-American history, *NPR Morning Edition*, Jan. 14. www.npr.org/templates/story/story.php?storyId=99259871 (accessed June 2012).
Stoller, P. 1989. *The taste of ethnographic things: the senses in anthropology*. Philadelphia: University of Pennsylvania Press.
Stoller, P. 1997. *Sensuous scholarship*. Philadelphia: University of Pennsylvania Press.
Sturken, M. 2007. *Tourists of history: memory, kitsch, and consumerism from Oklahoma City to Ground Zero*. Durham, NC and London: Duke University Press.
Thien, D. 2005. After or beyond feeling? A consideration of affect and emotion in geography. *Area*, 37(4): 450–6.
Thrift, N. 2008. *Non-representational theory: space/politics/affect*. London and New York: Routledge.
Tolia-Kelly, D.P. 2004. Locating processes of identification: studying the precipitates of re-memory through artefacts in the British Asian home. *Transactions of the Institute of British Geographers*, New Series, 29(3): 314–29.
Tolia-Kelly, D.P. 2006. Affect – an ethnocentric encounter? Exploring the 'universalist' imperative of emotional/affectual geographies. *Area*, 38(2): 213–17.
Tunbridge, J.E. and Ashworth, G.J. 1996. *Dissonant heritage: the management of the past as a resource in conflict*. Chichester: John Wiley and Sons.
Uzzell, D. and Ballantyne, R. 2008. Heritage that hurts: interpretation in a postmodern world, in G. Fairclough, R. Harrison, J.H. Jameson Jnr and J. Schofield (eds) *The heritage reader* (pp. 152–70). London: Routledge.
Waterton, E. and Watson, S. 2013. Framing theory: towards a critical imagination in heritage studies. *International Journal of Heritage Studies*, 19(6): 546–61.
Williams, P. 2007. *Memorial museums: the global rush to commemorate atrocities*. Oxford and New York: Berg.

Part I
Memories

2 Race and affect at the museum

The museum as a *theatre of pain*

Divya P. Tolia-Kelly

Museum displays effect and are forged through affective politics. When world cultures were framed in the nineteenth century, often underlying them were accounts of taxonomies of race where European bodies, affective capacities and sensibilities were defined as the best that humanity could achieve and offer. *Theatres of Pain* was originally a performance, an exploration of post-imperial affective politics at the twenty-first century museum, performed by Rosanna Raymond and myself at the inaugural conference of the Association of Critical Heritage Studies, Gothenburg, in 2012. Using this account, the exhibition space of the national museum is seen here to be experienced as a *theatre of pain*. The museum acts as a site of materializing the pain of epistemic violence, the rupture of genocide and the deadening of artefacts. The chapter examines the embodied experience of encountering these galleries and the effect of Tony Bennett's claim (2006) that the art museum becomes a mausoleum for the European eye, but which petrifies living cultures. The museum space from critical postcolonial perspectives is considered as a *theatre of pain*.

The central aim of this chapter is to consider 'affect' and emotion at the museum space. It outlines the value of thinking through theories of affect and emotion to understand the cultural politics and geographies of the museum. What contribution can affect and emotion make to thinking through heritage experience and, more importantly, what affects figure and shape specific encounters? Writers within the social sciences (Ahmed, 2004a, 2004b; Hemmings, 2005; Thien, 2005; Tolia-Kelly, 2006) have critiqued the occlusion of power within the conceptualizations of theories of affect. They have also argued that any 'universalist' account of experience risks ethnocentrism and homocentricism by default. Therefore at the heart of this research are critical perspectives from postcolonial theory and geopolitical critiques of exhibiting 'other' cultures at international art museums such as the British Museum. Work by Araeen (1987) and Sylvester (2009) has informed the research and framed the consideration of affect and emotion. The research poses the problematics of experiencing the gallery space by people who are from communities exhibited within the cabinets and galleries using critical theoretical frameworks (see also Golding, 2009). In particular, the research presents the experience of Māori visitors in their encounter with the exhibitions in the British Museum. Here, I propose thinking of the encounter with exhibits at the British

Museum beyond the usual registers associated with heritage sites such as awe, wonder, inspiration and the sublime in order to articulate how the museum space can be and is experienced as a *theatre of pain*. This conceptualization is inspired by historian Raphael Samuel's (1994) account of domestic collections as 'theatres of memory', of which museums are an example; they serve as formal sanctioned spaces of memory, but which are examined here as locations of pain and suffering. My argument is that cultural heritage is articulated through curatorial frameworks and logics that enable affective responses, circulations and capacities to disturb, give pain and sadden. For Māori visitors, including those from the London Māori community, Ngāti-Rānana,[1] pain, alienation and grief are the affective and emotional registers through which the Māori galleries are encountered. There are layers of this experience that are articulated in this chapter, including the geopolitics of the exhibition space as well as the 'voice' through which Māori culture is narrated. The Māori experience expresses discordancy between how Māori 'feel' their cultural heritage to be and the continuing resonances of imperial ways of framing, seeing and exhibiting 'other' cultures.

The chapter uses a postcolonial theoretical perspective on the embodied experience of the museum space, and outlines the ways in which the affective and emotional space of the museum is experienced by 'other' racialized communities. The visceral encounter within a spatial realm (in this case, the museum space) is considered alongside recent non-representational conceptualizations of affective atmospheres (Adey, 2008; Anderson, 2009; Kraftl and Adey, 2008; McCormack, 2008; Stephens, 2015), to extend thinking about the politics of 'other' cultures and race thinking. While thinking through these affective engagements in relation to the museum space (see also Waterton and Dittmer, 2014), I hope to outline the critical need to think cross-, trans-, intra- and inter-culturally. The chapter develops this argument through collaborative research conducted with the internationally renowned Polynesian artist, curator and writer Rosanna Raymond and Ngāti-Rānana at the British Museum. We consider the experience of the gallery, focusing on the Polynesian exhibition spaces (including Māori and Oceanic galleries) therein. Affect and emotion, overall, are examined through the research as valuable to the project of doing postcolonial readings of the museums exhibitions of Polynesian culture, but also of critiquing the remnants of imperial ways of seeing 'other worlds, peoples and places' (Said, 1979, 1993). They are however not limited to 'readings', but serve as a platform from which we can develop more inclusive post-imperial museum curatorship and stewardship that accommodate different spatial and temporal frameworks for heritage and history (see also Bohrer, 1994).

Affect at the museum

Scholars have engaged with affect and emotion at the museum in a number of ways. Thinking about affective presence through technologies of display goes beyond ethnographies that engage in postcolonial critiques (e.g. Boehner *et al.*, 2005; Coombes, 1994). Affective displays in museums may also have cumulative affects (see Dewan and Hackett, 2009) and are powerful in enabling critical pedagogy (Gregory and Witcomb, 2007; Witcomb, 2013). Alongside them are those thinking about the politics of affect in creating possibilities for new cosmopolitanisms (Schorch *et al.*, this volume) that challenge dominant accounts of cross-cultural encounters (Schorch, 2013, 2014). Affect works at various levels in the spaces of heritage, including thinking about the contribution of heritage spaces to producing and consolidating national identity and sensibilities at heritage sites (see Crang and Tolia-Kelly, 2010) or, indeed, garnering a particular atmosphere to enable engagement and understanding of a historical event such as an experience or logics of war (see Waterton and Dittmer, 2014; see also this volume). As Thrift (2004, p. 60) has argued, 'affect is a different kind of intelligence about the world ... emotions form a rich moral array through which and with which the world is thought and which can sense different things even though they cannot always be named'. It is thus important to reflect upon a space of heritage such as the museum as a space produced through affect and emotion. This account of the value of emotion and affect in these spaces is critical in acknowledging the power of affective circulations in shaping heritage encounters, producing alternative pedagogies, evoking counter-narratives and developing self-determined accounts of cultural heritage. There are layers of experience that are both emotional and affective, and it is important to discern between these two modes of thinking. Here, I will use *emotion* as an individualized response to the encounter or zone of contact, while *affect* is about the collectivities of emotional response that are not always identifiable, cognitive or, indeed, determinate. The affective, it is argued, is transpersonal, interpersonal and enables political events (Thrift, 2004) and responses, and contributes to geopolitics in the everyday (Pile, 2010). As such, McCormack (2008) and Bissell (2009) promote the idea of affective atmospheres as a way of conceptualizing a collective evocation and which are evoked in everyday life. This account of affective atmospheres is thus taken up within the social sciences to illustrate motivations for political action rooted in Marx ([1856] cited in Anderson 2009; see also Rovert C. Tucker, 1978), and translated into geopolitical economies (Thrift, 2004) and national sensibilities (Stephens, 2015).

Affective atmospheres are everywhere and inevitably are felt, and coalesce in the spaces of heritage, including landscapes, monuments and museums. It could be argued that heritage spaces are present as the consolidation of affective economies of the past. They are consolidated contours of how memory, affective charges and matter consolidate on the ground. Heritage spaces are material precipitates of affective memories at scales of nation and world. Materially, affective flows, responses and power can be inspired, can circulate and can shape the narratives, political meanings and life of heritage sites and spaces. Material, geopolitical sensibilities and the power of representation come into play when encountering museum spaces. It is also important to be mindful of the challenges made to the theorization of affect such that they include calls to avoid occluding power relations, geopolitics (Hemmings, 2005; Thien, 2005) and ethnocentrism (Tolia-Kelly, 2006). Affective atmospheres are embedded with memories that are figured and striated through geopolitical situations, national sensibilities and visceral relationships with the past. These fuel the affective circulations, relations and environments of heritage. They form the grammars of heritage spaces and landscapes and are simultaneously collective and contested. In this account, the articulation of affective atmospheres at museums and heritage sites is thus infused with differences and pluralities to which we are attuned and contribute. It is important that work on affective heritage should be mindful of contestations, authorized and occluded, that define the atmospheres in these spaces. Affective atmospheres are connected to place, space and site and as such 'are the shared ground from which subjective states and their attendant feelings and emotions emerge' (Anderson, 2009, p. 78). What is at stake here is the risk of thinking of experience, or collectivities of feeling, as emerging from a *singularized* 'shared' account of ground. At the museum, visitors are not of a single constituency, felt as occluded, 'outside' or indeed deliberately written out. It is inevitably not a singular plane, or indeed grounding in singular modes of feeling. Outlined below are ways in which museologists and heritage researchers have engaged with affect and emotion, thus situating the research. It is imperative to also remember that despite the recent affective turn in western academia, the affective atmosphere, as a framing logic of the museum space, is not novel for the majority population in the world, including colonized (Fanon, 1968), racialized (Hall, 1997a, b, 2005) and indeed oppressed people. Thus there are pluralities of orientation towards the logics of curatorship, display and narrative, based on the politics of identity, power and differences in sensibilities that come before, embedded in the encounter of embodied engagement.

Geopolitics at the museum

Baxandall (1991) has problematized museum representations by arguing that 'it is not possible to exhibit other cultures without putting a construction upon them' (p. 34). There is no value-free act of cultural representation. It is further complicated when W.E.B. Du Bois (1903) describes the problematic that many visitors and communities face when entering a space of 'national' or 'international' culture. Often racialized, postcolonial or indeed marginalized communities experience the heritage space in multiple ways or with a plurality to their orientation. A *doubleness* of sensibility or a striation of sensibilities, it could be argued, is at work; thus a presumed shared ground is fissured with co-constituencies and affectivities of a sensibility that is against the grain of expected receptions of museum narratives, politics and formats (see also Bennett, 1995, 2005). An appreciation of double-consciousness as it operates for racialized communities disrupts the idealized texture of 'shared ground', or indeed a notion of being part of collective constituencies that are easily felt and recognized. Constituencies of being part of a national sensibility sometimes are synthesized with those that *feel* part of the nation, or indeed at other moments are at an anti-thesis, coalescing with constituencies of the 'outsider', oppressed or unrecognized constituents of national citizenry.

Alienation from the space of the museum, and its dominant narrative, also creates affective responses that reflect emotions and affects that are compounded over time, and therefore 'affective atmospheres' do not necessarily do justice to the depth of subjugation, denial and violence experienced therein. This is simultaneously compounded by the very lack of voice, power or indeed righteousness of articulating one's own cultural story. This is what Spivak (1988) describes as the fall-out of alienation and imperial categorizations that resonate in museum accounts of 'other' cultures. This alienation is debilitating where there is a lack of self-knowledge and thus cathect (which is required to articulate and rectify the erasures of the palimpsests that have gone before). In terms of affective atmospheres at the museum, the lack of emotional/affective investment resulting from double-consciousness or indeed alienation challenges the usual articulations of 'other' cultures as sensual, feeling beings rather than intellectual, philosophical ones. The expected constellations of feelings that form the palate of response to the museum narrative implode with the articulation of the postcolonial politics of what it is to be racialized or 'other' in the space.

What is certain is that postcolonial affective encounters can disrupt homogenizing, occidentalist, imperial accounts of culture, and can help to co-produce museum spaces that are truly inclusive and anti-imperial in their taxonomies, narratives and hierarchies of culture represented within. Witcomb (2013, 2014) articulates in detail the value of affect in the museum encounter, in disrupting accounts of history that occlude Aboriginal, marginalized and 'other' voices, to provide political strategies to develop historical consciousness. Affect can be valuable in re-orientating the visitor, to embrace a critical engagement towards history that is conscientiously postcolonial, reinterpretive and plural without being imposed. Witcomb (2013) has argued that this critical historical consciousness can be developed beyond the usual nostalgia that is garnered within Australian national history narratives to include a feeling for counter-histories and memories (p. 255). For Witcomb, the essential point here is to activate affective registers to produce critical and inclusive sensibilities. Affect becomes part of new strategies of interpretation (p. 246). By eliciting feelings that translate into critical forms of thinking, thinking through 'affect' and 'emotion' can productively create the ground to address inequality and misrepresentation, and ultimately reconciliation between 'races', migrants and national citizens (p. 257). The feelings to be reconciled are initially located in feelings of being part of the British Empire for white Australians, and Indigenous people who do not see themselves as migrants. Two things happen in this account: Indigenous peoples are not so easily placed as 'other' to nation, and nostalgia is not embraced as an easy affect through which to co-opt white Australians to their 'pioneer' landscape. The possibilities for Australian museums are, in theory, better as they engage with 'indigenous peoples and their cultures and histories (as) part of the nation rather than as anthropological object' (p. 258). Witcomb's account outlines the problematics of revising representational frameworks in museums using affect to re-adjust accounts of the past and the lines of inclusion to national narratives and citizenry. The next section will look at affect as experienced by Māori at the British Museum.

Exhibiting Māori at the British Museum

The technologies of the museum have been the site of several critiques of the museum as a powerful space of discipline, as well as a tool of governmentality – a space through which citizens are *made* (Bennett, 2005). These critiques focus on the people flowing through as receivers and absorbers of ideas, values and 'ways of seeing' (Berger, 2008) other cultures, peoples and places. This section will focus on the museum display and cabinets, to highlight the work that the museum cabinet does to display Māori cultures. What emerges are the alignments of feelings that circulate as a result from Māori visitors, including Rosanna Raymond. This is distinct from superficial, immediate, sensory responses with objects in a proximal or a purely haptic register (Hetherington, 2003): these work at a different order.

The museum cabinet is the technology of an era of taxonomies and the display of exemplars of categories, types, genus and indeed 'races'. In the nineteenth century, there was an account of 'other' cultures as natural and separate; the cabinet was a technology through which difference, specificity and singular examples could be observed, and knowledge about them could be 'collected' through having observed them and 'ticking them off' a list. Representational practice was to give an object text and thus a biography which could be observed, understood and experienced as contact with that particular culture. The representation of the society or culture was thus deemed to have a place in the natural order of things in the world. Macdonald (1998) has argues that this naturalizing account of difference and identity is at the heart of the work of the museum case. The cabinet actively 'cases-in' an account of a culture which can be juxtaposed with another. Identities are essentialized and available to be possessed culturally, occularly and intellectually. By seeing them in the cabinet, they were at once knowable and positioned, for the power to gaze, deduce and categorize from a position of superiority. In this account of museum contact one never faced oneself in the cabinet: it was only ever 'others' that were culturally reified and objectified. Alpers (1991, p. 27) argues that the 'tendency to isolate something from its world, to offer it up for attentive looking and thus transform it into art like our own' is the *museum effect* that orchestrates a particular *way of seeing*. Artefacts are 'severed' from their original sites and situatedness and what is privileged is an enhanced sight of them. The visual supersedes the feel, texture and understanding in context. The reductive nature of a museum's way of seeing creates an art-object rather than positioning the artefact within grammars of everyday life. It is the curator that judges, defines through 'eye, taste and experience' (Alpers, 1991 p. 4) against a universal palate. What Alpers argues for is that to avoid misrepresentations museums need to enfranchise populations and give additional control to communities to shape representational practices that they have previously held to themselves. This is needed in addition to expanding the basis of 'expertise' and display design to include a variety of 'textures' and formats through which museums communicate the cultural meanings and values of other cultures.

To consider how these ways of seeing work, it is important to think about exhibitionary practices and communication of knowledge about Māori in the British Museum. Often a cultural representation focuses on a particular set of objects, so for Japanese collections it may be swords, for example. In the Māori galleries, jade is selected as a very important signifier of Māori heritage. The gaze onto the objects in the collections is examined further through the British Museum textbook *The Māori Collections of the British Musuem*. The front-cover image highlights a piece of jade with no signs of use, name or indeed temporal or spatial era. The object is without context or situation. It is removed from an organic account of its identity within a network of cultural values. This is often the case when engaging with a museum's way of seeing, but what happens when you embody the gaze that is curated for you to inhabit? When you look at an object, what do you see? In this example, the object, rather than having specific purpose, use, biography (Gell, 1998) or indeed an everyday life, becomes a representation of myth, reductive and supporting easily recognizable messages about the culture represented. The object becomes a metaphor, in this case for Māori society and culture and perhaps an account of Māori 'nation' or national culture. Furthermore, the object signifies the place of Māori in the universalizing aesthetic palate; it is seen as exemplifying the possibilities of Māori culture. In a post-imperial account the object articulates the Māori as embodying particular cultural capacities, grammars and vocabularies and therein their place in the network of human societies mapped out in the museum. The gaze with which we look mirrors the violences of the colonizer's value system, and positions Māori within the 'Great Chain of Being'. Until Māori are co-constructing the narrative, articulating the value and meanings of the objects, the imperial framework and narrative structure sticks, tenaciously. The act of looking onto an object can make known the culture of a peoples through a singular colonizing gaze. This captivates, assesses and fixes meanings and associations of culture within an epistemic framework that is powerful and which frames the artefact. This is an act of reducing, delimiting and re-colonizing the artefact. Code (2006) warns us about epistemic violences in the naming and categorizations of artefacts, which detextualise and sit inorganically in relation to the source of cultures that are represented. These misrepresentations are compounded by the ways in which we are conditioned to engage, observe and make knowable other cultures. There is an exoticizing and a separating that is supported by the architectures of display within museum spaces. Karp and Lavine (1991, p. 378) argue that 'no genre of museum has been able to escape the problems of exoticizing and assimilating inherent in exhibiting other cultures'. However, few have attended to the feelings and affective atmospheres that result. On looking at the object as a 'western' citizen, the grammars of being, looking and knowing through the museum gaze is now a habit, a rhythm with which we are familiar, and contestations are usually over aesthetics, space or indeed opportunity to gaze with the correct tools (Goodman, 1985). For Goodman, 'the museum has to contend with inexperience and ineptness in many viewers, a fixed and formidable environment for viewing, and usually with lack of any mobility or progression or time-value in the work itself' (1985, p. 56). However, Goodman is an important voice in drawing attention to the alternative values that viewers bring to the cabinet. He argues:

> Reverberations from a work may travel in cycles through our everyday environment, other works, and itself, again and again, with ever-changing effect. Works work by interacting with all our experience and all our cognitive processes in the continuing advancement of our understanding.
>
> (1985, p. 57)

The gaze, however, gains a different possibility when it is embodied by Māori themselves looking onto Māori *toanga*. Reverberations of sadness, pain and anger are felt at once. The *taonga* are not objects to Māori but gods, ancestors with biographies, spiritual power. The Māori experience on seeing *taonga* disrespectfully displayed, mislabelled or indeed exhibited is one of failing to responsibly take care of ancestral spirits. The experience of seeing the museum display results in the deadening, defiling and desecrating of their cultural ancestors.

Rosanna Raymond, my co-collaborator in this research, responds to these striations of pain, guilt and sadness in the poem below. Raymond was born in New Zealand, has self-defined as being of Samoan–English descent and currently lives and works in London. She was a founding member of the acclaimed art collective *Pacific Sisters*, and was co-curator and artistic director of the Pasifika Styles festival in Cambridge between 2006 and 2008. Her aesthetic practice centres on the aesthetics and positioning of the savage body, outside of accounts of international, modern, intellectual and philosophical art. Raymond, in her poem below, outlines her positioning as a result of continued misrepresentations of Māori art. Highlighted is the epistemic violence of labelling them as 'artefact' and the deadening of the cabinet, locked away from their true nature as enlivened and part of modernity, not pre-modernity.

The pain of being responsible is expressed in Rosanna Raymond's poem *The Silence of the Gods*:

> *A throng of gods*
> *Assembled in silence*
>
> *Accused of decadence*
> *Offered out of deference*
>
> *Emptied of resonance*
> *Collected for reference*
>
> *And now in idol consideration*
>
> *Engaged in your estrangement*
> *I gaze at you like a stranger*
>
> *Enjoying your sing song*
> *that fell on deaf ears*

I give you my name
And you give me your number

To revive you
To revere you

Raymond articulates the effect of W.E. Du Bois's (1903) double-consciousness. Double-consciousness encapsulates the museum experience of racialized, 'other' museum visitors – those doubly placed within and without narratives of nation, and national culture (Gilroy, 1990, 1993, 2013). It is the sense of always looking at one's self through the eyes of others. 'Other' in this account can include those that are represented, but not the architects of their representation, or those that are recognized as outside of a modern narrative of nation, or indeed are considered as outside modernity itself (Hall, 1997a; Gilroy, 1993). Du Bois is writing specifically through the African American subject position where a psychological space is created, mediated and promoted by a racist white American culture. What this advocates is that the space is produced through the rupture between the museum's account of a cultural story of being human and the experience of being human as 'other' to the dominant bodies represented.

Du Bois uses the metaphor of a veil. There is a veil, a barrier, a film between the narrative articulated on displays and the lived reality for racialized communities. There is a gap in the museum's own representation of your culture and the lived experience of your culture. There is a fissure between that is unreconciled. That gap is between the disenfranchised, displaced subject of the museum narrative and a wholly self-determined account of cultural identity. Du Bois (1903) thus argues that:

> After the Egyptian and the Indian, the Greek and the Roman, the Teuton and Mongolian, the Negro is the seventh son, born with a veil, and gifted with second sight in this American world – a world which yields him no true self-consciousness, but only lets him see himself through the revelation of the other world. It is a peculiar sensation, this double-consciousness, this sense of always looking at one's self through the eyes of others, of measuring one's soul by the tape of a world that looks on in amused contempt and pity. One ever feels his two-ness – an American, a Negro; two souls, two thoughts, two unreconciled strivings; two warring ideals.
>
> (1903, p. 2)

A century ago, the opportunity to self-determine the representation of black culture by black citizens in its fullest sense was at stake. Social and cultural oppression has resulted in the obscuring of the formation of a black African American 'structure of feeling' (Williams, 1975), that was owned, self-determined and articulated with power and freedom. More importantly for colonized subjects (see also Fanon, 1968), the voice of self-narrative was split into two, one articulating the inarticulatable and the other self-consciously reproducing the figure of the black American that was part of dominant ideologies. Spivak (1988) articulates and elaborates on this problematic. She argues that being the 'true' subaltern group, whose identity is its difference, there is no unrepresentable subaltern subject that can know and speak itself, thus raising the question: 'With what voice-consciousness can the subaltern speak?'(1988, p. 27). At stake here is an encounter with a narrated account of one's self as 'other', albeit in the texts of history, museums, art, philosophy and science, as an alienating one. At the museum, the seeing of your culture as 'other' is alienating because it is framed within an imperial taxonomy or fixity to an 'other' culture (Hall, 1997a; Spivak, 1988) which denies historical dynamism, heterogeneity and, most importantly, self-determination. The 'other' is articulated using the tools and voice of the historian, curator, anthropologist, who is always writing from a position of being outside the cultural realm of the other (Said, 1979, 1989). Even when subaltern voices take control or have opportunities to articulate, the tools available are not in synthesis with the subalterns' subject position. To self-determine or to define subaltern culture in a post-imperial world is to negotiate from a position of alienation, where, '[i]n the constitution of 'Other' in Europe, great care was taken to obliterate the textual ingredients with which such a subject could cathect, could occupy (invest) its itinerary' (1988, p. 24).

At the Māori cabinet there are powerful affective charges that emerge from this ground of doubleness, alienation and occluded accounts of Māori culture. At the museum, there is a joy and awe at seeing Māori *taonga*, being reunited with them and rekindling the relationship. This is experienced alongside the endurance of affective registers of grief, pain, loss and sadness that result from a feeling of guilt due to the failure to keep Māori ancestors safe, unviolated, undefiled, but most importantly, alive. Retaining life for Māori is about restoring the *Va*, the power of life (a Samoan concept) and spiritual connectedness in the present. Seeing Māori culture placed in a cabinet, in an alienating environment, without access to the contemporary life of the *Marae* is equivalent to seeing a body deadened through your own neglect. Moreover, Du Bois's conceptualization helps us understand the striations of affects including the pain of seeing Māori *taonga* as they sit as body-parts, without integrity or as part of contemporary connection with Māori family and heritage. For Māori, removal from community circulation is like burying them in an archaeological past that erodes the power and value of them. The objects only have value as part of embodied rituals and practices within communities, in context. Sitting outside of Māori life places their value out of reach and leaves their power deadened. Māori thus see themselves reflected through the cabinet, as artefact, as past. The co-constitution of modernity of which *toanga* are part is occluded in the grammars of the cabinet display. As well as the power and value of artefacts being deadened, the role of artefacts is reduced to the past. Their role in keeping ancestral knowledge as part of the present, as part of the contemporary ecological, political and spiritual network of regional, national and transnational nationhood that is *Māori*, is erased. It is as if they are of another country and not a means of articulating the contemporary world and Māori culture within it. Rosanna Raymond articulates this below:

Looking at *taonga* that is so familiar, yet very separated from its original place and purpose, can be a frustrating and painful process, especially if you feel connected to it, spiritually. I feel a strong bond to my ancestors when I meet artefacts. It is as if a direct line (*whakapapa*) opens up with my cultural heritage, the past becomes present.

(Interview, July 2010)

Time and space for Māori are turned on their head within the museum display. As Raymond argues, cultural heritage is a *live* relationship with the past; not one to be categorized, labelled and indeed displayed coldly outside of Māori stewardship. At the heart of this problematic is what Said (1979) argued in *Orientalism:*

[F]rom the beginning of Western speculation about the Orient, the one thing the Orient could not do was to represent itself. Evidence of the Orient was credible only after it had passed through and been made firm by the refining fire of the Orientalist's work.

(p. 283)

Taking this further, what is at stake here is not just representational politics, but an account of the affective logics of cultural collections that make sense of modern cultural truths, values and practices. To rip these apart is to produce a grief and discordancy with Māori belief systems and cultural memory itself.

Postcolonial affective atmospheres at the museum

The politics of art practice, it could be argued, is to make us feel. The politics of postcolonial expressive cultures have also incorporated the project of making us feel, in this case the positioning of the postcolonial subject (see Morrison *et al.*, 2013). As Pile (2011, p. 28) states, it is '[f]rom this perspective, rather than being the "victims" of race discourses, [that] bodies constantly threaten to destabilize processes of racialization'. Inhabiting and making the audience feel a space of embodied violation engenders a political transformation of empathy and thus a spirit of change. For artists such as Raymond, inhabiting the space of the violated enables the eradication of the structures repeating those violations, through sight, sound, memory and text. Raymond's project is to animate, to show the hauntings of eons of peoples who have been misrepresented and undermined, right down to their capacities to feel and think. Finding a place from which to simultaneously shatter mistruths and to articulate new ones is the project for Raymond; the site of the body becomes the counter-museum, counter-culture and counter-memory, all at once. The role of geographer–artist collaboration in this project has been to articulate, to enable, to make visible, to co-visualize the burdens of representation and the affective spaces of the museum as one of a *theatre of pain*. The aim is to situate the affective experience of epistemic violence that Māoris encounter in the twenty-first century, and map the cultural geographies of the new visualizations necessary for a redistribution of the sensible. Small modest steps ensue; in the space of a paper, a catalogue, an event that shifts the ground from beneath calcified layers of compounded memories and mistruths of imperial representation. It is important, however, to record the spaces of encounter through affective, representational and embodied accounts. The site of the visceral, therefore, is the site of contemporary reconciliation between postcolonial challenges to hierarchies, reductions and stratified accounts of 'other' cultures and the materialized cultural values expressed in museum displays.

Note

1 Ngāti Rānana London Māori Club aims to provide New Zealanders residing in the United Kingdom and others interested in Māori culture an environment to teach, learn and participate in Māori culture. The three guiding principles of Ngāti Rānana are whanaungatanga (togetherness), manaakitanga (looking after one another/hospitality) and kōtahitanga (unity). www.ngatiranana.co.uk/

Bibliography

Adey, P. 2008. Aeromobilities: geographies, subjects and vision. *Geography Compass*, 2(5): 1318–36.
Ahmed, S. 2004a. Affective economies. *Social Text*, 22: 114–39.
Ahmed, S. 2004b. Collective feelings: or the impressions left by others. *Theory Culture and Society*, 21: 25–42.
Alpers, S. 1991. The museum as a way of seeing, in I. Karp and S.D. Lavine (eds) *Exhibiting cultures: the poetics and politics of museum display* (pp. 25–31). Washington DC: Smithsonian Institution Press.
Anderson, B. 2009. Affective atmospheres. *Emotion, Space and Society*, 2(2): 77–81.
Araeen, R. 1987. From primitivism to ethnic arts. *Third Text*, 1(1): 6–25.
Baxandall, M. 1991. Exhibiting intention: some preconditions of the visual display of culturally purposeful objects, in I. Karp and S. Levine (eds) *Exhibiting cultures: the poetics and politics of museum display* (pp. 33–41). Washington DC: Smithsonian Institution Press.
Bennett, T. 1995. *The birth of the museum: history, theory, politics*. London: Routledge.
Bennett, T. 2005. Civic laboratories: museums, cultural objecthood and the governance of the social. *Cultural Studies*, 19(5): 521–47.
Bennett, T. 2006. Stored virtue: memory, the body and the evolutionary museum, in S. Radstone and K. Hodgkin (eds) *Memory cultures: memory, subjectivity and recognition* (pp. 40–54). New Brunswick and London: Transaction Publishers.
Berger, J. 2008. Ways of seeing. London: Penguin.
Boehner, K., DePaula, R., Dourish, P. and Sengers, P. 2005. Affect: from information to interaction. Proceedings of the 4th decennial conference on Critical computing: between sense and sensibility. www.dourish.com/publications/2005/cc2005-affect.pdf (accessed 22 October 2015).
Bohrer, F. 1994. The times and spaces of history: representation, Assyria, and the British Museum, in D. Sherman, and I. Rogoff (eds) *Museum culture: histories, discourses, spectacles* (pp. 197–222). London: Routledge.
Code, L. 2006. *Ecological thinking: the politics of epistemic location*. Oxford: Oxford University Press.

Coombes, A.E. 1994. The recalcitrant object: culture contact and the question of hybridity, in F. Barker, P. Hulme and M. Iverson (eds) *Colonial discourse/postcolonial theory* (pp. 89–114). Manchester: Manchester University Press.

Crang, M. and Tolia-Kelly, D.P. 2010. Nation, race and affect: senses and sensibilities at National Heritage sites. *Environment and Planning A*, 42(10): 2309–14.

Dewan, D. and Hackett, S. 2009. Cumulative affect: museum collections, photography and studio portraiture. *Photography and Culture*, 2(3): 337–48.

Du Bois, W.E. 1903. *The souls of black folk*. New York: Burghardt.

Fanon, F. 1968. *Black skin, white masks*. Translated by Charles Lam Markmann. New York: Grove Press.

Gell, A. 1998. *Art and agency: an anthropological theory*. Oxford: Oxford University Press.

Gilroy, P. 1990. Art of darkness: black art and the problem of belonging to England. *Third Text*, 4(10): 45–52.

Gilroy, P. 1993. *Small acts: thoughts on the politics of black cultures*. London: Serpents Tail.

Gilroy, P. 2013. *There ain't no black in the Union Jack*. London: Routledge.

Golding, V. 2009. *Learning at the museum: frontiers, identity, race and power*. Farnham: Ashgate.

Goodman, N. 1985. The end of the museum? *Journal of Aesthetic Education*, Special Issue: Art Museums and Education, 19: 53–62.

Gregory, K. and Witcomb, A. 2007. Beyond nostalgia: the role of affect in generating historical understanding at heritage sites, in S.J. Knell, S. MacLeod, and S. Watson (eds) *Museum Revolutions* (pp. 263–75). London: Routledge.

Hall, S. (ed) 1997a. *Representation: cultural representations and signifying practices Vol. 2*. Thousand Oaks and New Delhi: Sage.

Hall, S. 1997b. Who needs identity?, in S. Hall and P. du Gay (eds) *Questions of cultural identity* (pp. 15–30). New Delhi: Thousand Oaks and Sage.

Hall, S. 2005. Whose heritage? Un-settling 'the heritage', re-imaging the post-nation, in J. Littler and R. Naidoo (eds) *The politics of heritage: the legacy of 'race'* (pp. 21–31). London: Routledge.

Hemmings, C. 2005. Invoking affect: cultural theory and the ontological turn. *Cultural Studies*, 19(5): 548–67.

Hetherington, K. 2003. Spatial textures: place, touch, and praesentia. *Environment and Planning A*, 35(11): 1933–44.

Karp, I. and Lavine, S. (eds) 1991. *Exhibiting cultures: the poetics and politics of museum display*. Washington, DC: Smithsonian Institution Press.

Kraftl, P. and Adey, P. 2008. Architecture/affect/inhabitation: geographies of being-in buildings. *Annals of the Association of American Geographers*, 98(1): 213–31.

Macdonald, S. (ed) 1998. *The politics of display: museums, science, culture*. London: Routledge.

Marx, K. 1978 [1856]. Speech at the anniversary of the people's paper, in R.C. Tucker (ed) *The Marx-Engels Reader*, 2nd edition (pp. 577–8). London and New York: W.W. Norton & Co.

McCormack, D.P. 2008. Engineering affective atmospheres on the moving geographies of the 1897 Andrée expedition. *Cultural Geographies*, 15(4): 413–30.

Morrison, C., Johnston, L. and Longhurst, R. 2013. Critical geographies of love as spatial, relational and political. *Progress in Human Geography*, 37(4): 505–21.

Pile, S. 2010. Emotions and affect in recent human geography. *Transactions of the Institute of British Geographers*, 35: 5–20.

Pile, S. 2011. Skin, race and space: the clash of bodily schemas in Frantz Fanon's *Black Skins, White Masks* and Nella Larsen's *Passing*. *Cultural Geographies*, 18(1): 25–41.

Said, E.W. 1979. *Orientalism*. New York: Vintage.

Said, E.W. 1993. *Culture and imperialism*. London: Chatto and Windus.

Samuel, R. 1994. *Theatres of memory, volume 1: past and present in contemporary culture*. London: Verso.

Schorch, P. 2013. Contact zones, third spaces, and the act of interpretation. *Museum and Society*, 11(1): 68–81.

Schorch, P. 2014. Cultural feelings and the making of meaning. *International Journal of Heritage Studies*, 20(1): 22–35.

Spivak, G.C. 1988. Can the subaltern speak?, in C. Newson and L. Grossberg (eds) *Marxism and the interpretation of culture* (pp. 271–313). Champaign, IL: University of Illinois Press.

Stephens, A.C. 2015. The affective atmospheres of nationalism, *Cultural Geographies*. Online (accessed August 2015).

Sylvester, C. 2009. *Art/museums: international relations where we least expect it*. New York: Paradigm Publications.

Thien, D. 2005. After or beyond feeling? A consideration of affect and emotion in geography. *Area*, 37(4): 450–6.

Thrift, N. 2004. Intensities of feeling: towards a spatial politics of affect. *Geografiska Annaler: Series B, Human Geography*, 86(1): 57–78.

Tolia-Kelly, D.P. 2006. Affect – an ethnocentric encounter? Exploring the 'universalist' imperative of emotional/affectual geographies. *Area*, 38(2): 213–17.

Waterton, E and Dittmer, J. 2014. The museum as assemblage: bringing forth affect at the Australian War Memorial. *Museum Management and Curatorship*, 29(2): 122–39.

Williams, R. 1975. *The country and the city* (Vol. 423). New York: Oxford University Press.

Witcomb, A. 2013. Understanding the role of affect in producing a critical pedagogy for history museums. *Museum Management and Curatorship*, 28(3): 255–71.

Witcomb, A. 2014. 'Look, listen and feel': the First Peoples exhibition at the Bunjilaka Gallery, Melbourne Museum. *THEMA. La revue des Musées de la civilisation*, 1: 49–62.

3 Affecting the body

Cultures of militarism at the Australian War Memorial

Jason Dittmer and Emma Waterton

The machine-gun turret of the Second World War bomber, "Tail End Charlie," sat in the middle of the exhibition hall, its tiny doors open as if inviting the museum-goer to crawl inside (see Figure 3.1). This invitation was of course virtual rather than actual—not least because very few of the museum-goers viewing it with us appeared small enough to fold their bodies inside the turret. Nevertheless, the virtual invitation drew on previous experiences of tight spaces—the backseat of a coupe, an economy-class airline seat, and so on—to inspire an imagined corporeal link between us and the airmen who flew so many missions in these exposed, dangerous, and unheated spaces.

It was only as we drew closer that we noticed the bullet holes riddling one side of the turret, our embodied sense of its tight spaces nagging our conscious minds that the turret's occupant was unlikely to have survived. Those bullet holes, blasted through steel and unchanged in the years since, triggered an imagined moment of death—our own—overlaying that of the unknown occupant. The affective homology between our bodies and those of the men seventy years ago who manned these turrets, when brought into relation with the now-cold violence of the bullet holes and their twisted edges, produced an event through which a multiplicity of lines of flight unspooled in and through us. Our critical analysis of the exhibition came to a momentary halt as we imagined ourselves as soldiers and felt grateful that we had not experienced the trauma of war.

This chapter builds on our memories of that encounter by considering the cultures of militarism displayed at the Australia War Memorial (henceforth "Memorial") in Canberra, Australia, as enmeshings of the social and somatic. Derived from embodied interactions between people and wider discourses and technologies, both serendipitous and planned, we move to understand the museum experience as provoked by far more complex and performative processes of engagement than traditional museology literature might have it (though see work by Witcomb, 2013a, b, 2012). Indeed, our intention is to understand the Memorial as a place of affect and effect. To make our case, we start from the assumption, borrowed from past work in this area (e.g. Dittmer, 2013), that larger narratives and contexts of conflict are often elided or bypassed if there is no clear way to link combatants to a morally unambiguous and superior position *vis-à-vis* their enemies. In these circumstances, those producing cultures of militarism often take to a new strategy

Figure 3.1 "Tail End Charlie".

(intentionally or not), drawing from the affective and emotive realm in order to buttress support for their stance (see Muzaini and Yeoh, 2005). In practice, of course, nothing is as clear-cut as this binary; geopolitical narratives are always mediated in affective worlds that shape their reception. Instead, our case study is of interest precisely because of the competing interests that shape its affective and discursive becoming.

At the heart of our exploration lies the Memorial, a key heritage site within the Australian context, founded in 1941 in Canberra. Its functions, as outlined in the *Australian War Memorial Act of 1980*, are "to maintain and develop the national memorial," to "develop and maintain […] a national collection of historical material," to "exhibit […] historical material from the memorial collection," and to "conduct, arrange for and assist in research pertaining to Australian military history." The site is therefore multiplicitous—at the same time "a shrine, a world-class museum, and an extensive archive" (Australian War Memorial, 2013, n.p.). Leaving aside the archive for the moment, it is worth considering the tension between the Memorial's roles as a shrine and a museum of military exploits. The latter might be understood as part of the quintessential state-centered culture of militarism; the former, however, points to the losses of war—the bodies broken, the limbs lost, the families fragmented. At times these aims are congruent; at others they seem to work at cross-purposes. The Memorial is a particularly interesting heritage site because of the way that Australian narratives of national identity are predicated not as much on military prowess but on military futility on behalf of others. Famously, it is British disregard for Australian lives at Gallipoli in the First World War that is often posited as sparking the emergence of a new nation—distinct from the British Empire—the "depth and soul" of which, popular memory suggests, was consolidated during the Kokoda campaign in the Second World War (former Australian Prime Minister Paul Keating, cited in Nelson, 1997). While the emergence of an independent Australia and national spirit is much more complex than those stories capture alone, what is relevant here is how cultures of militarism are at once central to Australian collective memories of "self" and yet also strangely dissonant. Therefore, the shift to an affective register enables the Memorial to encompass a specifically Australian sense of war as both a locus of national pride and a scourge whose costs must be counted.

To commence our explorations of the social and somatic affordances of the Memorial, we begin with a review of the ways in which heritage has been taken up by cultural and political geography, before turning to the literature on embodied nationalism, tracing the connections between the individual corporeal body and the larger body politic. Our rationale for beginning with the geographical literature emerges out of the hesitancy with which theories of affect have been dealt with in the heritage field (as identified by the conference sessions[1] from which this chapter emerges). While there is a rich itinerary of sources that point to the centrality of emotion, affect, and feeling, there is little in that corner of the literature that deals explicitly with, *and names*, theories of affect (though see Byrne, 2013; Vincent, 2014; Waterton, 2014, 2015; Waterton and Watson, 2014). For example, David Uzzell, borrowing from Robert Abelson's concept of "hot cognitions," is clearly invoking a sense of "affect" by using the term "hot interpretation" to point to the passions of being human, arguing that in tandem with generating interesting and enjoyable information, heritage sites need also to shock, to move, to be cathartic (1989, p. 46). But the most instructive attempts to rearticulate heritage as an affective encounter have emerged from geography (see Crang and Tolia-Kelly, 2010; Tolia-Kelly, 2006), and so it is to the overlaps between geography and heritage that we turn in this chapter. In particular, we draw on the literature in critical geopolitics that has examined visuality and affect, arguing that the Memorial serves as an archive (though not the archive referred to in the Act above) of technologies meant to inspire empathic identification with fighting men and women in a heritage setting. This archive enables us to examine the changing nature of cultures of militarism and consider the way that they shape museum-goers' experiences. After a discussion of our methods, we go on to consider the various affective technologies found in the Memorial itself, from simple dioramas to vast multimedia experiences. Like the open doors of the turret that began this chapter, these technologies invite museum-goers in, positioning them in ways that afford affective encounters with long-dead men and women. Finally we conclude by sketching out an agenda for the renewed consideration of heritage in relation to the "affective turn."

A melding of geography, heritage and affect

Heritage and geography

We start with a review of the literature that considers the relationship between heritage and geography as it has developed over the past quarter-century. This is an engagement that has been riven from its beginnings, with authors frequently inciting binaries in the literature: heritage as a conservative and radical concept, seemingly simultaneously (Hardy, 1988), for instance, or heritage envisaged as nationalist nostalgia and a postmodern pluralism (Johnson, 1996). Both binaries implicitly adopt a normative stance on "good" (e.g. tourism that celebrates difference) and "bad" (e.g. celebrations of past wars) heritage, but equally they both consider heritage primarily as a matter of discourse and narrative. This is the way heritage has usually been considered within geography. Hardy, for instance (1988, p. 333; see also Graham *et al.*, 2000; Lowenthal, 1985), offers what can be taken to be a commonplace formulation of heritage as:

> a value-loaded concept, embracing (and often obscuring) differences of interpretation that are dependent on key variables, such as class, gender and locality; and with the concept itself locked into wider frameworks of dominant and subversive ideologies (where the idea of heritage can be seen either to reinforce or to challenge existing patterns of power).

This early framing of heritage as a matter of meaning and interpretation is entirely congruent with the emergence of post-structuralism as a strong theme running through human geography in the 1990s (Crang, 2003), and is one that has been rehearsed in much of the literature since. Key texts in this regard would be David Lowenthal's *The Past is a Foreign Country* (1985) and Graham *et al.*'s *A Geography of Heritage: Power, Culture and Economy* (2000), both of which continue to be well-cited resources in current heritage literature. As these volumes attest, geographers in the 1990s and afterward tended to deal with heritage through the lenses of landscape and space, with each reworked by the post-structural turn (Cosgrove, 1985; Daniels, 1993). The resultant appreciation of the political nature of landscape that emerged at that time simultaneously prompted a similar questioning of heritage, particularly monuments and their contestation:

> Memorials influence how people remember and interpret the past, in part, because of the common impression that they are impartial recorders of history. Their location in public space, their weighty presence, and the enormous amounts of financial and political capital such installations require imbue them with an air of authority and permanence. […] Further, their apparent permanence suggests the possibility of anchoring a fleeting moment in time to an immovable place. Composed of seemingly elemental substances—water, stone, and metal—memorials cultivate the appearance that the true past is and will remain within reach.
> (Dwyer and Alderman, 2008, pp. 167–8)

From there, an influential literature started to emerge that considered together the concepts of landscape, heritage, and spaces of remembrance (see for example Johnson, 1995, 2015; Tolia-Kelly, 2004). Dwyer and Alderman (2008) usefully consider three approaches to studying the nexus of these: memorial landscapes as text, as arena, and as performance. The first of these (as text) considers memorials as semiotic contributions to the landscape, written by the artists/politicians that produce them and read by visitors. Past work in this vein includes Azeryahu's 1996 study of street naming, Alderman's 2010 study of a slavery memorial's caption in Savannah, Georgia, and Johnson's 1996 study of Strokestown House, the Irish "heritage tourism" destination. The second approach (memorial as arena) de-emphasizes the memorial itself and focuses on the politics surrounding it, such as its location or its scale. Key studies of this type include Leib's 2002 study of the contest surrounding the location of a statue of Arthur Ashe on Richmond's (Confederate) Monument Avenue; Alderman's 2003 consideration of the scalar politics involved in the designation of Martin Luther King, Jr Boulevards; and Johnson's 1995 analysis of the politics of nationalism, gender, and post-coloniality in Irish memorial-making (see also Johnson, 2015).

Closest to our own conceptual framework is Dwyer and Alderman's third approach—memorials as performance. These studies note that "the memorial landscape is constituted, shaped, and made important through the bodily performance and display of collective memories" (Dwyer and Alderman, 2008, pp. 173–4). This strand of research points in particular to the agency of visitors to memorials, as they can ignore memorials, or desecrate them, or perform rituals at them (Cresswell, 1996). Several studies have, for instance, found performative aspects to memorial landscapes such as the carnivalesque renaming of Basque streets (Raento and Watson, 2000) and the everyday vocalization of words such as New Zealand/Aotearoa, which are simultaneously ethnic memorializations (Kearns and Berg, 2002).

Our approach differs from these, however, in that it sees the subjectivity and agency of both memorializers and memorial visitors as more-than-human in nature. With this, the oft-noted "elemental" materiality of memorials (see the Dwyer and Alderman block quote above) enters into assemblage with the bodies and tools of those who seek to inscribe meaning (memorial as text), to contest geographies (memorial as arena), or to perform space (memorial as performance). People contest memorials, but equally memorials enable new spaces and politics to emerge. While affirming this "heritage" of geographic research on memorials as materializations of geographic identities, narratives, and discourses, we point toward a different sensibility. Memorials have long since been described as "sites of memory"—or *lieux de mémoire* (Nora, 1989)—though more recent authors have bolstered this conceptualization with notions of memory as an embodied process. As Hoelscher and Alderman (2004, p. 350) put it, for example, "[t]hrough bodily repetition and the intensification of everyday acts that otherwise remain submerged in the mundane order of things, performances like rituals, festivals, pageants, public dramas and civic ceremonies serve as a chief way in which societies remember." In this we could not agree more; however, this definition leaps from the individual body to the body politic—a commitment that is verbalized in the range of names deployed for this kind of memory: "collective memory," "social memory," "public memory," "historical memory," "popular memory," and "cultural memory" (this list is given in Hoelscher and Alderman, 2004, p. 348). Instead we propose to consider memory as the somatic residue of specific events, an emergent result of bodies brought into assemblage with a range of objects, bodies, and discourses. Indeed, as Dwyer and Alderman (2008, p. 166) argue, this "'performance' metaphor recognizes the important role that bodily enactments, commemorative rituals, and cultural displays occupy in constituting and bringing meaning to memorials, suggesting that the body itself is a site of memory."

Empirically, the study most closely related to ours is perhaps that of Crouch and Parker (2003, p. 399), who, in their article "Digging up Utopia?", consider the non-representational aspects of heritage (in particular, archaeological digs), but do so from within a framework of a rational actor. Essentially, they argue that "[p]ractice includes active bodily engagement, wandering round, knowing 'with both feet'. We express ourselves through the landscape we construct in the process of doing." Rather than seeing the landscape as produced through performances of heritage, we emphasize the mutual constitution of both subject *and* heritage through their constant becoming together. In other words, memorials do work upon our bodies, and larger "collective memories" (*et al.*) are an aspect of our bodies in assemblage (Protevi, 2009). Such an approach has resonated in heritage studies where, for instance, Kearney (2009, p. 213) has argued that "the depository of these intangible cultural expressions [i.e. collective memory] is the human mind and body, ancestors, and homelands; all of which become instruments for its enactment, or literally its embodiment." Kearney's research revealed the critical importance of bodily senses such as smell to the production of embodied heritage and memory. Other work in the field that has contributed to the laying of a careful foundation for a new theory of signification for heritage has come from scholars such as Crang and Tolia-Kelly (2010) and Waterton and Watson (2014). These, like Kearney, are seeking out a fuller account of "affect," based on the assumption that heritage sites circulate with—and evoke—strong emotions and feelings that resist representation. As Waterton and Watson (2014, p. 76) put it:

> This is because our abilities to respond to the semiotic landscapes that surround us emerge in large part from the ways in which we have already reconciled previous experiences, moments, knowledges and events. Indeed, it is through bodily remembering that our engagements with heritage spill out beyond representation, with memories *being remembered* and moving through our bodies, where they are expressed once again and come to affect ourselves, other bodies and other representations.

Similarly, our study turns to the affective techniques through which the body is remade in minute ways that resonate with others in the body politic to produce collective memory.

Visuality and affect

As noted above, we are not the first to propose such a move within geographies of heritage; for instance, Crang and Tolia-Kelly (2010, p. 2316) have argued that

> Heritage sites differentially enable and arrest the circulation of objects, people, emotions, and ideas. They are at least in significant part also intended more or less successfully to fix, stabilize, and store both things and categories. [...] Work on affect in geography has attended rather less to the fixities and intransigencies, to relations that are fetishised and reified, to performance as repetition. [...] Heritage sites allow us to look at such moments [of viscosity].

While we would argue that such indeterminacy is a crucial aspect of the non-representational turn in human geography, and have attempted to consider that aspect of this study elsewhere (Waterton and Dittmer, 2014), we agree that it is quite possible to lose sight of the patterns that emerge within a possibility space. We also share Crang and Tolia-Kelly's interest in national heritage and "the differentiated affective energies created by relationships between geography (site, situation, and spaces), places (how they are encountered, experienced, and felt), the body (race, citizenship, and positioning), and the 'heritage' apparatus (exhibits, taxonomies, and conservation)" (2010, p. 2316). In our attempts to extend Crang and Tolia-Kelly's observations, we turn to a consideration of the technologies deployed in order to attempt to pin down affective circulations of particular types and embed them in particular somatic experiences. For this, we rely on Thrift's (2004, pp. 64, 67) description of "a tendency towards the greater and greater engineering of affect" in contemporary society, which, he argues, is driven by tendencies to liberal choice in more spheres of life and heightened mediatization of politics, as well as the discovery of spaces and temporalities of the body upon which a "microbiopolitics" can be enacted. None of this is new, as Thrift himself notes, but it has intensified in recent decades, and it is these drivers of affective politics that we argue are at work in the Memorial.

The rise of liberal choice, for instance, comes into view when recent developments in museum studies and practices are considered; after much postcolonial criticism of the nationalist narratives often ensconced in Western museums, many now seek to present multiple narratives and allow visitors and associated community groups a stake in their composition. Biopower, then, becomes a replacement for the previous monolithic narrative made available through heritage institutions. Like Thrift, Anderson (2012, p. 29) sees the contemporary relation between affect and biopower as a product of recent neoliberal drives:

> If productive forces are to be "generated", made to "grow", and be "|ordered", then the contingencies of life must be known, assayed, sorted and intervened on. But contingency must never be fully eliminated, even if it could be. To do so would be to also eliminate the circulations and interdependencies that supposedly constitute the "freedom" of individuals and commerce in liberal-democracies. In short, choice must be available so that several degrees of freedom are possible; but as much as possible the affective ground on which choices are made can be narrowed to harness the productive forces of life.

If a liberal notion of choice has become common within museum practices, the mediatization of politics (another trend identified by Thrift) is also apparent in the rise of "the screen" as a medium through which heritage content is presented (Witcomb, 2007). Those charged with planning and designing museum spaces have moved to bolster the traditional textual panels and audio-visual tours we are accustomed to finding within their walls with increasingly varied and sophisticated multimedia interpretation strategies, many of which are aimed at triggering an affective pedagogy (Witcomb, 2013a). Thus, while lighting, logistics, volume and visitor "flow" have for quite some time been considered powerful elements in creating museum atmospheres, the highly mediated world in which we live is also being called upon to "trigger and diffuse" affect (Thrift, 2008, p. 254). Examples of this include the incorporation of hologram technology, which has seen, for instance, the insertion of "solid", 3D holavisions into the museumscape, as with "Shane Warne" at the National Sports Museum in Melbourne, Australia and the "ghostly" apparitions of several historic characters at the Abraham Lincoln Presidential Library and Museum in Springfield, Illinois. Here, perhaps more than anywhere, we begin to see evidence of Thrift's (2004, p. 58) assertion that affect is becoming more easily and actively engineered, "more akin to the networks of pipes and cables that are of such importance in providing the basic mechanics and root textures of urban life."

In a similar vein, Carter and McCormack (2010), drawing on Deleuze's work on cinema (2005a [1985], 2005b [1986]), have argued that the cinema can also serve as a conduit through which affects circulate, particularly in the post-9/11 era of geopolitics. With reference to the film *United 93*, they argue that

> [t]he real significance of the film is its attempt to suggest what it might have been like to have *felt* these events. *Feeling is the primary affective logic of the film*: feeling that circulates within and between the bodies of those on the aircraft; feeling which, due to the skill with which the film is constructed, cannot but be felt by the viewer—however critical.
> (Carter and McCormack, 2010, p. 111)

In earlier work, and with reference to the film *Black Hawk Down*, Carter and McCormack (2006, pp. 239–40) argued that that film resonated with various discourses at work in the post-9/11 moment in ways supportive of intervention. Drawing on the insight of Colls (2012), we would add that the sites of encounter between the film's affects and the discursive frame of 9/11 were audience members' bodies, themselves differentiated in myriad ways that refract experience and produce unexpected outcomes. Nevertheless, the emergence of a pattern in responses indicates a key, and often overlooked, component of war films—the fact that they valorize an intensely emotional form of comradeship and brotherhood emerging from the more-than-human affects of battle. This was crucial to how the film resonated in the immediate wake of 9/11. Through the felt intensity of its own cinematic logics, *Black Hawk Down* had the potential to become re-territorialized on one of the key somatic markers of the post-9/11 world, the military body. Through the military body's immersion in the affective event of intervention, and its capacity to affect and be affected by other bodies in this field, it becomes a site for the redemption and revalorization of militarism as a heroic experience. The heroes of the film are simultaneously individuals—acting, intervening, suffering—and a group of bodies—sharing, relating, collaborating; this complex relation was mirrored in the audiences, who were simultaneously individuals—watching, thinking, consuming—and there in Mogadishu, with the soldiers. There is no need of a longer narrative to make the audience identify with the soldiers, only the felt connection enabled by the screen.

While there are clear differences between the usage of screens in the Memorial and in Hollywood cinema, and certainly in their ability to disseminate widely, there are nevertheless some key similarities in the way that they are used to attempt affective engineering, such as their ability to amplify and circulate affects, as well as to resonate with or disrupt already extant affects. Further, the screens in the Memorial and in the cinema can be seen to reinforce one another, circulating affects that are modulated differently but that nonetheless resonate with each other; they each help render the other sensible to the viewing bodies. Thrift's final two trends—the rise of microbiopolitics and the increased ability to engineer urban spaces—are the subject of our empirical study. We argue that the assemblage of the Memorial is indicative of techno-cultural shifts in both areas, and is a particularly salient site for examination because the institution is uniquely cognizant of its own "heritage" and therefore has preserved early attempts (the First World War galleries in particular—currently under redevelopment) to engineer affect in the way described by Carter and McCormack, alongside newer attempts (Second World War and later galleries).

Ethnographic visuality

Before moving on to explore those moments in the Memorial where the affective and emotive realms are piqued, we first need to detail our method of data collection and the mode of analysis through which we came to understand these dynamics. For this, our research drew inspiration from developments within "audience studies" that advocate a move away from traditional interviews and surveys toward ethnographic approaches that might capture the "'texture' of places" (see Rose, 2007; Dittmer and Gray, 2010). Like a number of other studies geared around affect, the project drew upon autoethnography in order to explore our own experiences. In this we found ourselves aligned with numerous scholars who emphasize narrative as a means of making sense of human experience and advocate for a research process in which social scientists turn "an ethnographic eye on themselves and their own lived experiences" (Bochner, 2012, p. 156; see also Ellis and Rawicki, 2013; Stewart, 2013). To improve upon our rigor as narrators, we supplemented our auto-ethnographic inquiry with close observations of Memorial technologies and their "becoming" with other visitors. We were using our own bodies as "instruments of research," to borrow from Longhurst *et al.* (2008, p. 215), but also observations of *other* bodies moving through the same space in a sensory ethnographic practice conducted over a six-day period in August 2012.

During our time at the Memorial, we used photography/video recording and audio recording and took detailed fieldwork notes in order to record the experiences unfolding within and around us. We took detailed and close photographs of museum exhibits, their various lighting strategies, and textual accompaniments, as well as panoramic snapshots that captured fuller pictures of entire galleries and the chronologies of their script (culminating in more than 700 photos). We captured sound-recordings from those exhibitions reliant on audio, and took in-depth notes regarding those that engaged with our bodies in ways that expanded beyond the visual and audible. Tone, image, rhythm, movement, pulses, vibrations, lighting, texture, and so forth were all considered in our field-notes for the role they might play in affecting the sensuous bodies of ourselves and visitors. This approach enabled us to engage with a mode of inquiry that assumed a world of becoming (Connolly, 2010) rather than a stable heritage site to be described and analyzed. We ourselves were a part of the museum assemblage during this time; as such, a continual part of our method was a reflexive concern with the role we played in shaping encounters, not only in terms of how we engaged with technologies but also with other museum-goers, as bodies, in the spaces of the museum—or, as

McCormack (2008, p. 2) puts it, "becoming affected and inflected by encounters with and within distinctive kinds of thinking space."

The rather recent turn in research methodology on which we draw here is paired with a more traditional concern with the history of museum practices. From its origin, the Memorial was charged to "commemorate the sacrifice of those Australians who have died in war" (Australian War Memorial, 2013, n.p.). As described in the introduction, this is a double-edged sword with respect to cultures of militarism, and by using a range of affective technologies to promote connection through space and time with Australian armed forces, a wide range of subjectivity-shaping intensities and empathies can emerge. We are particularly interested with how changes in affective technology since the Memorial's founding have attempted to fix those outcomes through what Anderson (2012) refers to as the application of biopower. However, most accounts of biopower ignore the role of virtual pasts in the techno-somatic assemblages that enable its applications; we hope to remedy this absence of the past in our accounts of the present.

Affective technologies

In the following sections we illustrate the developing nature of museum interpretation and evidence those touch-points we felt as being crucial to our co-production with the Memorial of experiencing subjects. We move from a consideration of dioramas, to audio-visual installations, to haptic advances, to simulations, in a bid to appreciate the incorporation of affective technologies into the museumscape.

Rescaling war: the Memorial dioramas

The use of dioramas as part of the interpretative repertoire drawn upon within the spaces of museums can be traced to the mid-nineteenth century; since then, they have been used to illustrate a range of cultures, habitats, and other natural histories. Their incorporation into the Memorial occurred in tandem with its official opening in 1941, though many of the dioramas on display today pre-date that—some by almost twenty years—as indeed does the impetus to include them. The following excerpt from a letter penned long before the Memorial was to open—by Charles Edwin Woodrow Bean, a war correspondent and official historian, to John Linton Treloar, Captain (though later reaching the rank of Lieutenant Colonel) in command of the Australian War Records Section—reveals something of the line of thinking behind their inception:

> A model of a sunken road, with figures and dug out entrances, is normally employed only as a sort of three dimensional map of the place it represents to explain it to your brain. But I think it could be made to explain it to your sensibilities as well—to give you the impression of the utter fatigue, or the danger, the feverish unreality which comes over everyday landscapes during battle times ... not a sort of Noah's Ark model, but a real picture, with

the atmosphere, the gradations of shade and colour, *the feeling of the scene*, created by an artist.

(letter from C.E.W. Bean to John Treloar, 14 May 1918, cited in Rutherford, 2004, emphasis added)

Crucial, then, to these early dioramas is the *feeling of the scene*, an attention to the replication of affective atmospheres emanating from the battlefield. And indeed, the traditional format of dioramas tends to de-emphasize human bodies in favor of the larger scene; the size of bodies is literally reduced to the point where it becomes materially difficult to render them with nuance. The materials of diorama then tend to contradict the logic of Deleuze and Guattari's (1987) concept of faciality, in which the screen is filled with a face that conveys pure affect. This traditional emphasis on atmosphere rather than bodies (despite the sheer number of the latter) is borne out in the early dioramas of the Memorial.

While the *First World War* galleries were closed to visitors in June 2013 for redevelopment,[2] those who moved with the "flow" engineered into the building's design prior to their closure would first have encountered the *Lone Pine* diorama in the Gallipoli gallery, which at the time of our fieldwork formed part of the larger *First World War* galleries. From there, they would move into the Western Front gallery, which contained a more fulsome example of the Memorial's collection of reduced-scale models, including those depicting *Pozieres 1916*, *Somme Winter 1916–1917*, *Bullecourt 1917*, *Ypres 1917*, *Dernancourt 1918*, and *Mont St Quentin 1918*, as well as a series entitled *The Evacuation of Wounded* (1917). A second series could be found in the Palestine gallery, comprising a sequence of nine diorama scenes that track the transportation of supplies in Palestine in 1916. All of these examples had a similar design insomuch as they combine painting with sculpture, with a composite of plaster, wood and wire used to fashion an "earthly" setting onto which painted lead and wax figures are inserted and set against a painted backdrop. In these terms, the dioramas were seemingly reduced to a simple assemblage of manufactured components, but they were of course simultaneously performing aesthetically, as pieces of visual art created to communicate trauma and the experience of loss. In them, we can trace something of a "conventional" semiotics of war representation that is recognizable to the viewer because of its links with a broader cultural assemblage around war. The diorama, then, is a representative, narrative artwork that, according to both C.E.W. Bean (excerpt above) and accompanying exhibition text found online and at the Memorial itself, has been designed to directly release the "devastation" and "danger of battle," thereby creating "an emotional account of the experiences of Australians who fought in the First World War."[3] This is the representational force of the diorama—and it has an affective intensity that might "grab" some visitors, though not all.

As an interpretative practice, dioramas ask us as visitors to follow their narrative, enter into it, and imagine our own bodies treading wearily across the muddy landscapes of their depiction. Their material components, however, are organized in ways that encourage that engagement at the visual level only, with visitors assuming a role of observer: nothing about the material components of the diorama and its display translates into something that is accessible to the ear, to touch,

Affecting the body 59

to taste, or to smell. With the *Lone Pine* diorama, for example, a feverish chaos permeates the diorama, with soldiers running, falling, and crawling in several directions at once—with Anzac soldiers caught mid-movement above ground and Turkish soldiers below (Figure 3.2).

It is a provocative piece, with the eye drawn to a central figure leaning backwards, arm stretched high above his head as if waving in greeting—cut down in the sprint across "no-man's land" or summoning his troops? Standing in front of that particular diorama we were moved to confusion—we were not able to quite fathom where they should all be heading—but that was the only intensity we were moved to share with them. Certainly we were engaged with, or pressed to think about, the people those figurines represented, but there was a qualitative difference between our miscomprehensions and the experience of utter confusion that advance undoubtedly produced. The thresholds of engagement were limited—we could not slide easily "into" the dioramas and their narratives—which in return may limit their potential affects. There are of course other material and immaterial influences at play in this example, including personal memories and connections that could be triggered by the affective affordances of the dioramas, along with a range of other cultural or historical energies. So too does the wider assemblage, within which the dioramas are placed, play a role. The larger exhibition hall, for example, does not radiate as a space of angst; it was quiet, respectful, even church-like, with the architecture of the hall staging an environment for contemplation but not so much from a personal and emotional perspective. The field-notes we gathered there did not flow with thoughts about atmosphere or reflect on spaces infused with moods and feelings. In part this is a consequence of the overall space of the hall, which at the

Figure 3.2 The *Lone Pine* diorama.

time of our fieldwork was light, clean, and ordered, with large objects fixed into the center of the room around which were featured the dioramas, information boards, and small sculptures raised upon plinths (Figure 3.3).

Each diorama was accompanied by extensive textual explanation, which afforded visitors the chance to read about the battle in question. This textualism shaped the space, as visitors moved quietly, with heads bowed over the text accompanying each display, occasionally raising their eyes to the smattering of flags and paintings that adorned each wall. In an audio recording of the room, made as part of a sound survey of the Memorial, the air conditioning is audible.

As Insley (2008) has argued, the evolution of museum interpretation can in fact be traced through an analysis of diorama developments made in a bid to update and refresh displays. Of particular interest to us, then, is the Memorial's Kapyong 1951 diorama, completed more recently in 2007 and found in the *Conflicts 1945 to Today* galleries, specifically the Korean Gallery. In contrast to those found in the *First World War* gallery, the Kapyong diorama has been strengthened by the use of digital technology used to "recreate the sights and sounds of the battlefield."[4] Situated in a small, darkened space sandwiched between two "corridors" that wend their way through the gallery, the diorama is recessed into the wall and fronted by low glass shelving that displays related artefacts and objects. At times, the room is bathed in a bluish-purple light; at others it is far darker. Sounds of the battlefield and other audio affects are piped into the space, along with recollections voiced by Corporal Ray Parry, a soldier involved in the battle of Kapyong. What makes this diorama particularly unique, however, is not simply that it has been laced together with an audio

Figure 3.3 The *First World War* galleries, prior to their redevelopment.

Affecting the body 61

accompaniment; rather, it is the fact that the faces of the figures employed within it are sculpted as individual portraits that resemble the faces of four soldiers who *lived* the scene depicted (Figure 3.4). Here, we see the innovative use of the diorama format in ways that incorporate Deleuze's faciality, and build in the capacity for embodied empathy.

Figure 3.4 The Kapyong faces: in photographs and the *Kapyong* diorama.

62 *Jason Dittmer and Emma Waterton*

Reading this information, both authors were physically alerted to the fact that this was something we cared about, a realization that was distributed throughout our bodies with the widening of eyes, a slight pulling back from the display and a muscular tensing. This, for us at least, triggered something of a connection between ourselves and the people remembered, and between us and the darkened spaces of the exhibit. Indeed, we were being asked to pay attention to real people: Their faces and their voices. But in reflecting on this more recent diorama, equipped with the sound contribution of Ray Parry's voice, we were also moved to reflect on the role played by "unsound" in the *First World War* galleries. Was this "unsound" not equally powerful at compelling a response from visitors: to talk in hushed tones and moderate their behavior by virtue of the cues commanded by the museumscape itself? Further attention, it seemed to us, was required for a consideration of the role played by sound and the audio, to ascertain if sound might be differently affective to vision.

The sounding Memorial

Our next example looks more closely at the role played by audio-visual technologies in eliciting affective intensities within museum settings, and in this can be considered alongside a number of cogent critiques examining the relationships between technology and affect (see Witcomb, 2013a for discussion of technology and affect in the museum context). In addition to visually focused stimuli, in this section we are also struck by the role played by sound in concert with the visual: How does sound contribute to an immersive atmosphere in ways that moderate mood and motivate a series of affects? To further develop a line of inquiry into the role played by audio-visual strategies in enabling the transmission of affect, we focus on the *Menin Gate* exhibition—a very small exhibition space tinctured with both sound and atmosphere. This exhibition revolves entirely around one visual and one audio focus deployed in concert: the *Menin Gate at* Midnight,[5] painted by Will Longstaff, and Schubert's *Unfinished Symphony*, recorded by the Anima Eterna Symphony Orchestra and delivered via a hidden sound system in the spaces of the exhibition. In addition to the painting itself, two large panels of interpretive text adorn the room, along with a small panel providing specific exhibition detail and a metal bar that carefully partitions visitors from the painting itself. Apart from these intrusions, the exhibition space is empty. It is a dark space. The walls are painted black and there is only a limited source of lighting filtered into the room via a few spotlights on the ceiling, some directed at the painting and others at its interpretation. The painting itself depicts the Menin Gate Memorial—a memorial to the missing, who are absent yet present—which is located in Ypres, Belgium. The structural presence of the Memorial, which stands against a darkened sky, seems to dominate the painting; that is, until one's eye flickers to the harvested cornfields in the foreground, moves past the small clusters of "poppies" and realizes that the painting's textures are in fact evocative reminders of fallen soldiers, their ghostly forms marching across the scene (Figure 3.5).

The familiarity of the cornfield is ruptured as we are forced to wonder if those are poppies in the foreground after all. The volume and intensity of the accompanying

Figure 3.5 The ghostly figures of the *Menin Gate at Midnight*.

symphony are uneven in their presentation—at times quite sedate, at others reverberating and loud. Visitors passing nearby, as we were on one occasion, can be drawn toward the little darkened gallery at those particularly loud moments, when the sound is amplified beyond the immediate confines of its display, causing regular eddies in the patterns of movement within the Memorial.

In this part of the museum, it is the painting and the symphony, in concert with the darkened atmospheric space in which they are fused, which stand out and, in so doing, create a certain mood for visitors. We might form an impression, but it is one, to borrow from Ahmed (2014), that depends almost entirely upon the way in which those sights and sounds *impress* upon us or leave their mark: "*we need to remember the 'press' in an impression*" (Ahmed, 2014, p. 6, emphasis in original). This is amplified when the small room is shared with other visitors. The haunting sounds and narrow beams of illuminations, alighting on the ghostly figures of the painting, intersect with those human bodies as they move and pause in the room. The resultant atmosphere, then, is co-produced, to borrow from Sørenson (2015, p. 66), through "the co-presence of various bodies (humans, things, architecture)."

The technological simplicity of the Menin Gate chamber stood in contrast to some of the more ornate cinematic exhibits. For example, the "Anzac Hall" is given over entirely to three sound-and-light shows: *Striking by Night* (a filmic recreation of a bomber mission over Berlin), *Sydney under Attack* (a detailed account of Sydney's response to a Japanese mini-submarine in its harbor), and *Over the Front: the Great War in the Air* (a cinematic portrayal of biplane warfare). Each sound-and-light show is paired with at least one artefact: a Lancaster bomber, a German U-boat, and several biplanes, respectively (see Figure 3.6).

64 *Jason Dittmer and Emma Waterton*

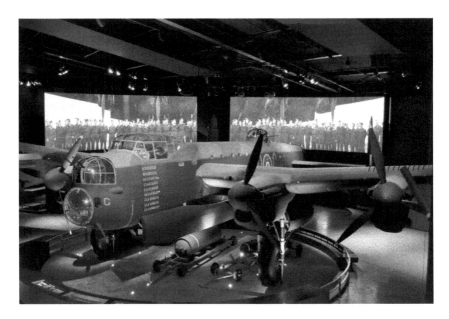

Figure 3.6 "Anzac Hall," *Striking By Night* sound and light show.

This juxtaposition enables viewers simultaneously to orient their body in relation to the physical artefacts, getting a sense of their size and relative vulnerability, and to take in a narrative—drenched in life-and-death drama—in which those artefacts feature. By relating the artefact, the film, and the embodied listener/viewer, the Anzac Hall managed to conjure a powerful array of emergent forces. For instance, it was difficult to listen to the wartime hunt for a Japanese midget submarine in Sydney's harbor (narrated entirely from the Australian perspective) without making visual reference to the submarine hung diagonally in the space of the hall, its smooth hydrodynamics ruptured by the depth charges that sank it. The virtual bodies of its occupants were called forth, their fate tacit but certain.

Indeed, the sound-and-light displays of the Anzac Hall all attempted to produce some feeling of what it was like to participate in war, effectively linking the listener/viewer to the combatants of the past. *Over the Front* was particularly adept at this, following one biplane pilot into a large aerial dogfight. The film is shown in a range of split-screen orientations, fragmenting the scene. The protagonist's perspective is juxtaposed with those of other pilots looking at him and other fixed third-person perspectives of the battle; as the biplanes swirl around one another the variety of perspectives both bewilders and disorients while also maintaining the protagonist as a focal point. The battle proceeds, although the first notes of an epic dirge begin to hint that this is not a comfortable war story. Indeed, we watched the audience transfixed as they struggled to assimilate all the perspectives simultaneously on display. The music swells, and the central protagonist's plane accumulates too much damage. The pilot bails out of his plane, but it is clear that he is too badly wounded to survive. In the audience, a mother turns her young son away from the screen and whispers something inaudible to him as she leads him away.

Over the Front is an example of how the visual field can be shaped to both engender the disorientation and confusion of war, while also predisposing the audience to identify with (in this case) a pilot who gave all. This process of subjectification parallels other cinematic attempts to have the audience identify with a protagonist, a common feature of narrative films. However, unlike most films with which the audience is familiar, this film ends in the protagonist's death. The virtual presence of the war film genre in the Anzac Hall, structuring listener/viewer expectations of *Over the Front*, both heightens the affective intensity when the outcome is clear and enables unexpected moments, such as the young son's exposure to the cost of war, when militarism is undercut through somatic identification with the dead. Deviant moments such as this were not witnessed in relation to the dioramas discussed earlier; their elevated perspective on a military landscape (with the exception of Kapyong) precluded identification with any particular soldier, and their use of narrative either focused on processes (such as logistics) or on the overall outcome of a battle.

Haptic advances

The Memorial is arguably at the forefront of museological practice that attempts to incorporate technologically assisted haptic simulations and multimedia modules into the visitor experience. While many museums (including the Memorial) have focused on generating opportunities for visitors to touch artefacts (or replicas of artefacts) in order to provide another form of visitor access to objects, the Memorial has pioneered passive forms of haptic experience that work to position visitors as military subjects during wartime. In short, these technologies do not revolve around the visitor touching, but rather the visitor being touched, buffeted, or otherwise made to *feel*. This addition of the haptic, as Paterson (2006, p. 695) has argued, "enhances the experience of the user, and is commonly expressed in terms of immersion or presence." Though Paterson is talking primarily about the incorporation of haptic technologies into video-games, his observations about a sense of presence are worth reflecting upon in the museum context, too. In his 2006 article, "Feel the presence: Technologies of touch and distance," Paterson argues that:

> Like mimesis, haptics becomes a form of production, to enhance operations, to provide richer user experience, even to promote experimentation through free-flowing play and creativity. But, primarily, haptics creates a whole set of forces and corresponding sensations, a fusion of feelings that are generated and retro-engineered from the perspective of the user, not imposed by programmers or coders themselves, in order to recreate the right "feeling".
>
> (p. 698)

A simple version of this sort of process was found in the Second World War galleries: an audio-visual display (*Zombie Territory*) narrates the challenges of battling against Japanese "Kamikaze" pilots in the Pacific Theatre, drawing the eye of those walking through the gallery and causing many to stop as images of flaming planes hurtled across the screen. Unnoticed by many, however, was the fact that they were now standing upon a small simulation of a metal ship's deck, complete with non-slip texturing and a motor underneath to simulate the vibration caused by the ship's engines. The floor and the video are synced, so that the floor only vibrates when the video is showing. This particular haptic advance is a subtle thing; many walked over the "ship's deck" without even noticing it was there, particularly when the video didn't work to arrest their movement. One group of teenagers, however, focused entirely on the floor, ignoring the video and instead calling out for their friends to join them on the thrumming deck; they identified "the best" spots on the deck. As they eventually moved off, one teenage girl announced: "That was weird." This interaction was unusual in its focus on the deck. More often, the floor remained a source of background affect, patterning the experience of those watching *Zombie Territory*. Crucially, the deck locates the viewer on the ships under Kamikaze attack, tacitly reinforcing an assumed Australian (or allied) subject position and thereby using affect to position the museum-goer in relation to a particular subjectivity.

There are other instances in which the Memorial uses haptic technologies to circulate affective experience. Elsewhere, we have explored the light and sound installations "Royal Air Force Bomber Command" and "Dust Off," both of which dabble with the haptic (see Waterton and Dittmer, 2014). It is worth revisiting the latter, "Dust Off," which is one of two light and sound shows currently installed in the Vietnam War gallery, the other being "Heliborne Assault." Both revolve around an Iroquois helicopter and events of Operation Bribie, which took place in February 1967 (see Figure 3.7).

Figure 3.7 "Dust Off" in the Vietnam War gallery.

In addition to the incorporation of black-and-white video footage, loud machine-gun fire, flashing lights, and physical objects—all of which charge the atmosphere with a sense of rapid movement—the "Dust Off" installation also relies in part on the sensibilities of the museum-goer's body. First, this is achieved by the sheer volume of accompanying sounds—the machine-gun fire, the rain, and sounds of a helicopter in flight—which at times prompted us and those around us to flinch, grimace, or cover our ears. It was LOUD. But positioned in the primary spaces of this sensuous, interpretive experience, there is a second assault on the museum-goer's body, which will at some point in the "show" be subjected to a very real buffeting sensation. This is produced by overhead fans, spinning at full pelt, which have been designed to simulate the downward gusts of helicopter rotor blades as they generate lift for flight. In that moment, the museum-goer literally *feels* history on their skin, in their hair, in their eyes, on their body. It strikes at them, caressing them, and moves them. Temperatures change and our bodies cool; we are less steady and kinesthetically sense motion; and the history on display no longer looks real, it *feels* real (after Paterson, 2006). In those moments, the spaces between the museum and museum-goer shrink: the former "ends," or so it would seem, right there upon the skin of the latter. Certainly, there is no longer a clear separation between the two. Instead, the bodies of museum-goers become less an elusive element of the interpretative experience and more a crucial part of its mechanism, through which they are being asked to configure meaning. Indeed, they are part of the narrative process through which they, as museum-goers, go about disclosing the world of warfare: seeing other visitors respond to the fans (holding onto hats, staring upwards, etc.) helps to complete the scene. The video around which the haptic simulations revolve runs for just over six minutes and the fans/simulated blades are working only for a portion of that time, yet their motion nevertheless seems to stretch out the experience in ways that exceed the visual and audible.

There are other attempts at immersion at the Memorial that extend the haptic advances outlaid above in that they move beyond "touch" and "sensation" toward mimicking a sense of being *in* the spaces of warfare. Here, attempts are made to further collapse the "distance" associated with representations of war by introducing simulations into the proximate space of the museum-goer. Ellington *et al.* (1998, pp. 2–3) describe simulations as "ongoing representations of real situations," with pedagogical value for the inculcation of embodied memory that might be applied in future, anticipated events (Anderson and Adey, 2011). As such, they involve the production of topological spaces in which the simulation's scenario and rules are operative. In Game Studies, this space is defined by the "magic circle" (Klabbers, 2009), which separates the "game" from everyday life in both space and time. Of course, given that at a minimum the bodies of those participating in the simulation cross over into and out of the magic circle, it is clear that the magic circle is leaky, or, worse, impossible to define in topographical space (Castronova, 2005). It is more helpful to consider simulations as assemblages of objects, bodies, and the discourses that code them, with the "simulation" as an emergent effect of their particular topological constellation. These simulations are therefore both set apart from "our world" and yet also fully part of it, as the elements of the assemblage function multiply.

68 *Jason Dittmer and Emma Waterton*

What distinguishes this sort of interpretive strategy from the previous haptic examples is the degree of perceived agency afforded the visitor. Of course, in the haptic examples the passive relation to the sense of touch in no way minimizes the agency of the visitor's body, which is always fully implicated in the production of the exhibition. Nevertheless, simulations offer a more interactive experience, with museum-goers offered the chance to interact with various elements of the simulation. One example is the bridge of the HMAS Brisbane, an Australian guided-missile destroyer that served in Vietnam and the first Gulf War (see Figure 3.8).

After decommissioning, the bridge of the ship was removed and installed in the Memorial, with the windows replaced by computer screens on which a naval scenario unfolds. Plexiglas screens off the front of the bridge from museum-goers, but otherwise visitors are allowed to move about the bridge, touch whatever interests them, and otherwise imagine themselves underway. The sensing body of the visitor is wholly enclosed within the simulated space of a ship's bridge at sea.

This immersion was heightened in an area of the Memorial called the "Discovery Zone." This area, aimed at visiting school groups and other children, offered a complete sensory experience that allowed visitors to "see, touch, listen, and smell their way through five distinct environments that replicate the living conditions during five periods of conflict since Australia's Federation."[6] These include a trench on the Western Front, a Cold War-era submarine, a helicopter from the Vietnam War, an Australian backyard barbeque during the Second World War, and a temporary building housing Australian peacekeepers "somewhere in an emerging nation" (see Figure 3.9).

Figure 3.8 The Bridge, HMAS Brisbane.

Figure 3.9 The Discovery Zone helicopter.

These spaces not only attempt to simulate life in other places and times, but also enable embodied movement within them. Visitors are able to look through the submarine periscope, take the stick of a helicopter, and otherwise actively participate in the scenarios being replicated. It is notable that, excepting the helicopter, each of these spaces is a domestic space (of a sort), a place where Australians *lived* during wartime; children are thus encouraged to compare their experience of these spaces with their own living conditions. This emphasis on the domestic privileges a somatic comparison organized around such embodied sensibilities as comfort, warmth, dampness, and spatial cramping. Given the permeability of the "magic circle," children's experience of "climb[ing], jump[ing], crawl[ing], touch[ing] and explor[ing]"[7] in the Discovery Zone uses their bodies as a vector through which the peacetime domestic is brought into topological relation with the wartime domestic. The presumed outcome of such a comparison is an affective orientation of gratitude to those who suffered thus, given the narrative of sacrifice that suffuses the Memorial. This somatic predisposition leaves itself on the bodies of the children in the Discovery Zone, enabling this trace to return with the school groups to their various home towns, helping to construct a national body politic marked by the simulation-spaces of the Memorial.

Conclusions

In this chapter, we have introduced and explored a range of affective experiences found at the Australian War Memorial in Canberra. In them, we have attempted to articulate the ways in which contemporary museum practice might engage in the provocation of affect through sensorial forms of experience. We began with reflections on early attempts at interpretation found at the Memorial, the dioramas, before moving on to explore the incorporation of light, sound, and movement. To close, we introduced what we see as moves to incorporate a more immersive strategy, through which museum-goers are invited to "feel the presence" of war, to borrow from Paterson (2006). In reflecting on those examples, we want to avoid any teleological argument regarding technology and the ability of affective technologies to produce specific political subjects in a top-down fashion. We also want to be clear about our understanding of how affect and its potential contagion works: not all visitors to the Memorial will share our feelings and responses to, or even descriptions of, its atmosphere. We are, after all, precisely the sort of museum-goers that undoubtedly underlie visions of the assumed audience to which the Memorial "speaks" and makes comfortable: white, middle-class, educated, Western, English-speaking, and so forth. In other words, we are represented, invited, and not forced to feel the pain of exclusion (the mechanisms of which are a topic for another chapter). To that effect, our research recorded many moments of excess in which museum-goers performed other subjectivities when confronted with these technologies; subjects continue to be nomadic and excessive of disciplinary power. Our purpose was to document, in terms of museological practice, the relationship between heritage and affective technologies as encountered through the experiences we felt compelled to "pick up" or "enter into" as we moved through the Memorial. By and large, we have been building an argument that suggests that the historical shift from dioramas, to audio-visual technologies, to haptic technologies reflects an ongoing concern of curators and historians with conveying the experiences of war via techniques of the body. The shift from a static, purely visual technology (diorama) to moving pictures with sound (as in the Anzac Hall), and the subsequent augmentation of those audio-visual experiences with haptic technologies, reflects a multi-modal intensification of these techniques of the body as the desired outcome—a museum-going body marked with traces of militarism by the experience of the Memorial—becomes more important.

The wider culture of militarism has been under sustained assault as postcolonial critique becomes mainstream, as financial crisis undercuts military spending, and as the human cost of recent intervention haunts contemporary foreign policy debates. Given that the conventional justifications for permanent war no longer convince, or at least require buttressing, spaces such as the Memorial become crucial sites for the production of militarized bodies. This is the case not necessarily in terms of the recruitment of new soldiers (although high-school "cadet" programs make up a significant number of the overall visitors), but rather through the somatic patterning of visitors' bodies through experience of "war," of the past, and of the Memorial. The production of embodied predispositions toward Australian military bodies, as strong individuals, and empathy toward them, is crucial to the maintenance of contemporary militarism.

This is, however, a schizophrenic militarism, as manifested in the split purpose of the Memorial. On the one hand, it is a museum celebrating military exploits; on the other it is a sacred memorial to their sacrifice. The affective predisposition engendered by the techno-cultural assemblage of the museum elements exists in tension with the memorial elements' emphasis on loss and sacrifice. The emergent outcome—a somatic trace that predisposes subjects to support contemporary and future intervention or to oppose them—is not determined by the Memorial itself, but rather by its interaction with the pre-existing political subjects who enter into assemblage with it, if only for a while.

Notes

1 "Affecting Heritage: revisiting the geographies and politics of heritage through affect and emotion," double session at the 4th *International and Interdisciplinary Conference on Emotional Geographies*, 1–3 July 2013, hosted by the University of Groningen, The Netherlands; "Affecting Heritage: revisiting the geographies and politics of heritage through affect and emotion," double session at the RGS-IBG Annual International Conference, 28–30 August 2013, hosted by Imperial College London.
2 The First World War galleries were closed for redevelopment as part of the nationwide 1914–18 centenary commemorations, and were officially reopened on 22 February 2015. Information provided about the new First World War galleries on the Memorial's website confirms that the "iconic dioramas will remain an integral part of the galleries": www.awm.gov.au/1914-1918/first-world-war-galleries/ (accessed 22 October 2014).
3 See the Australian War Memorial online encyclopedia entry for "Dioramas": www.awm.gov.au/encyclopedia/dioramas/ (accessed 22 October 2014).
4 Quoted on the Australian War Memorial webpages: www.awm.gov.au/visit/post-1945-galleries/korea/ (accessed 22 October 2014).
5 The *Menin Gate at Midnight* painting, a permanent fixture at the Memorial since its opening in 1941, was on loan to the Canadian War Museum in Ottawa for a First World War centenary exhibition from September 2014, returning in mid-2015 (www.awm.gov.au/media/releases/memorial-farewells-menin-gate-lions-and-painting-overseas-loan/).
6 This extract and next are quoted on the Australian War Memorial webpages: www.awm.gov.au/education/schools/discovery-zone/ (accessed 22 October 2014).
7 Quoted on the Australian War Memorial webpages: www.awm.gov.au/visit/discovery-zone/ (accessed 22 October 2014).

Bibliography

Ahmed, S. 2014. *The cultural politics of emotion*. Edinburgh: Edinburgh University Press.
Alderman, D.H. 2010. Surrogation and the politics of remembering slavery in Savannah, Georgia. *Journal of Historical Geography*, 36(1): 90–101.
Anderson, B. 2012. Affect and biopower: towards a politics of life. *Transactions of the Institute of British Geographers*, 37(1): 28–43.
Anderson, B. and Adey, P. 2011. Affect and security: exercising emergency in UK civil contingencies. *Environment and Planning D: Society and Space*, 29: 1092–109.

Australian War Memorial. 2013. About the Australian War Memorial. www.awm.gov.au/about/ (accessed 11 March 2013).
Azaryahu, M. 1996. The power of commemorative street names. *Environment and Planning D: Society and Space*, 14(3): 311–30.
Bochner, A.P. 2012. On first person narrative scholarship: autoethnography as acts of meaning. *Narrative Inquiry*, 22(1): 155–16.
Byrne, D. 2013. Love and loss in the 1960s. *International Journal of Heritage Studies*, 19(6): 596–609.
Carter, S. and McCormack, D.P. 2006. Film, geopolitics and the affective logics of intervention. *Political Geography*, 25(2): 228–45.
Castronova, E. 2005. *Synthetic worlds: the business and culture of online games*. Chicago: University of Chicago Press.
Colls, R. 2012. Feminism, bodily difference and non-representational geographies. *Transactions of the Institute of British Geographers*, 37(3): 430–45.
Connolly, W. 2010. *A world of becoming*. Durham, NC: Duke University Press.
Cosgrove, D. 1985. *Social formation and symbolic landscape*. Totowa: Barnes & Noble Books.
Crang, M. 2003. Qualitative methods: touchy, feely, looky, see? *Progress in Human Geography*, 27(4): 494–504.
Crang, M. and Tolia-Kelly, D. 2010. Nation, race and affect: senses and sensibilities at national heritage sites. *Environment and Planning A*, 42(10): 2315–31.
Cresswell, T. 1996. *In place/out of place: geography, ideology and transgression*. Minneapolis: University of Minnesota Press.
Crouch, D. and Parker, G. 2003. 'Digging up' Utopia? Space, practice and land use heritage. *Geoforum*, 34(3): 395–408.
Daniels, S. 1993. *Fields of vision: landscape imagery and national identity in England and the United States*. Princeton: Princeton University Press.
Deleuze, G. 2005a [1985]. *Cinema 1: the movement-image*. London: Athlone.
Deleuze, G. 2005b [1986]. *Cinema 2: the time-image*. London: Athlone.
Deleuze, G. and Guattari, F. 1987. *A thousand plateaus*. London: Continuum.
Dittmer, J. 2013. *Captain America and the nationalist superhero: metaphors, narratives, and geopolitics*. Philadelphia: Temple University Press.
Dittmer, J. and Gray, N. 2010. Popular geopolitics 2.0: towards new methodologies of the everyday. *Geography Compass*, 11(4): 1664–77.
Dwyer, O.J. and Alderman, D.H. 2008. Memorial landscapes: analytical questions and metaphors. *Geoforum*, 73(3): 165–78.
Ellington, H., Gordon, M. and Fowlie, J. 1998. *Using games and simulations in the classroom*. London: Kogan Page.
Ellis, C. and Rawicki, J. 2013. Collaborative witnessing of survival during the Holocaust: an exemplar of relational autoethnography. *Qualitative Inquiry*, 19(5): 366–80.
Graham, B., Ashworth, G. and Tunbridge, J. 2000. *A geography of heritage: power, culture and economy*. London: Arnold Press.
Hardy, D. 1988. Historical geography and heritage studies. *Area*, 20(4): 333–8.
Hoelscher, S. and Alderman, D. 2004. Memory and place: geographies of a critical relationship. *Social and Cultural Geography*, 5(3): 347–55.
Insley, J. 2008. Little landscapes: dioramas in museum displays. *Endeavour*, 32(1): 27–31.
Johnson, N. 1995. Cast in stone: monuments, geography, and nationalism. *Environment and Planning D*, 13(1): 51–65.

Johnson, N. 1996. Where geography and history meet: heritage tourism and the Big House in Ireland. *Annals of the Association of American Geographers*, 86(3): 551–66.
Johnson, N. 2015. Heritage and geography, in E. Waterton and S. Watson (eds) *The Palgrave handbook of contemporary heritage research* (pp. 159–70). Basingstoke: Palgrave Macmillan.
Kearney, A. 2009. Homeland emotion: an emotional geography of heritage and homeland, *International Journal of Heritage Studies*, 15(2–3): 209–22.
Kearns, R.A. and Berg, L.D. 2002. Proclaiming place: towards a geography of placename pronunciation. *Social and Cultural Geography*, 3(3): 283–302.
Klabbers, J.H.G. 2009. *The magic circle: principles of gaming and simulation*. Rotterdam: Sense Publishers.
Leib, J.I. 2002. Separate times, shared spaces: Arthur Ashe, Monument Avenue and the politics of Richmond, Virginia's symbolic landscape. *Cultural Geographies*, 9(3): 286–312.
Longhurst, R., Ho, E. and Johnston, L. 2008. Using 'the body' as an 'instrument of research': Kimch'i and pavlova. *Area*, 40(2): 208–17.
Lowenthal, D. 1985. *The past is a foreign country*. Cambridge: Cambridge University Press.
McCormack, D.P. 2008. Thinking-spaces for research-creation. *Inflexions*, 1(1). www.senselab.ca/inflexions/Inflexions%20Issue%20One%20McCormack%20final%20word%20version.doc.pdf (accessed 19 June 2013).
Muzaini, H. and Yeoh, B.S.A. 2005. War landscapes as 'battlefields' of collective memories: reading the reflections at Bukit Chandu, Singapore. *Cultural Geographies*, 12(3): 345–65.
Nelson, H. 1997. Gallipoli, Kokoda and the making of national identity. *Journal of Australian Studies*, 21(53): 157–69.
Nora, P. 1989. Between memory and history: *les lieux de memoire. Representations*, 26: 7–24.
Paterson, M. 2006. Feel the presence: technologies of touch and distance. *Environment and Planning D: Society and Space*, 24: 691–708.
Protevi, J. 2009. *Political affect: between the social and the somatic*. Minneapolis: University of Minnesota Press.
Raento, P. and Watson, C.J. 2000. Gernika, Guernica, Guernica? The contested meanings of a Basque place. *Political Geography*, 19: 707–36.
Rose, G. 2007. *Visual methodologies*, 2nd edition. Thousand Oaks: Sage.
Rutherford, R. 2004. Teaching the terrain: First World War battlefields at the Australian War Memorial, *The Free Library*. Australian Map Circle. www.thefreelibrary.com/Teaching+the+terrain%3a+First+World+War+battlefields+at+the+Australian...-a013429296323 (accessed 23 October 2014).
Sørenson, T. (2015) More than a feeling: towards an archaeology of atmosphere. *Emotion, Space and Society*, 15: 64–73.
Stewart, K. 2013. Regionality. *Geographical Review*, 103(2): 275–84.
Thrift, N. 2004. Intensities of feeling: towards a spatial politics of affect. *Geografiska Annaler: Series B, Human Geography*, 86(1): 57–78.
Thrift, N. 2008. *Non-representational theories*. London: Routledge.
Tolia-Kelly, D.P. 2004. Landscape, race and memory: biographical mapping of the routes of British Asian landscape values. *Landscape Research*, 29(3): 277–92.
Tolia-Kelly, D.P. 2006. Affect – an ethnocentric encounter? Exploring the 'universalist' imperative of emotional/affectual geographies. *Area*, 38(2): 213–17.

Uzzell, D. 1989. The hot interpretation of war and conflict, in D.L. Uzzell (ed) *Heritage interpretation: the natural and built environment* (pp. 33–47). London: Belhaven Press.

Vincent, R. 2014. Experiments with bodies in social space: towards a contemporary understanding of place-based identities at the social history museum. *Museum Management and Curatorship*, 29(4): 368–90.

Waterton, E. 2014. A more-than-representational understanding of heritage? The 'past' and the politics of affect. *Geography Compass*, 8(11): 823–33.

Waterton, E. 2015. Visuality and its affects: some new directions for Australian heritage tourism. *History Compass*, 13(2): 51–63.

Waterton, E. and Dittmer, J. 2014. The museum as assemblage: bringing forth affect at the Australian War Memorial. *Museum Management and Curatorship*, 29(2): 122–39.

Waterton, E. and Watson, S. 2014. *The semiotics of heritage tourism*. Bristol: Channel View Publications.

Witcomb, A. 2007. The materiality of virtual technologies: a new approach to thinking about the impact of multimedia in museums, in F. Cameron and S. Kenderdine (eds) *Theorizing digital cultural heritage: A critical discourse* (pp. 35–48). Cambridge, MA: MIT Press.

Witcomb, A. 2012. On memory, affect and atonement: the Long Tan memorial cross(es). *Historic Environment*, 24(3): 35–42.

Witcomb, A. 2013a. Understanding the role of affect in producing a critical pedagogy for history museums. *Museum Management and Curatorship*, 28(3): 255–71.

Witcomb, A. 2013b. Using immersive and interactive approaches to interpreting traumatic experiences for tourists: potentials and limitations, in R. Staiff, R. Bushell and S. Watson (eds) *Heritage and tourism: place, encounter, engagement* (pp. 152–70). London: Routledge.

4 Affect and the politics of testimony in Holocaust museums

Steven Cooke and Donna-Lee Frieze

In 2010, the Jewish Holocaust Centre (JHC) in Melbourne unveiled its new permanent exhibition, replacing one that had remained, mostly unchanged, for the past twenty years since a major redevelopment in 1990. The former exhibition had received many plaudits from visitors and reviewers for its homespun, intimate aesthetics and display techniques, largely based on photographs (Light, 2002). Central to the JHC's role as a site of mourning *and* education, the exhibition included the use of personal testimony from Melbourne's Holocaust survivors, both in the exhibition displays and through the survivors who ran the museum and shared their stories with individuals and groups. A continuing anxiety over the thirty-year history of the JHC has been the passing of Holocaust survivors. These survivor guides were central to the discourse of a "living museum," seen as giving the organization its uniqueness compared to other Holocaust institutions as well as other museums generally. Oral survivor testimony was perceived as a key aspect of the museum's pedagogic potential: The affective encounter with survivors telling their stories while the visitor was viewing the exhibition was identified as having a transformative function, particularly for school-age students who comprised the majority of the visitors. The exhibition redevelopment in 2010 was, in part, a manifestation of that anxiety, with the urgency to incorporate survivor video-testimony increasing as the survivors aged and their memories faded. However, replacing a much-loved exhibition was fraught with difficulties, as the survivors were still very much part of the museum decision-making process. As the JHC had gradually moved from a survivor-volunteer based place of mourning to a professionally run museum with paid employees, there was a need to preserve the voices of the survivors who had been guides at the museum since its opening. Approaching a time when the survivors are not bodily present to share their stories, how might their testimonies still have transformative potential and inform interpretive techniques?

Building on the work of Walter Benjamin, Simon *et al*. (2001, p. 285) question the "how and what" of Holocaust remembrance, which they argue should be "interminably current," akin to James E. Young's (1993) suggestion that the most appropriate way to remember the Holocaust is to continually debate this act of remembering. The anxiety over the passing of the survivor generation is coupled with the "affective turn" in the humanities and social sciences to form a series of issues about the response of Holocaust museums and the relationship between memory, testimony, pedagogy, and affect in heritage, particularly what Simon *et al.* (2000, p. 7) term "remembrance as critical learning." This raises a number of questions. What is the role of the survivor guides within Holocaust education at the JHC? How does the JHC approach the transition from live guides to their video-testimonies within the context of affect? What are the differences between video-testimony and oral testimony in the museum context? How might a "politics of listening" (Witcomb, 2014) and affect in the museum open up new forms of civic engagement? We argue that these questions are relevant not only to the specific case of Holocaust heritage sites—whether they are the sites of destruction or monuments and memorials at significant geographical distances from places of atrocity, such as the JHC—but also, given the increased use of narrative and affective techniques within museums and heritage sites, to other heritage places and landscapes.

The work of testimony in the museum

The critical examination of the past in the present, particularly the growth in interest in dark or dissonant heritage (Tunbridge and Ashworth, 1996), including "places of pain and shame" (Logan and Reeves, 2009), has been a theme of cultural heritage and museum studies in recent years. Arising from a number of disciplines, including human geography (see Crang and Tolia-Kelly, 2010), anthropology, tourism studies, cultural studies, and sociology, this work has foregrounded the politics of the past in the present and the way that difficult histories are represented, understood, and performed at such sites. This brings our attention to the relationship between people and place and much work has been done on the pedagogic potential of visiting Holocaust sites, whether the sites of the atrocity (Charlesworth, 1994, 1998; Bastel *et al.*, 2010; Maitles and Cowan, 2008) or museums (Smith, 2007). Smith (2007, p. 273) states that a key factor in the learning experience in museums is being away from the everyday, entering a world "which lies outside one's own experience," creating transition and informing the visitor. This "numinous" experience—the performance of being there, the immediacy of place—provides an affective experience that leads to further action and behavioural changes (Latham, 2013). New methodologies are being developed to understand this dynamic (see Schorch, 2013), but further research is needed to understand the role of emotional and affective registers from the perspective of museum visitors: A point to which we return in the conclusion.

The idea of a civic role for museums is not new and has a long history (Hein, 2012; see also Bennett, 1995). However, more recent work has discussed new approaches that explore the role of affect in museums and heritage, covering subjects such as interpretation (Staiff, 2014), heritage tourism (Waterton and Watson, 2014), pedagogy (Cooke and Frieze, 2015b; Witcomb, 2013), and exhibition design (Witcomb, 2010, 2014). Within museum studies, the work of Andrea Witcomb has been particularly influential, articulating a movement away from a pedagogy of walking (Bennett, 1995) to a "pedagogy of feeling," particularly the way in which affective "responses open up possibilities for interpretation that engage with the politics of representation and identity formation" (Gregory and Witcomb, 2007, p. 262). A component of this is the use of personal narratives in the museum (Kelly, 2010; Henning, 2006), which allows multiple voices and, Marzia Minore argues, has the potential to reduce "the historical and cognitive gap between the museum and the public" (Minore, 2012, p. 144). This may invoke "a greater emotional involvement" (Minore, 2012, p. 144) on the part of the visitor. The argument here is that the inclusion of personal stories results in the democratization of the museum spaces, through the decentering of the museum's authority.

This process of democratization is more complex in Holocaust museums, where personal narratives, Holocaust testimony, have an additional function, with testimonies not only used as evidence of a crime but also to evoke questions regarding survival, dehumanization, physical and emotional violence, group destruction, and living with deep trauma. This power ascribed to written and oral testimony in understanding historic and momentous events is evident in other contexts too. For example, historian Patrick Lindsay, writing about the historiography of the First World War, argues that "those who were there speak with the greatest authority" (Lindsay, 2013, p. 10; see also Ziino, 2010). This authority, as testified by victims of atrocities, is seen as a moral and effective pedagogic approach. As Simon *et al.* (2000, p. 4) claim:

> This is a hope that anxiously attends to a horrific past in expectation of the promise that, by investing attention in narratives that sustain moral lessons, there will be a better tomorrow. It is a tomorrow fully cognizant of the warning that forgetting could lead to a return to the horrors of history.

As Simon *et al.* further argue, however, this repetitive "moralizing pedagogy ... provides no guarantee of a redeemed society" (2000, p. 4), although the alternative—not to eulogize on the lessons of hatred—suggests a society that is deemed to have forgotten, and hence, not learnt. For survivors and educators of atrocity history, remembrance through testimony is resistance of a violent future, or at least the hope of it. The authors utilize Walter Benjamin's argument that the storyteller will have a narrative that contains, openly or covertly, something useful for the receiver—what he refers to as "counsel." According to Benjamin, counsel can take three forms: a moral, practical advice, or a proverb. These three aspects of counsel are grounded in morality, not ethics, a point not specified by Benjamin:

> In every case the storyteller is a man who has counsel for his readers. But if today "having counsel" is beginning to have an old-fashioned ring, this is because the communicability of experience is decreasing. In consequence we have no counsel either for ourselves or for others. After all, counsel is less an answer to a question than a proposal concerning the continuation of a story which is just unfolding. To seek this counsel one would first have to be able

> to tell the story. (Quite apart from the fact that a man is receptive to counsel only to the extent that he allows his situation to speak). Counsel woven into the fabric of real life is wisdom.
>
> (Benjamin, 1999, loc. 1317)

Writing in the 1930s, after the First World War and before the Holocaust, Benjamin may have been more pessimistic about the promises of counsel. However, in the age of testimony (Felman, 1992), he may have found a renewed optimism. Indeed, Simon (2000, 2005) has adopted Benjamin's concept of counsel through testimony and its affective potential. Even though Simon (2000) borrows Emmanuel Levinas' concept of the Other (a philosophy of the ethical), Benjamin and Simon are specifically referring to *moral* components of counsel, which, according to both, have power and reciprocity through affect.

One way of understanding this power is through the work on affect and memory in Holocaust museums by such authors as Andrea Witcomb and Jennifer Hansen-Glucklich, who build on Benjamin's notion of "aura," the authentic presence, which Benjamin argues "withers in the age of mechanical reproduction" (in Hansen-Glucklich, 2014, p. 125). Although, as Hansen-Glucklich points out, there are differences between Benjamin's conception of the aura of art and that of Holocaust artefacts—and also, we would argue, between artefacts and Holocaust survivors—the notion of aura does prove a useful starting point to think through the impact of survivor testimony. This is because

> by emphasizing the authenticity and aura of the objects and thereby forging an affective connection based on empathy and identification … [s]uch exhibits seek to transform passive spectators into involved and concerned Holocaust witnesses.
>
> (Hansen-Glucklich, 2014, p. 142)

Holocaust museums need to be understood within the context of changes in museum practice more generally, particularly the move from objects as sites of authority to objects as support for broader narratives, "from provenance value to representative value" (Hansen-Glucklich, 2014, p. 123). Hansen-Glucklich goes on to argue that Holocaust museums have not generally followed this trend, relying "on the presence of authentic artefacts to overcome the psychological and cognitive distance that time, space, and experience engender between Holocaust victims and museum visitors" (2014, p. 124). While we would dispute Hansen-Glucklich's broader claim that Holocaust museums have not responded to changes in museum theory in the way in which they present artefacts as "authentic relics" rather than as "de-sacrilized" objects (Hansen-Glucklich, 2014, p. 123), we concur that they have a particular relationship to concepts of "authenticity." This is particularly the case with the JHC survivor guides' oral and video testimony, where their authority is paramount. This is because part of their rationale is a response to Holocaust denial, affirming the veracity of what took place.

The role of survivor guides within education at the JHC

The JHC was formed in 1984 by members of the Federation of Polish Jews and the Kadimah (a Yiddish cultural centre) in Melbourne. The majority were Polish Jewish Holocaust survivors who became involved in political organizations after migrating to Melbourne in the 1940s–50s. Importantly, many of those involved in the founding were Yiddish speakers, and this lingua franca would have a major impact on the first exhibitions and, later, decision-making regarding the form and content of the later 2010 exhibition. The role of Holocaust survivors in shaping the narratives in the exhibition space and collections is integral to understanding the work of the JHC.

Even though the JHC officially opened in March 1984, the founders were part of an ongoing memorialization process that had been active among survivors since the 1940s. Numerous authors have examined the development of Holocaust memory in Australia, identifying a number of external factors, such as the trial of Adolf Eichmann (Ritter, 2007) or increasing anti-Semitism (Berman, 2001), as drivers for the changes to commemorative practices and emergence of permanent Holocaust museums in the 1980s. Melbourne also has the largest survivor population per capital outside of Israel, many of whom achieved positions of prominence in the Australian Jewish community and had been involved in communal memorial activities, including exhibition, in the postwar years. As we have argued elsewhere, the JHC should be seen within the context of the incremental development of Holocaust memorialization in Australia rather than as a break with what went before (Cooke and Frieze, 2015a). The JHC was a material manifestation of the intangible culture of Holocaust memory. The homespun feel of the original museum was a comfort to the originators and the many survivor volunteers who began working at the Centre:

> It is more like a mausoleum to inter the remains of those of our dead who have no remains—our collective, often nameless dead without graves, whose bodies were released in the air where they still hover around us like unappeased spirits.
>
> (Harry Redner in Jewish Holocaust Centre, 1984)

The memorial rationale of the JHC from its inception in the mid-1980s was also coupled with education. The two were intimately connected: keeping alive the memory of loved ones who perished in the Holocaust meant fighting denial, ignorance, anti-Semitism, and racism. Students and other visitors could learn about the Holocaust through the microcosm of narratives from the survivors who helped create the JHC. It was asserted that the power of the survivor story told by the survivor face-to-face with the visitor would have a life-changing or cathartic reaction from the listener, and would provoke Benjamin's counsel and Simon's affective potential.

The survivor guides were integral to the discourse of a "living museum," linking testimony with commemorative events, but also to providing eyewitness testimony that would support (and thus verify) other primary accounts. For example, when the then Director of Tourism in Victoria, Don Dunstan, visited the JHC in 1986, the former Premier of South Australia lauded the JHC as a "live museum rather than a static exhibition [which] creates the interest and shows the devotion of all concerned" (Cylich, 1986, n.p.). The role of the survivor guides' testimony was reinforced in 1987 when the JHC won a special award in the Victorian section of the Westpac Museum of the Year Awards with the citation: "To the JHC's voluntary staff for having the courage to be the living exhibits of an horrific time. This is a unique museum" (Anon, 1987, p. 3). The survivors were living history, seen as the lifeblood of the JHC, without which it was "an inanimate shell waiting animation" (see also Cooke and Frieze, 2015a). The relationship between the (student) visitor and the survivor was considered key to the visit's transformative potential. Volunteer curator Saba Feniger, who developed the long-standing 1990 exhibition, wrote:

> unlike any other, the Holocaust Museum can never be impersonal or cold: the informative aspect is exceeded by a humane one. I want to emphasize that the visitor's emotional reaction is not simply a result of looking at horrific photographs. The approach is an intellectual and spiritual one, it creates empathy of one being with another.
>
> (Feniger, 1996, p. 13)

Thus, the role of survivors and their face-to-face encounter with visitors and the oral history that apparently binds them has been crucial and critical to the unique experience of the museum since it began. As the survivors pass, this face-to-face encounter is being replaced with video-testimony, in the hope that this counsel will continue. The question becomes: What are the potential consequences of this change from live testimony to recorded and non-interactive narratives?

Although visitors to the JHC were many and experiences varied, perhaps unsurprisingly for a small, volunteer-run museum, little in the way of visitor research was undertaken. Instead, the "visitor" was understood as Hansen Glucklick's "ideal visitor": "Sympathetic to the core values and beliefs of the context culture and [who] responds viscerally to the evocative forms produced in its spaces" (2010, p. 210). This was particularly the case with the focus on school education. The early compulsion to "never forgive and never forget" was complicated by the mission of tolerance and the dangers of racism within contemporary society. Educating secondary school students was seen as central to achieving these aims. The role of survivor guides thus became the basis for a moral pedagogical paradigm. Due to their unique status as witness to the Holocaust, their expertise was perceived as being unique and authoritative, a status still in evidence in 2011 when the President of the JHC could state:

> Sometimes I take for granted the extraordinary gift with which I have been presented—the privilege of hearing Holocaust survivors tell their stories. Recently I attended a function at which Holocaust survivor Kitia Altman was the speaker. She was to speak for twenty minutes and asked me to stop her when her allotted time was up. Kitia spoke for forty minutes and I would not

have wanted to stop her. What she teaches us is so powerful, and the audience was spellbound. Her message is not filled with hate or despair, but rather with hope—hope for the future.

(Rockman, 2011, p. 3)

This statement, made when many survivors had already passed away, illustrates the continuing anxiety over the role of survivors and survivor testimony in the museum. The JHC would have to transition, with difficulty, from live oral history to truncated video-testimony. Given the ways in which testimony has been used within the historiography of the Holocaust and in museums, the change from the responsiveness, the intimacy, and even the contradictory narratives of speaking with Holocaust survivors to the linear narrative of highly edited video-testimonies could have the potential to reduce "the subject matter to a simplistic morality tale; one shorn of its specific historic context, which particularizes when, where, and who was affected" (Kushner, 2006, p. 292).

The authority that accompanies testimony, in its power, has affective pedagogic potential, and is seen as particularly effective when the experience with survivors is "first hand":

What the survivor does when he or she enters the classroom is put flesh back on the bodies of those victims, allowing students to see them not for what was done to them but for who they were in the fullness of the lives they led.

(Barbara Appelbaum in Michael Berenbaum, 2008, p. 40)

This was seen by the JHC as a powerful means to moral and ethical change in the broader community. The corporeal engagement and proximity (Waterton and Watson, 2014 p. 25) to a survivor was thought to bridge a gap between the then (the Holocaust) and the now (contemporary Melbourne). Hansen-Glucklich (2014, p. 132) suggests that Holocaust artefacts provide a "concrete presence in time and space ... [which] bridges the gap between viewers and victims and thereby evokes empathy and transforms viewers into witnesses." The physical proximity to survivors in the exhibition worked in the same way. The survivor guides who had been in Auschwitz further bridged this gap, by displaying their tattoos (Josem, 2010, p. 7). This encounter was the prompt for a temporary photographic exhibition at the JHC in 2010, by Andrew Harris:

... because the moment when survivors show visitors their tattoos is so powerful that students often remark that "this made it real". The Holocaust itself can be such an overwhelming topic that seeing this tangible marker brings it home to them.

(Josem, 2011, p. 18)

This was reinforced by an advertising campaign for the JHC through the History You Can't Erase poster (see Figure 4.1) at the same time, which used the indelible mark left on some survivors by their experiences as a counterpoint to Holocaust denial.

Figure 4.1 History You Can't Erase poster.

Testimony in the exhibitions

The former exhibition comprised mainly text, with a small number of objects, one diorama, and a model of the Treblinka death camp by Holocaust survivor Chaim Sztajer, which stood in contrast to the documentary style of the rest of the exhibition that included many graphic images. As Young suggests, Holocaust survivors in general preferred literal representations, as these worked as proof of their experiences, particularly in the face of Holocaust denial (in Witcomb, 2010, p. 43). The model also connoted the combating of denial as there were very few survivors of this death camp. The model of Treblinka was an artistic, albeit intensely personal representation, which Witcomb argues creates a close connection between the past and the present because of the act of its creation. The presence of the model maker, who could often be found close to the model until his death in 2008, allowed the visitor "a glimpse into ongoing grief": The Holocaust was not something that was past, but an ongoing trauma in the present (Witcomb, 2010, p. 47). The exhibition included very few artefacts, although what were collected or donated were given prominence. However, reproductions of images were the main mode of display. The tolerance and human rights agenda was seen as working through an affective response on the part of the visitors and the survivor. Feniger (1996, pp. 13–14) considered that the museum "provoke[d] deep emotions … for each one of this museum's exhibits … although inanimate, has a soul." This relationship between the visitor and the survivor guides was seen as the major point of differentiation between the JHC and other museums. As Feniger (1996, p. 14) wrote: "They are a real and authentic part of the historical period that we present. The importance of their role is beyond comparison. The visitor's face-to-face encounter with a survivor, though sometimes awe-inspiring, is also reassuring." The act of creating the exhibition was an act of witness: photographs were "an instrument of survival," with each photograph or document which was donated having a story "with strong emotional associations" (Feniger, 1996, p. 14).

This was particularly the case when the photographs were of survivors' families, which often interrupted the set chronological narrative of the genocide. Such disruptions opened up a space for critical reflection by the museum visitor, which foregrounded the intimate connections between the Holocaust and Melbourne through the stories of those survivors, overcoming the conceptual and imaginative distance of an event too often seen as "over there" and "back then" (Cooke *et al.* 2014).

With education as a central remit of the JHC and "never again" a continuing refrain, the attempt to communicate the lessons of the Holocaust for contemporary society, particularly to school children, and primarily through the live testimony of Holocaust survivors, was a key theme in the redeveloped exhibition. The removal of those disruptions and the display of more normative Holocaust narrative raises the issue of how the visitor might now "learn from attempts to face the traces of lives lived in times and places other than one's own" (Simon *et al.*, 2001 p.286).

The planning for the redevelopment of the exhibition space started in 2007, and its explicit pedagogic function was optimized by the education and curatorial departments at the JHC working together. The development and research of the exhibition was supported in part by a grant from the Conference on Jewish Material Claims against Germany and this allowed the JHC to employ exhibition firm Lilford Smith for conceptual help. The main themes of the 1990–2010 exhibition would remain. However, the number of photographs, particularly those with graphic content, would be reduced, revealing more of the white walls. This would mean that the exhibits would have "more impact" and the text would be easier to read. Martin Lilford, the exhibition designer, said:

> The museum has been designed to guide visitors through the Holocaust in the order events unfolded. Angular walls serve to break up the journey and help to reinforce thematic changes. The simplicity and consistency of the museum's design aims to assist visitors to focus on the content, providing clarity and room for reflection. Images, documents, artefacts, audiovisuals and the written word have been woven together to engage and inform. The display is punctuated by artworks that bring direct expression and emotion to the indescribable. Ultimately the museum serves to educate and provide a message of hope.
>
> (Anon, 2009, p. 14)

The *Treblinka* model remained a unique contribution to the JHC, but new technologies allowed the visitor to "tour" the model, lighting up different areas with a computer explaining the details. Jayne Josem, curator and head of collections, identified the now interactive model as a "chilling insight into the way the Nazis went about murdering the Jews and, being a model, it is more tangible, more real, than words on a page" (Josem, 2010a, p. 7).

Role of oral testimony in the museum

Given the importance placed on the relationship between the visitor and the survivor guides, a key part of the planning for the new museum was the incorporation of the testimonies of survivor guides:

> the real imperative was the need to keep the Holocaust survivors' voices alive in the museum, with most survivors now in their eighties. Survivors have been the lifeblood of the Centre, volunteering in all areas, but most significantly as museum guides, talking to visitors about their experiences. This feature has been crucial to the Centre's success, so the challenge was to ensure that visitors to the museum would continue to learn about the Holocaust directly from the stories of survivors.
>
> (Anon, 2012, p. 6)

This encounter between visitor and survivor guide was understood as an affective event. Josem, who led the redevelopment of the exhibition in 2010, argued that the encounter with survivor guides would provide an "affective jolt" (Cooke and Frieze, 2015b):

> We hope these nonchalant students will learn just what people are capable of, if left unchecked. We want their visit to be a "wake-up". In saying this we must acknowledge that not all of our young visitors are disinterested. Some are indeed motivated and interested, but the majority needs to *be jolted* into a sense of awareness of the horrors of the Holocaust and of racism and prejudice.
>
> (Josem, 2010a, p.7, emphasis added)

For Josem, again the challenges are always present:

> Every day busloads of students traipse through this Centre. They saunter in, teenagers with attitude, gelled-up hair, piercings everywhere, tattoos, pants slung low in the case of boys, skirts too high in the case of girls. They seem impossibly tall. They are wired up—mobiles, texting, tweeting, iPod shuffling. They are thinking about friendship issues, boyfriend troubles, sport, TV, YouTube and Facebook. Many have never met a Jew. Our job here is to penetrate through their digital armoury and get them to focus on and understand why they have come. *We want them to listen, hear, learn, engage, react, respond and remember*. It is a tall order, every day of every week. These students, these pierced, gelled giants that wander through daily, mostly arrive at the Centre indifferent to Jews, indifferent to the Holocaust. *They arrive indifferent, but after hearing from our survivors and walking through the museum, they leave different*. We rebuilt the museum to ensure that we continue to build on the wonderful foundations the founding survivors established for us and to ensure that we continue to engage the younger visitors so that they learn from history. And we rebuilt the museum because history has taught us never to take anything for granted, even our seemingly secure democracy.
>
> (Josem, 2010b, p. 7, emphasis added)

The change from live oral history from survivors to survivors' stories meant the shifting process included a focus on the pedagogical impacts, but changed how memorialization practices were made available to visitors at the JHC. One approach, given the large collection of unedited video-testimonies from the Melbourne survivors, was to use condensed versions of these video-testimonies in the exhibition.

Video-testimony

After Claude Lanzmann's 1985 nine-hour documentary, *Shoah*, and despite faded memories and the contradictory and problematic adherence to historical accuracy, Holocaust video-testimonies slowly became valued by scholars and historians specifically for their showcasing of nuanced memories, but also for their value of filling in the gaps of historical knowledge. The eminent historian Christopher Browning admits that because little knowledge of the camps in Starachowice in Poland existed, he was fortunate to find video-testimonies in the Shoah Visual

History Foundation (a bank of approximately 52,000 testimonies) that provided specific information on the camps (Blumenstyk, 2005). In addition to this example of the historical value of testimonies, scholars have begun to appreciate the worth of a single testimony, which contains a unique story and the unraveling of an inimitable life.

The grassroots organization Holocaust Survivors Film Project Inc. began collecting video-testimonies at Yale University in 1979. This led to the pioneering work of Geoffrey Hartman's Fortunoff Video Archive for Holocaust Testimonies in the 1980s and has spurred groundbreaking work into the value of survivor testimonies from such eminent scholars as Lawrence Langer, Dori Laub, Shoshana Felman, and Hartman himself, who produced analyses of testimonies within psychoanalytic and literary studies. They initiated the early steps in formulating the scholarly importance of testimony in "the era of the witness"—the eyewitnesses to the Holocaust, and to the viewers of the testimonies: The witnesses to the eyewitnesses.

In the early 1990s, Langer, a literary scholar, first analyzed written testimony and then video testimony. His prize-winning *Holocaust Testimonies: The Ruins of Memory* reflected the worldwide desire to capture, understand, and listen to the voices of Holocaust survivors, who were rapidly aging. At first, the analysis of video-testimonies received a lukewarm reception. Unlike written testimonies, scholars felt that nuanced and disorganized memory in survivor testimonies was unreliable. Langer, Laub, and Felman's focus was traumatic memory that appeared "to operate in a historical vacuum" (Maclean *et al.*, 2008, p. 21), ignoring the value of a survivor's religious, cultural, and national background, including the survivor's class and political beliefs, that impinged upon the video-testimony narrative and influenced how a life is remembered, retold, and recalled. Although literary techniques were acknowledged in written testimonies, the early analyses of video-testimonies also ignored such filmic techniques as framing, panning, zooming, and the varying mise-en-scene of each video testimony.

By these standards, the JHC was a little late in coming to the centrality of video-testimonies in broader Holocaust narratives, although the Centre certainly understood the importance of testimony within its community from its inception in 1984. In other words, the JHC understood the need of its unique and highly concentrated Holocaust survivor community to bear witness. In addition, the JHC was concerned from its inception with survivor attrition and what would happen to the legacy of the survivor community once they had passed on. The need for testimony in the new museum was not only about pedagogy, but also about the continuation of the memorial function of the JHC.

Video-testimony was privileged over audio and written testimony in the redeveloped museum for several reasons. By the 1990s (when the JHC began to record survivor stories on video), video-testimonies were beginning to become a popular medium for survivor stories. The sheer volume of video-testimony meant that the JHC had a large bank to choose from. The Centre thus had a dedicated bank of video-testimonies that were seen as a valuable resource to share. Placing the video-testimonies within the museum space would release them from the "hidden" area of the JHC where they were recorded and stored. The JHC also viewed video-testimony as the most "authentic" replacement for the real survivor guides: The combination of body language, gestures, and facial expressions with audio intonations and tones would provide the most affective experience for the visitor. Written testimonies, by their nature, omit the breaths, pauses, junctures of speech, interruptions, and, most importantly, the silences.

Storypods

The centrality of the survivor story and the limited time spent at the museum meant that the video-testimonies—some of which ran for many hours—had to be condensed. The quality of the interview, the complexity of the narrative, and the "complete" life narrative of a testimony was edited into snippets of areas of the testimony narrative that would make the most impact. The JHC designed and curated these sections into "Storypods," interactive displays that would "feature the inspiring stories of Holocaust Survivors who volunteer as museum guides" at the JHC (JHC, n.d.). These would not be used only in the museum space but would be adapted:

> for delivery over the internet via the Centre's existing website and by DVD, and to develop engaging audio-visual curricular materials for teachers and students. These materials will use the unique Storypod technology with its oral history, written documents, historical photographs and timelines, and eyewitness testimonies of Holocaust survivors to engage students in learning experiences. Through this innovative curricular program, we aim to engage students with sophisticated multimedia material, gaining their interest by being both informative and intellectually stimulating.
>
> (Fineberg, 2011, p. 3)

For the JHC the concept of the Storypods was the perfect "in place of" for the museum, with the Director of Education at the JHC confident that "in the future [the Storypods] will mean that the survivors' voices will never be silent" (Civins, 2011, p. 5). Similarly, Josem assured survivors that the transition from live oral history, to video-testimony, and finally to the truncated Storypods would be seamless.

However, there was healthy caution and skepticism. The Director of Education added to his statement by asking: "But will people remember that the images speaking from a glass screen are those of real people" (Civins, 2011, p. 5). While Holocaust museums are understood as agents of change (Grenier, 2010), this change to *Storypods* will inevitably impact upon pedagogy, effect, and affect. As Witcomb (2010) has argued, there is a difference between seeing a survivor in person recounting a story and seeing and hearing visual and aural testimony. Levinas (cited in Levinas and Malka, 1984, p. 48, emphasis in original) illustrates this, arguing that

> [i]f you conceive of the face as the object of a photographer [or film maker], of course you are dealing with an object like any other object. But if you *encounter* the face, responsibility arises in the strangeness of the [O]ther.

If the video-testimony does not allow an encounter, then there is no possibility of the ethical. The image, then, for Levinas does not constitute an extant object, not in this world of the ethical and not in the represented narrative. It is the corporeality of the encounter with the Other that binds the Levinasian ethical response; it is the face-to-face. One can appreciate the Other through their very alterity or wholeness as a person and not via an aesthetic appreciation of their features or voice or their very inability to interact. What Levinas neglects to acknowledge is that representation—for not all of it can be illegitimate—is a presentation of what is *possible* in the affective and ethical encounter. While we dispute Levinas' claim that *all* representation is incapable of affect and the ethical, we are cautious about these possibilities in truncated *Storypods*.

Conclusion

The continuing focus on the affective potential of museum exhibitions, particularly the role of personal narratives in creating testimony, has posed specific difficulties for Holocaust, genocide, and human rights museums in particular due to the ethical need to affect the other and the moral obligation to counsel. The hidden archive of testimony at the JHC was a process of witness, a memorial function, but without an explicit pedagogical focus. It was disconnected from the survivor guides' face-to-face encounter with the visitors to the museum. The Testimonies Department was primarily organized by a Holocaust survivor, Phillip Maisel, who contacted the survivors, arranged the interviews, filmed each testimony, and was often the interviewer. Thus, the purpose of this testimony was to engage in a monologue or an interview in a safe environment (always filmed in a private room at the JHC, unlike the Shoah Foundation's video-testimonies, which are filmed in survivors' homes), with neither moral nor ethical obligations except to honour their own memories. The survivors were not impeded by time constraints: the testimony could last half an hour or several hours. Thus, the function of this type of testimony was antithetical to the face-to-face testimony in the museum, where there were time constraints, ethical encounters, moral obligations, interactive challenges, and pedagogical and affective concerns. The former's purpose was private; the latter, public.

The unedited video-testimonies that become *Storypods* may lose their affective potential when they are geographically relocated within the JHC from the Testimonies Department to the museum displays and to a standalone smartphone app. Many of the video-testimonies were re-filmed to correspond to and focus on objects, in particular photographs hidden or recovered after the Holocaust. The re-filming takes place with a black backdrop, separating them from not only the JHC but also a defined space and place: in effect these testimonies can become anyplace. Further, these highly curated testimonies have the potential to dislocate experiences from the before and after. Does this stop, in Benjamin's terms (1999), the story from continuing? We would follow Kushner in arguing that "choosing confusion over smoothness in the representation of life story testimony is to do greater justice to the way the Holocaust was actually experienced on an everyday level" (Kushner, 2006, p. 292). In intently focusing on keeping the survivor stories alive, the result is perhaps a testimony that is devoid of affect and Levinas' ethical encountering. It may, however, retain Simon's reconceptualization of counsel: a moral imperative that is steeped in "learning lessons." As our theoretical understandings of the work of affect in museums and heritage sites develop, more work is needed to understand how the visitor engages with this process and how this might result in longer-term attitudinal and behavioural change.

With the affective turn in museums, personal narratives are seen as a key way of democratizing the institution and where the encounter with "narrative fragments" (Minore, 2012, p. 146) allows the visitor to construct their own experiences. The use of new media facilitates this process. However, if a defining element of the post-museum is the discursive space for conversation (Lindauer, 2007, p. 305), how can technology facilitate this without flattening the complex geographies and histories of the Holocaust and of those who experienced it?

Bibliography

Anon. 1987. Holocaust centre, Jewish museum win special prizes. *The Australian Jewish Times*, May 14: 3.

Anon. 2009. Museum upgrade: state of play. *Centre News*, 31(2): 15.

Anon. 2012. Jewish Holocaust Centre celebrates prestigious MAGNA award and record number of visitors. *Centre News*, 34(1): 6.

Bastel, H., Matzka, C. and Miklas, H. 2010. Holocaust education in Austria: a (hi)story of complexity and ambivalence. *Prospects*, 40: 57–73.

Benjamin, W. 1999. The storyteller: reflections on the works of Nikolai Leskov, in H. Arandt (ed) and H. Zorn (trans) *Illuminations*. London: Pimlico (Kindle Edition).

Bennett, T. 1995. *The birth of the museum: history, theory, politics*. London: Routledge.

Berenbaum, M. 2008. Survivors as teachers, in B. Zuckerman, Z. Garber, J. Schoenberg and L. Ansell (eds) *The impact of the Holocaust in America: the Jewish role in American life. An annual review*, 6 (pp. 31–62). West Layfayette: Purdie University Press.

Berman, J. 2001. *Holocaust remembrance in Australian Jewish communities, 1945–2000*. Crawley: University of Western Australian Press.

Blumenstyk, G. 2005. Holocaust stories move to academe (survivors of the Shoah Visual History Foundation). *The Chronicle of Higher Education*, 52(11): 5; A2 and USC 'Shoah Foundation Institute', http://dornsife.usc.edu/vhi/# (accessed 1 May 2015).

Charlesworth, A. 1994. Teaching the Holocaust through landscape study: the Liverpool experience. *Immigrants and Minorities*, 13(1): 65–76.

Charlesworth, A. 1998. 'Children, do something different': reflections on running a Holocaust field trip. *Journal of Holocaust Education*, 7(1 & 2): 126–32.

Civins, Z. 2011. Education. *Centre News*, 22(2): 5.

Cooke, S. and Frieze, D. 2015a. *The interior of our memories: a history of Melbourne's Jewish Holocaust Centre*. Melbourne: Hybrid.

Cooke, S. and Frieze, D. 2015b. Imagination, performance and affect: a critical pedagogy of the Holocaust? *Holocaust Studies: A Journal of Culture and History*, 21(3): 157–71.

Cooke, S., Alba, A. and Frieze, D. 2014. Community museums and the creation of a 'sense of place': Holocaust museums in Australia. *Recollections*, 9(1). http://recollections.nma.gov.au/issues/volume_9_number_1/papers/community_museums (accessed 8 May 2015).

Crang, M. and Tolia-Kelly, D.P. 2010. Nation, race and affect: senses and sensibilities at National Heritage sites. *Environment and planning A*, 42(10): 2315–31.

Cylich, S. 1986. Dunstan visits centre. *Australian Jewish News*, 18 May, n.p.

Felman, S. 1992. Camus' *The Plague*, or a monument to witnessing, in S. Felman and D. Laub (eds) *Testimony: Crises of witnessing in literature, psychoanalysis, and history* (pp. 93–119). New York and London: Routledge.

Feniger, S. 1996. 'Museums' and the Jewish Holocaust Museum. *Centre News*, 13(1): 13–4.

Fineberg, W. 2011. Director's cut. *Centre News*, 33(2): 3.

Gregory, K. and Witcomb, A. 2007. Beyond nostalgia: the role of affect in generating historical understanding at heritage sites, in S. Knell, S. Macleod and S. Watson, S. (eds) *Museum revolutions: how museums change and are changing* (pp. 263–75). London and New York: Routledge.

Grenier, R.S. 2010. Moments of discomfort and conflict: Holocaust museums as agents of change. *Advances in Developing Human Resources*, 12(5): 573–86.

Hansen-Glucklich, J. 2014. *Holocaust memory reframed: museums and challenges of representation*. Piscataway: Rutgers University Press.

Hein, G.E. 2012. *Progressive museum practice: John Dewey and democracy*. Walnut Creek: Left Coast Press.

Henning, M. 2006. *Museums, media and cultural theory*. Maidenhead: Open University Press.

Jewish Holocaust Centre. 1984. Commemorative booklet produced to mark the opening of the JHC.

Jewish Holocaust Centre. n.d. *Storypods*. www.jhc.org.au/storypods/ (accessed 8 July 2015).

Josem, J. 2010a. A labour of love. *Centre News*, 31(1): 6–7.

Josem, J. 2010b. Keeping the survivors' voices alive. *Centre News*, 32(2): 7.

Josem, J. 2011. Marked: Holocaust survivors and their tattoo. *Centre News*, 33(1): 18.

Kelly, L. 2010. Engaging museum visitors in difficult topics through socio-cultural learning and narrative, in F. Cameron and L. Kelly (eds) *Hot topics, public culture, museums* (pp. 194–210). London: Cambridge Scholars Publishing.

Kushner, T. 2006. Holocaust testimony, ethics, and the problem of representation. *Poetics Today*, 27(2): 275–95.

Latham, K.F. 2013. Numinous experiences with museum objects. *Visitor Studies*, 16: 3–20.

Laub, D. 1995. Truth and testimony: the process and the struggle, in C. Caruth (ed) *Trauma: explorations in memory* (pp. 61–75). Baltimore: Johns Hopkins University Press.

Levinas, E. and Malka, S. 1984. Interview with Salomon Malka, in J. Robins (ed) *Is it righteous to be? Interviews with Emanuel Levinas* (pp. 93–102). Stanford: Stanford University Press.

Light, H. 2002. A home of Jewish culture and civilisation: Rabbi Lubofsky and the Jewish Museum of Australia, in Anna Blay (ed) *Eshkolot: essays in memory of Rabbi Ronald Lubofsky* (pp. 4–14). Melbourne: Hybrid.

Lindauer, M.A. 2007. Critical pedagogy and exhibition development: a conceptual first step, in Simon J. Knell, Suzanne Macleod and Sheila Watson (eds) *Museum revolutions: how museums change and are changing* (pp. 303–14). London and New York: Routledge.

Lindsay, P. 2013. *Spirit of Gallipoli. The birth of the Anzac legend*. Richmond, Victoria: Hardie Grant Books.

Logan, W. and Reeves, K. (eds) 2009. *Places of pain and shame: dealing with 'difficult heritage'*. London and New York: Routledge.

Maclean, P., Abramovich, D. and Langfield, M. 2008. Remembering afresh? Videotestimonies held in the Jewish Holocaust and Research Centre (JHMRC) in Melbourne, in P. Maclean, M. Langfield and D. Abramovich (eds) *Testifying to the Holocaust* (pp. 3–28). Sydney: Australian Association of Jewish Studies.

Maitles, H. and Cowan, P. 2008. Why are we learning this? Does studying the Holocaust encourage better citizenship values? Preliminary findings from Scotland. *Genocide Studies and Prevention*, 3(3): 341–52.

Maitles, H. and Cowan, P. 2012. 'It reminded me of what really matters': teacher responses to the lessons from Auschwitz project. *Educational Review*, 64(2): 131–43.

Minore, M. 2012. Narrative museum, museum of voices: displaying rural culture in the Museuo della Mezzadria Senese', Italy, in J. Fritsch (ed) *Museum gallery interpretation and material culture* (pp. 136–49). New York and London: Routledge.

Ritter, D. 2007. Distant reverberations: Australian responses to the trial of Adolf Eichmann, in T. Lawson and J. Jordan (eds) *The Memory of the Holocaust in Australia* (pp. 51–73). London: Valentine Mitchel.

Rockman, P. 2011. From the president. *Centre News*, 33(1): 3.

Schorch, P. 2013. Contact zones, third spaces, and the act of interpretation. *Museum and Society*, 11(1): 68–81.

Simon, R.I. 2000. The paradoxical practice of zakhor: memories of 'what has never been my fault or my deed', in R.I. Simon, S. Rosenberg and C. Eppert (eds) *Between hope and despair: pedagogy and the remembrance of historical trauma* (pp. 9–25). Lanham, Boulder, New York and Oxford: Rowman and Littlefield Publishers.

Simon, R.I. 2005. *The touch of the past: remembrance, learning and ethics*. New York and Basingstoke: Palgrave Macmillan.

Simon, R.I., Rosenberg, S. and Eppert, C. 2000. Introduction – between hope and despair: the pedagogical encounter of historical remembrance', in R.I. Simon, S. Rosenberg and C. Eppert (eds) *Between hope and despair: pedagogy and the remembrance of historical trauma* (pp. 1–8). Lanham, Boulder, New York and Oxford: Rowman and Littlefield Publishers.

Simon, R.I., Eppert, C., Clamen, M. and Beres, L. 2001. Witness as study: the difficult inheritance of testimony. *The Review of Education/Pedagogy Cultural Studies*, 22(4): 285–322.

Smith, S.D. 2007. Teaching about the Holocaust, in M. Goldenberg and R.L. Millen (eds) *Testimony, tensions, and Tikkun: teaching the Holocaust in colleges and universities* (pp. 271–83). Pastora Goldner Series in Post-Holocaust Studies, Seattle: University of Washington Press.

Staiff, R. 2014. *Re-imagining heritage interpretation: enchanting the past-future*. Farnham: Ashgate.

Tunbridge, J.E. and Ashworth, G.J. 1996. *Dissonant heritage: the management of the past as a resource in conflict*. Chichester: John Wiley and Sons.

Waterton, E. and Watson, S. 2014. *The semiotics of heritage tourism*. Bristol: Channel View Publications.

Witcomb, A. 2010. Remembering the dead by affecting the living: the case of a miniature model of Treblinka, in S.H. Dudley (ed) *Museum materialities: Objects, engagements, interpretations* (pp. 39–52). New York: Routledge.

Witcomb, A. 2013. Understanding the role of affect in producing a critical pedagogy for history museums. *Museum Management and Curatorship*, 28(3): 255–71.

Witcomb, A. 2014. 'Look, listen and feel': the first peoples exhibition at the Bunjilaka Gallery, Melbourne Museum. *THEMA: La revue des Musees de la civilization*, 1: 49–62.

Young, J.E. 1993. *The texture of memory: Holocaust memorials and meaning*. New Haven and London: Yale University Press.

Ziino, B. 2010. 'A lasting gift to his descendants': family memory and the great war in Australia. *History and Memory*, 22(2): 125–46.

5 Museum canopies and affective cosmopolitanism
Cultivating cross-cultural landscapes for ethical embodied responses

Philipp Schorch, Emma Waterton and Steve Watson

Museums as 'cosmopolitan canopies'

> The existence of the canopy allows such people, whose reference point often remains their own social class or ethnic group, a chance to encounter others and so work toward a more cosmopolitan appreciation of difference.
> (Anderson, 2004, p. 28)

In his article 'The cosmopolitan canopy', Elijah Anderson (2004)[1] describes contemporary urban landscapes as those strongly affected by the forces of globalization, migration and industrialization. In Anderson's terms, public spaces in the United States have inevitably become racially, ethnically and socially more diverse; at the same time, those markers of difference have simultaneously contributed to the division of cityscapes into ethnic neighbourhoods and the resultant separation of social groups. This line of thinking reflects Mike Featherstone's (2002)[2] comments on the significance of the city in cosmopolitan dispositions, Ulrich Beck's (2002) concept of cosmopolitanization as a kind of internalized globalization *within* the nation-state and Saskia Sassen's (2000, p. 153) characterizations of the city as a contested space where wealthy elites and low-income others jostle for space, each transnational in character but embedded and competing in specific places. The existence of Anderson's 'cosmopolitan canopies', however, enables people who are often confined to their ethnic group or social class to '*encounter others*' and thus potentially develop a '*cosmopolitan appreciation of difference*' (2004, p. 28; our emphasis). Anderson, (2004, p. 28) goes on to identify such settings or 'canopies' within the urban context of Philadelphia in the USA, including areas such as 'the Reading Terminal, Rittenhouse Square, Thirtieth Street Station, the Whole Foods Market, and sporting events'; surprisingly, museums do not feature on his list.

In large part, this surprise comes about because museums have for quite some time been imagined as inherently cross-cultural landscapes that can potentially facilitate the development of a cosmopolitan ethics, a characteristic that is reflected in the wider literature (see Kreps, 2003, 2011; Schorch, 2013a, 2014a). Furthermore, it is important for scholars to approach 'cosmopolitanism' as a concrete lived experience rather than an abstract normative ideal, something our research seeks to advance by focusing on specific urban settings and by considering cosmopolitanism through theories of atmospheres (Anderson, 2004), encounters (Delanty, 2011), performances (Woodward and Skrbis, 2012), practices (Kendall *et al*., 2009) and interpretive meanings (Schorch, 2014a). Again, museums are not explicitly noted in most of these theoretical discussions and empirical investigations, despite their frequent appearance as anchor points and hubs of activity in urban cultural quarters. At the same time, there have been only 'limited incursions' (Mason, 2013, p. 42) of cosmopolitan thinking into the fields of museum and heritage studies themselves, though among these there are a handful of useful instances in which it does make an appearance (Daugbjerg, 2009; Mason, 2013; Schorch, 2013a, 2014a; Staiff, 2014). In one example, Sharon Macdonald (2013) investigates the *Memorylands* of contemporary Europe and detects evolving forms of cosmopolitan heritage and memory which do not simply override other frames of reference and forms of attachment such as the nation (see also Daugbjerg and Fibiger, 2011). Macdonald (2013, p. 173) goes on to argue that 'memorial forms in the cityscape' such as museums 'become important stimuli for bringing interlocutors together' to engage in interactions across cultural boundaries. Likewise, Russell Staiff (2014) uses K.A. Appiah's book *Cosmopolitanism: Ethics in a World of Strangers* (2006) to unravel issues of universalism and cultural relativism in the sphere of heritage interpretation. He concludes that emphasizing 'commonalities' – things shared between cultural groups – while fraught with risks and the implications of unequal power relations, should at least facilitate conversations across cultural differences and boundaries, offering a way to negotiate the borders or limitations of interpretation at particular heritage sites (Staiff, 2014, p. 157).

Affective cosmopolitanism in the context of the museum

For the purposes of this chapter, we are keen to explore the extent to which aspects of this cosmopolitan debate resonate with affect and emotion as embodied performativities evoked in, and by, particular settings such as museums. The value of such an approach lies in the way it brings together theoretical insights with a 'detailed attention to the political, economic and cultural geographies of specific "everyday practices" …' (Nash, 2000, p. 662). Accordingly, our understanding of these performativities commences with the capacity for affecting and being affected as developed in Spinozan–Deleuzian terms (Deleuze, 1988; Deleuze and Guattari, 1994), and in the sense adopted by non-representational theorists as pre-cognitive, pre-personal dimensions of experience. Mapped across the museum as a space of cross-cultural encounter, we might consider such experiences as profoundly affective in that they prompt and set in motion embodied engagements and acts of making meaning (Schorch, 2014b). What is of interest to us, however, is the way in which this is consonant with the social dynamics identified by Anderson within the cosmopolitan canopy, and how this translates into the potentialities for affecting the way people engage in and with museums. This brings an extended lens to the non-representational literature, which has a tendency to overlook situated accounts of cosmopolitan encounters (see Tolia-Kelly, 2006).

Indeed, in order to consider the nature and scope of such cross-cultural encounters, and how museums might moderate or facilitate them, we need to go beyond the stripped-down pre-personal, autonomous notion of embodied affect that is apparent in some of the non-representational literature informed by Spinoza's ethics and its Deleuzian readings. This is because of the difficulty we have with assuming that affect can somehow be separated from human meaning-making; indeed, we are more interested in a theory of affect that encompasses the subject and subjective responses expressed *inseparably* as emotion, cognition and the construction of meaning. It is thus worth reinforcing at this point that although we take a lead from non-representational theories of affect, we have not adopted the hard boundary definitions that distinguish affect and emotion that are apparent, for example, in Steve Pile's (2010) interpretation of their significance in geography. Instead, we feel more comfortable with Liz Bondi and Joyce Davidson's (2011) willingness to live with the inherent messiness of these concepts, and Deborah Thien's (2005, p. 453) view that a focus on affect alone occludes the emotional landscapes and inter-subjective processes that constitute daily life and social and cultural experience (see also Harding and Pibram, 2002).

In drawing to mind what we might term an *affective cosmopolitanism*, our purpose in this chapter is to probe at the relationship between the affective and the subjective, emotional and cognitive (or non-representational and representational), particularly in relation to the kinds of engagement associated with the cosmopolitan canopy. Though she does not use precisely our terms and topics, we have been influenced by Leila Dawney's (2013, p. 629) suggestion that '[t]here is a need to develop tools for thinking about the way in which the affective and subjective registers operate through each other and are constitutive of each other' (see Dawney, 2013, pp. 629–31, for a summary of this debate). Purposefully problematizing the relationship between affect and the reflective subject is for us key to understanding the ways in which meanings are constructed and engaged across cultural differences. This does not mean that we have adopted the easy way out of conflating affect and emotion (which is, in any case, insupportable); rather, we are interested in the affective–subjective dynamic in forming an agenda for future research. In a more general context, Dawney describes this dynamic as an 'oscillation' between the two registers and, as such, a site for the 'social production of experience' (2013, p. 632). To return briefly to the question of theory, it is therefore the *more-than-representational* domain identified by Hayden Lorimer (2005) that locates our own thinking.

In terms of our agenda in this chapter, it seems obvious that the simple duality of 'visitor' and cultural 'museum display' is unavoidably and actively mediated by the agency of the museum, in much the same way that Gerard Delanty's (2011) 'third party' facilitates many cultural encounters. As Delanty (2011, p. 644) goes on to argue, '[i]t is also increasingly the case that many cultural encounters are occurring against the wider context of world culture and democratization, which serve as forms of mediation'. It seems reasonable to suppose, therefore, that where the museum is acting as a third party to encounters between visitors or between visitors and cultural displays, it is doing so not just through its narrative and other representational practices, but also through its design, spatial affordances and *the affective potentialities that are then created*, both deliberately on the part of the museum and in the nexus of what the visitor also brings to the engagement.

This engagement, and particularly the affective potentialities thereby afforded, is the basis of our claim that museums act as cosmopolitan canopies, that is, as settings which allow for actually existing cross-cultural encounters and a potentially cosmopolitan condition that can only can only emerge through the practice of meaning-making and the 'act of interpretation' (Schorch, 2013a). These are the interpretive practices across cultural differences through which a cosmopolitan encounter can be established, navigated and nurtured. Understanding how these practices take place paves the way to understanding how different cultural actors engage in the process of cultural and potentially cosmopolitan world-making (Schorch, 2014a). As such, we propose that such encounters provide a framework for investigating the ways in which cross-cultural experiences are modulated by affordances that begin with registers of affect. This broadly reflects Dawney's (2013) concept of 'interruption', where a given situation – for example, a museum display or exhibition – might stimulate the body's capacity to be affected in some way (see Tolia-Kelly, this volume). In a museum context, this could be any number of provocations, from a visceral reaction to an image or narrative that might, in turn, contagiously affect others and/or rise up into, or oscillate with, emotional responses and cognitive understandings (see Waterton and Dittmer, 2014). We might see expressions, for example, of joy or sadness, pleasure or discomfort, identification, empathy, alienation, hostility, boredom and so on and so forth, all of which can be represented to a certain degree through language in social milieus.

But let us not forget what a visitor brings in terms of an assemblage of personal and cultural subjectivities, such as their past experiences, schooling and cultural beliefs, all of which operate in tandem with our temperaments and dispositions. As Sara Ahmed (2010, p. 41) notes, '[the] moods we arrive with do affect what happens: which is not to say we always keep our moods'. The museum does not, therefore, etch its presence on a blank sheet. Antecedents of a style of thinking that engages with affective responses can be found in David Uzzell's (1989, p. 46) 'hot interpretation', which is a term he used to foreground our humanness, arguing that heritage sites have at times the power to shock, move and be cathartic. The work of Gaynor Bagnall (2003) and her influential study at two heritage sites, the Museum of Science and Industry in Manchester and Wigan Pier (both in the UK), similarly challenge the conventional view that museum-goers are passive and uncritical consumers of 'heritage', arguing that, to the contrary, what is evident are performances that demonstrate a 'complexity and diversity' in respect of the visitors' engagements that is registered as much in emotion and imagination as it is in cognition (2003, p. 87). Such thinking has provided momentum to a growing field of study that, unlike conventional museum studies (see Kirchberg and Tröndle, 2012), acknowledges the agency of the visitor and the dynamics of interaction between the visitor and display as the core of visitor experiences (Latham, 2007; Schorch, 2015a, 2015b; Soren, 2009; Witcomb, 2013).

More recently, Lisa Costello (2013) has extended this approach to those museums that serve a particular memorializing function, in her case the Jewish

Museum Berlin, which focusses on Jewish culture throughout European and German history and the Holocaust as central themes. In Costello's work, set against the affordances of visceral affect and consequent emotional registers, there is a stark dissonance of cultural perspectives, about which moral judgements are invited that are consonant with contemporary attitudes towards tolerance, responsibility and the idea of a universal moral lesson. And yet, there are so many facets of experience and meaning made possible; interlocking, competing for attention and diverse in the way they engender engagement: 'The design of the space allows audiences an array of responses that are both intellectual and physical, encouraging a negotiation of multiple narratives of collective memory with each visit' (2013, pp. 5–6). The museum thus actively engages its visitors in the process of making meaning within its spaces, transforming them from bystanders into active witnesses by asking them to think (or rethink as is more often the case) about the events portrayed, and to link these thoughts, experiences and performances with the present (2013, p. 18).

In adopting a theory of *cosmopolitan affect*, therefore, it is possible to see visitors' agency as operating in the co-production of meaning at a more-than-representational level: meaning is conceptualized as generated, explored and shared in all manner of ways, drawn as it is from memories and preconceptions, the narratives of overarching discourses and not least the somatic nature of engagement and the emotional. Staiff (2014, pp. 46–69) explores the somatic and embodied nature of heritage by employing a historicist and dialogical perspective on the work of writers and artists to provide insights into the way that the 'bodily experience of a heritage place or object or landscape' can be described, suggesting that 'the body is the locus of experience: memories, referencing, emotions, imagination, knowledge, dreaming, temporal/spatial mobility and being are all bodily' (2014, p. 47). His emphasis on the embodied experience of heritage is reflective of Waterton and Watson's (2014) concern with an embodied semiotic of heritage engagement that goes beyond the visual and representational and into the sensory world of affect, where places and objects constitute semiotic landscapes that conjure intensities of experience in which the past is both immanent and yet fluid and contingent in its meaning. What this approach also suggests is that cosmopolitan affect may be afforded across, between and within cultural entities – in a cross-cultural landscape – so that what is intensely felt in one subject, group, community or polity may be less significant, or not significant at all, to another, but is yet susceptible to ethical interpretation and, therefore, a higher level of affordance in formalized heritage spaces such as museums.

Cross-cultural encounters in the museum: the Museum of New Zealand Te Papa Tongarewa (Te Papa) and the Immigration Museum Melbourne (IMM)

Our aim in setting down an initial understanding of what we have termed an affective cosmopolitanism has been to prepare the ground for supposing that museums might constitute 'cross-cultural landscapes'. As revealed in the previous section, key to this theorization is the assumed potential of museums to engage their visitor-audiences through their embodied and sensory capacities, their emotions and the emotional–cognitive assemblages that they bring with them. These 'terms of engagement', it seems to us, provide a framework for the kinds of cross-cultural encounters that correspond with Anderson's notion of the cosmopolitan canopy and all that implies in terms of a humanizing cosmopolitan ethics. We will consider the nature and content of this fusion of ideas in the context of two museums that have a clear purpose in representing cross-cultural relations: the Museum of New Zealand Te Papa Tongarewa (Te Papa) and the Immigration Museum Melbourne (IMM). Drawing on empirical investigations of global visitors' experiences in both museums, the following analysis aims to illustrate how each operates as a cosmopolitan canopy within the respective cityscapes of Wellington and Melbourne, by facilitating cross-cultural encounters and engagements that are entangled with travel practices, thus affording spaces that evoke embodied, affective and emotional responses. To think about them in Anderson's (2004, p. 24) terms, both museums allow visitors to 'encounter people who are strangers to them, not just as individuals but also as representatives of groups they "know" only in the abstract. The canopy can thus be a profoundly humanizing experience'. Based on the research findings, we argue that Te Papa and the IMM put into practice a form of museological intervention, an interruption, to use Dawney's term, which, through the humanized, multi-sensory performativity of displays, provokes *at once* critical cosmopolitan and embodied responses *through* visitors' interpretive engagements. Cosmopolitanism thus emerges as a critical faculty (Delanty, 2012) and, to borrow from Mica Nava (2006), 'structures of feeling' (following from Raymond Williams, 1977) that emerge *through* the 'cosmohermeneutics' of cross-cultural encounters, entangling self and other through visitors' interpretive dialectics of reflexivity and empathy (Schorch, 2014a). At the same time, the research findings suggest that the biographies of visitors intertwine with interpretive engagements and with exhibitions (Schorch, 2015c). There is, then, no 'cosmopolitan Te Papa' or 'cosmopolitan IMM' in a totalizing sense; rather, there are particular cross-cultural negotiations, framed by affective and emotional registers, in specific contexts that might lead to intercultural literacy and ethical positions or ethnocentric misreadings and indifferent tolerance, among other potential responses.

Background: Te Papa and the IMM

Te Papa

Te Papa, which opened in 1998, considers itself a bicultural organization based on the principle of partnership enshrined in the Treaty of Waitangi, signed in 1840 between the British Crown and Māori. The Treaty is widely regarded as the founding document of Aotearoa New Zealand, and after decades of negligence it has gained constitution-like status in recent years. Today, concrete policies and practices such as Mana Taonga (living spiritual and cultural links between material treasures and people) and Mātauranga Māori (Māori knowledge) ensure Māori participation and involvement in the museum (Hakiwai, 2006; McCarthy, 2007, 2011; Schorch and Hakiwai, 2014; Schorch *et al.*, 2016; H. Smith, 2006). Importantly, Māori input into exhibition developments is not confined to Māori galleries, but adds a Māori and thus bicultural dimension to social and natural history as well as art galleries within the museum. The Treaty of Waitangi thereby assumes the central position within Te Papa's spatial layout and thematic composition by forming the main part of the *Signs of a Nation/Ngā Tohu Kotahitanga* exhibition and standing in a wedge-shaped space underneath a high cathedral-like ceiling (see Figure 5.1).

This space divides the museum into two sides: one devoted to Māori themes and the other to British settlers and other more recent immigrants from Asia and the Pacific region. Equally, though, there is an intention to draw these poles and their often conflicting histories together towards a common future. Te Papa thus houses a variety of cultural differences in their material, discursive and spatial manifestations under a common 'canopy'. The visitor study upon which this chapter is based involved interviews with visitors from Canada, the USA and Australia, and aimed at eliciting how tourists from other Anglo settler nations with similar but different postcolonial realities and Indigenous populations responded to Te Papa's explicit bicultural approach (for detailed research design see Schorch, 2015c).

The IMM

The IMM was also founded in 1998 and was Australia's second migration museum, after the South Australian Migration and Settlement Museum in Adelaide, established in 1986. Born out of an initiative by the state of Victoria and specifically devoted to the topic of immigration, the IMM has assumed a specific political position by constructing immigration as an integral part of Australia's history, as is obvious in the words of the museum's patron, the governor of Victoria at the time, who stresses that 'the story of immigration is essentially the story of all non-Indigenous Australians' (IMM, 1998, p. iv). This inclusive founding principle has been translated into museum practices and collection policies, and is reflected in the interrelated permanent galleries and temporary exhibitions. While the latter

Figure 5.1 Signs of a Nation/Ngā Tohu Kotahitanga exhibition within Te Papa.

Source: Te Papa.

100 *Philipp Schorch et al.*

are dedicated to particular communities, the former present critical approaches that place individual experiences within the socio-political and historical contexts of migration, thus providing an analysis of the host society as much as a history of migrants themselves (Witcomb, 2009). Thus, these galleries deal with the history of immigration policies and their impacts on those affected by them, the reasons for migration, and the experiences of migrants in Australia (see Figure 5.2).

The IMM, then, in ways similar to Te Papa, attempts to offer a spatial 'canopy' under which different cultural perspectives can interact across a common sphere. Furthermore, the IMM has strengthened its critical edge by tackling contemporary issues such as racism in the latest exhibition, *Identity: Yours, Mine, Ours* (Schorch, 2015a; Schorch *et al.*, 2015; Witcomb, 2013). In the case of the IMM, the visitor study, which incorporated interviews with Australian individuals and pairs of adults, set out to examine how visitors experience or engage with the representation of migration at the museum (for detailed research design see Schorch, 2014a).

Cross-cultural encounters and cosmopolitan engagements

Recent approaches to museum visitor studies have generated a nuanced understanding of what exhibitions might achieve by using qualitative methods to investigate visitor experiences through an analytical lens of 'encounter' and 'engagement' (Macdonald, 2002; Sandell, 2007; Smith, 2011; Schorch, 2014a, 2015c). In the

Figure 5.2 Leaving Home long-term exhibition, Immigration Museum, Melbourne.
Source: Museum Victoria. Photographer: Jon Augier.

cosmopolitan studies literature, a related focus on 'encounter' (Delanty, 2011) has emerged and can be drawn on to analyse the empirical realities of cross-cultural encounters and potentially cosmopolitan engagements in specific settings, such as museums (Schorch, 2014a). Moreover, museums and their practices of collecting and displaying the 'other' might offer a range of 'opportunities for encounters beyond the self' (Mason, 2013, pp. 44–5), especially in museums devoted to biculturalism (Te Papa) and migration (IMM) and the associated movement between cultural worlds of meaning. In short, we are interested in the ways in which a cross-cultural encounter can *become* a cosmopolitan engagement *through* the 'act of interpretation' (Schorch, 2013a, 2014a) and associated 'self-transformation in light of the encounter with the other' (Delanty, 2011, p. 642). Based on the two visitor studies at Te Papa and the IMM, we show how such museum encounters and interpretive engagements proceed through the cosmopolitan power of individual objects, the cosmopolitan agency of photographs and the cosmopolitan faces and stories of tour guides.

The cosmopolitan power of objects

Recent scholarship has expanded on the 'material turn' in the humanities and social sciences and the position of 'museums in the material world' (Knell, 2007) by emphasizing the material nature of museum experiences (Witcomb, 2010) and associated constructions of meanings (Schorch, 2014b). While taonga or Māori treasures, for example, have been turned into 'objects of ethnography' (Kirshenblatt-Gimblett, 1991), it is vital that, rather than seeing them merely as the products of social relations and knowledge, one considers their active mediation of those relations and knowledge through their own material and social agency (Gell, 1998; Henare *et al*., 2007; Latour, 2005). Thus, objects do not only reflect or embody external realities but also exert their own influence and *enact* relationships. Following Alfred Gell (1998, p. 6), the analytical lens should thus be geared towards the 'practical mediatory role' of objects 'in the social process' by zooming in on what materialities *do* rather than what they represent (see also Chua and Elliott, 2013). Such awareness should not, however, be reduced to an object-centred focus. Rather, meaning arises out of the interpretive space *in-between* objects and people, and vice versa (Schorch, 2015c).

A clear sense of this sort of mutual constitution throughout the processes of meaning-making was articulated by one participant, Bruce, from the USA, in the interview after his visit to Te Papa. Through his observations we can gain insight into how his museum experience arose out of a multi-sensory, embodied 'object-subject interaction' (Dudley, 2009), a process of active and mutual engagement between self and the physical world. This has similarities with arguments recently developed by Rosalyn Diprose (2011), who has explored the role of buildings, as non-human agents, in gathering affect. For Diprose (2011, p. 6), a 'building' is of course an ordinary thing; nonetheless, it carries a capacity to assemble and arrange 'atmosphere, wind, light, wood, stone, vegetation, as well as the flesh and sensibilities of its occupants and of those living beings that it leaves outside'.

The reflections Bruce offers on *Te Hau ki Turanga*, a communal Māori meeting house, echo these sentiments. He encountered this whare (building) in the *Mana Whenua* exhibition (see Figure 5.3), which explores and celebrates Māori as tangata whenua (original people) of Aotearoa New Zealand, and remarked:

> We took off our shoes and walked into the little house and kind of looked at all that. It always amazes me how cultures retain information. I mean writing is a cool thing but it takes a lot of being able to stay in one place and have a fairly complex society for it so it pops up ... the sort of non-character ways of retaining information that the hut embodies ... that the pylons are the ancestors and by looking at them you can recall your history, I mean it's an interesting memory device. I suppose in computer terms it's a very lossy way of doing things losing information over time, but it keeps at least the highlights for you and it keeps them really present. I guess one of the things that came up while I was looking at those structures was that these are very connected societies that have a very close connection to predecessors and to the community.

Through this exchange it becomes apparent that *Te Hau ki Turanga* exercises a form of agency that opens up an imaginative world, enabling Bruce to develop

Figure 5.3 *Te Hau ki Turanga* in the *Mana Whenua* exhibition.
Source: Te Papa.

an insight into the expressions of community and genealogy that are materialized in the object. This demonstrates the mobility of affect and its flows between human and non-human entities. These insights into the cultural other, however, are always mediated through the interpretive community of the self (Schorch, 2013a), in this case exemplified through the functioning of a computer. For Susan, also from the USA, a multi-sensory, embodied engagement with a canoe in the same exhibition framed her encounter with cultural differences and provoked the interpretive construction of cross-cultural meanings (Schorch, 2014b):

> And then looking at the canoe and seeing how small of a canoe that is, how wide it is and trying to imagine a six-foot man sitting in that cross-legged or even hunched down, being able to feel that and like 'that's crazy'. You know, I wouldn't be able to experience that if it was set up behind glass and like looking at it. I wouldn't actually be able to tell the depth I feel. And that not just me personally, but you just, you can almost feel yourself stepping into the canoe when it's set up in the middle of the floor like that and when you are able to walk into the building…

As Bruce and Susan's narrations indicate, those encounters between themselves and objects – and thus between self and other – began with an embodied engagement and *became* a cosmopolitan engagement through the interpretive transactions taking place between both poles and their ultimate entanglement. That is, the encounter facilitated a process of engagement from the affective to the cognitive that traces a variety of affordances, from the design of the space and the exhibition within it through to an emotional and cognitively framed interpretive engagement expressed in the interpretive performances of reflexivity and imagination. It is this crucial point of engagement that hermeneutically produces an entanglement of self and other and thus a cosmopolitan moment across differences and commonalities (Schorch, 2014a). For Bruce and Susan, particular taonga or Māori treasures *became* a 'medium of intimacy', as Classen and Howes (2006, p. 200, 202) would argue, and facilitated a 'corporal encounter' which allowed them to connect to both 'sensory as well as social biographies' of a carved object. For Māori people, taonga possess a life-force or mauri (Hakiwai and Diamond, 2015). They *are* ancestors and therefore *are* people and instantiate relations (Henare, 2007), which collapses the common dichotomy between subject and object. At the heart of this simultaneous, mutual constitution of human and non-human actors lies interpretive practice. Importantly, then, the biography of an object should not be equated with, or reduced to, its socio-cultural life trajectory, but should rather be understood through the biographies of relationships *enacted* as through taonga, in this case including Bruce and Susan. It is therefore more precise to speak of webs of biographies that *emerge* through the inextricable entanglements of human life and the material world (Schorch, forthcoming).

Susan's follow-up interview, which was conducted via phone six months after her visit to Te Papa, offers evidence of the long-term impact of this multidimensional interplay, which lives on in her memory as a 'felt presence':

The displays and exhibits that I really remember were the Māori displays and the, I don't remember what it was called, not like a temple but a meeting room where they perform their meetings?! And you were able to take off your shoes and enter in and just kind of sit there and soak it all in. You feel the presence and everything and like all the beautiful carvings and it's nice being able to touch everything and just look at the different, the very beautiful intricate details on the carvings.

For Susan, a 'felt presence' seems to imprint on her memory more profoundly than factual information such as the name of the 'Māori displays'. This 'felt presence', or 'eerie sense', as another visitor from Australia put it in his follow-up interview, is a clearly embodied engagement or act of meaning-making that is performed across cultural boundaries, which, throughout a process that moves from and between the affective to the cognitive, shapes a cosmopolitan entanglement of self and other, through the interpretive movements of reflexivity and imagination. This means that there is no cosmopolitan object or subject *per se*, but instead there emerge potentially cosmopolitan moments which erupt from processes that begin with affective–subjective engagements and proceed through the simultaneous, mutual constitution of objects and subjects across cultural differences.

The cosmopolitan agency of photographs

There is a growing body of literature within and beyond heritage studies which calls our attention to dimensions such as 'senses', 'feelings', 'emotions', 'affect' and 'embodiment' to gain a more nuanced view of the human experience (Crouch, 2015; Dudley, 2009; Edwards *et al.*, 2006; Gregg and Seigworth, 2010; Gregory and Witcomb, 2007; L. Smith, 2006; Thrift, 2008; Waterton, 2014; Witcomb, 2010). However, while these perspectives importantly allude to the so-called 'non-representational' dimensions of experiences, they often do so at the expense of power, situatedness and biography (Thrift, 2008) and language (DeLanda, 2006) (see detailed discussions in Schorch, 2013b, 2014b). As a specific consequence of these debates, 'there is a tendency for images to be treated as visual or non-verbal, which creates a false contrast with language' (Hughes-Freeland, 2004, p. 209; though see detailed discussions in Waterton and Watson, 2014 against this line of thinking). The visitor study at the Immigration Museum Melbourne included the *Leaving Dublin* temporary exhibition (see Figure 5.4), which photographically captures the current generation of Irish migrants to Australia, exposing this 'contrast' as 'false' through an empirical interrogation of interpretive processes.

Turning to an interview with Paul, who migrated to Australia from Hungary after World War II, we can observe the agency of *Leaving Dublin* in facilitating an encounter between viewer and exhibition through engagement with faces and stories: 'I like the exhibition. The photographs were fantastic, very evocative and artistically … I mean I'm no photographer, but I was struck by just how wonderful the photographs were and just related some of this to my own experience.' The exhibit provides at once a window to the other and a mirror to the self. That is,

Figure 5.4 Leaving Dublin: Photographs by David Monahan touring exhibition 2012–13, Immigration Museum, Melbourne.

Source: Museum Victoria. Photographer: Benjamin Healley.

by engaging with the 'evocative … photographs' depicting other migrants, Paul was able to relate 'some of this' to his 'own experience' of having been a 'refugee' and 'migrant', which he elaborated on during his interview when shifting his reflections from the photos to the 'stories':

> The individual stories were quite touching, bringing up all these things of fear and loss and leaving a community and realizing that to have a decent life, this was again a theme in the exhibition, people need to somehow take roots in a new community which may be quite strange and forbidden even.

Mirror and window, or self and other, become entangled through the interplay of embodied narratives and narrative embodiment. That is, Paul's 'own experience' is embodied in the 'photographs' of fellow migrants' faces, while 'the individual stories' embody the 'fear and loss and leaving'. 'These are the thoughts that come to me', Paul concludes, 'by looking at a picture or photographs and hearing particular stories'. The simultaneous presence of embodied narratives through faces and narrative embodiment through stories humanizes the museum encounter and entangles the 'experiences' of self and other (Schorch, 2014a). It becomes apparent that 'evocative … pictures' of bodies and their 'touching … stories' are irretrievably intertwined dimensions of Paul's interpretive engagement, thus pointing to the mutual constitution of affective registers and interpretive strategies such as stories and their subjective consequences arising through the performed practices of a lived experience (Schorch, 2014b).

The interpretive interplay between 'picture' and 'story', or embodied narrative and narrative embodiment, assists in opening the encounter between exhibition and viewer to empathetic and reflexive engagements. For Julia, who was born and raised in Australia, 'the photographs' of 'them', the faces of the protagonists, embody a 'kind of symbol' that hints at a potentially happy end to their stories:

> What I liked about the photographs was the darkness, but in most of them there was light shining through at some point. Something was illuminated and quite bright gold light, which I guess relieved that sense of sadness and, you know, the pain of saying goodbye with this kind of symbol of something new, maybe in the distance but that was going to come to them. I hope it did for them.

Lisa, who recently emigrated with her boyfriend Kyle from Ireland to Australia, shares Julia's empathetic identification with the experiences of the other and in the process reflects upon the self:

> Parts of the exhibition were related to Dublin, Ireland, which I found particularly enjoyable. It was good to hear the stories of fellow immigrants and see that we are not alone … young people coming over for work and to start a family just because it's difficult to do at home in Ireland at the moment, and that would be part of why we came over, with a view to starting a new life for ourselves. And just to see that people had done that before and it's the same emotions and missing family and the same kind of struggles.

Lisa's encounter with the exhibition becomes an interpretive engagement that is both reflexive and empathetic, the latter of which, as Andrea Witcomb (2009, p. 64) argues, is a 'prerequisite for dialogue, for the recognition of commonalities'. Indeed, empathy requires a shared symbolic terrain or 'common sphere' (Dilthey, 1976), so that the hermeneutic negotiations of cultural differences can lead to understandings (Schorch, 2013a). The interpretive dialectics of empathy and reflexivity, then, create commonalities across differences, thus entangling self and other through the 'cosmohermeneutics' (Schorch, 2014a) of cross-cultural museum encounters. These actual experiences and their narrative expression allude to the *material-in-the-verbal* through the bodies and flesh of narrative characters. Importantly, then, 'language' should be seen evocatively rather than through a structural or representational lens. Claims like 'language is not life; it gives life orders' (Deleuze and Guattari, 1987, p. 84) derive from such a limiting structural and almost mechanical view on linguistics which does not capture the infinite and fluid world of interpretive and imaginative engagements (see detailed discussion in Schorch, 2013b). Rather, we should recognize that 'images, *like words*, evoke worlds' (Hughes-Freeland, 2004, p. 209). That is, images, language and words, like objects, never only *represent* external realities, but *become* meaningful through their performative effect and embodied affect on readers and viewers.

Humanizing cosmopolitanism

A culture cannot speak or engage in encounter and dialogue; it depends on the face and story of a cultural actor. It follows that there cannot evolve a cross-cultural dialogue between totalized collective entities, but only an interpersonal dialogue among cultural human beings (Schorch, 2013a, 2014a). A practice of affective cosmopolitanism, then, requires the humanizing of cross-cultural encounters as the basis of interpretive engagements so that a potentially cosmopolitan 'self-transformation in light of the encounter with the other' (Delanty, 2011, p. 642) can occur. Returning to the interview with Bruce, we can observe that a tour guide at Te Papa lends a face with a story to a cultural group:

> One of the cool things was that, according to the tour guide, it was basically presented by the Māori not by, you know, a bunch of white guys saying what we present of the Māori, which made a lot more tellable and believable and didn't have this sort of stench of imperialism on it. So it made it a lot easier to sort of, because if somebody is telling about themselves rather than somebody telling about somebody else, we call that hearsay in the law.

Through the medium of a live Māori presence embodied in the tour guide, Bruce engages with another cultural world at Te Papa after initial reluctance fuelled by his 'experience with native culture in the United States'. Through this humanized, interpretive mediation, Bruce recognizes the self-representation of Māori in the wider exhibition spaces, which seems to offer some remedy for the 'stench of imperialism' which he associates with many 'presentations of non-dominant cultures'. The tour intervention and the associated humanization of culture through the face and story of the tour guide, then, open up the potentialities of the museum encounter, readying it for a mutually negotiated cross-cultural dialogue and facilitating embodied, cross-cultural forms of emotion- and meaning-making (Schorch, 2013a, 2015a, 2015b). Strikingly, Bruce departs from the specificity of the situation to assume a wider moral stance. He talks about 'non-present cultures' in general and links their alien representation to the 'hearsay' concept 'in the law', his own professional field. Bruce's interview thus attests to the interpretive process through which visitors narrate their biographies into the museum experience, and the museum experience into their biographies (Schorch, 2015a, 2015b), as well as to the moral quality of this interpretive dialectics.

The study at the IMM that includes the *Getting In* gallery, which tracks changes in Australia's immigration policies through history, also illustrates this point. Part of this gallery is an interview booth, an interactive touch screen that embodies the viewer in the position of both an immigration officer and a visa applicant. The display thus requires the viewer to physically and imaginatively assume the roles and perspectives of various others. For Angela, who was born and raised in Australia, such an affective–subjective framing of an empathetic identification evolved into an emotionally modulated reflexive and critical examination of socio-political contexts through the experiences of 'someone from Iraq' and her own life. Being confronted with a face and 'story behind' a 'poor policy' is, according to her, 'absolutely appalling' but 'doesn't surprise' her, since she is 'working with people who do those interviews' in immigration detention health services in Western Australia. The interview booth at the IMM and her professional life, full of 'direct experience working with immigrants and asylum seekers', enable Angela to understand the concrete 'story' of 'someone' behind an abstract 'policy'. This humanization of migration shapes an interpersonal rather than an abstract encounter, and renders possible a moral and emotional relationship between self and other, thus shaping a discursive cosmopolitan space. That is, the 'reflexive condition' of cosmopolitanism as a 'mode of critique' (Delanty, 2012) can only emerge through the practice of meaning-making and the 'act of interpretation' (Schorch, 2013a, 2014a). A cosmopolitan 'fusion of horizons' (Gadamer, quoted in Ricoeur, 1991), through which the perspective of the other is being incorporated into the broadened horizon of the self (Delanty, 2012; Held, 2002), can only be hermeneutically achieved through the concrete interpretive performances by individuals rather than the abstract merging of collective entities. In other words, a culture cannot reflect, critique or transform itself. Instead, it requires the embodied and affective potential, and the reflexive, critical and transformative faculty, of cultural actors (Schorch, 2014a). The development of a 'cosmopolitan appreciation of difference', then, depends on the 'profoundly humanizing experience' (Anderson, 2004) afforded by a process that begins with embodied encounters and is resolved in moments of meaning-making and change, the very essence of which can be captured by an affective cosmopolitanism provoked in and through the cross-cultural museum canopies of Te Papa and the IMM.

Conclusions

Drawing on two studies conducted with global visitors to Te Papa and Australian visitors to the IMM, this chapter has offered an interpretive exploration of museum experiences as embodied, interpretive engagements with cultural differences, something we have described as *cosmopolitan affect*. We have argued that Te Papa and the IMM to varying degrees put into practice a form of museological intervention that is both affective and cosmopolitan. We suggested three features through which both museums facilitate particular forms of cross-cultural encounters and thus provoke affective, cosmopolitan engagements: the cosmopolitan power of individual objects, the cosmopolitan agency of photographs and the cosmopolitan faces and stories of tour guides. By deploying humanized cultural perspectives and multi-sensory performative displays, each museum *enacts* rather than represents or teaches cultural difference. Moreover, both museums not only enact cultural plurality but also build bridges across these pluralities. The required conversation across cultural differences occurs through the performativity rather than representational function of displays on the one hand and visitors' interpretive dialectics of reflexivity and empathy on the other; both sides become entangled through the interpretive space that is 'in-between' viewer and display or between self and other, thus creating a shared sphere of affordances within the 'canopy' of the museum. To capture this, we have proposed what we have termed an *affective cosmopolitanism*, which we have illustrated as being enacted in and through the cross-cultural museum canopies of Te Papa and the IMM. This is *at once* an empirical and a normative concept; an embodied, social practice *and* a philosophical ideal that not only cultivates cross-cultural landscapes for intercultural literacy and ethical responses, but also complicates ethnocentric misreadings and the indifferent tolerance of others. As an exploration of the working of affect in heritage, we have explicitly situated ourselves – with certain qualifications – in the non-representational or rather more-than-representational domain of theory as we feel that this provides a more accurate understanding of the interplay of the structured sequences of museum engagements, the role of interpretation, the dissonant or uncontrolled affordances of the museum as a landscape for cross-cultural encounters and the capacity of visitors to be affected by them. Based on our empirical enquiry and theoretical reasoning, we have offered affective cosmopolitanism as an analytical category and lens which captures and illuminates the affective–subjective dynamic of museum and other cross-cultural experiences as embodied encounters in structured spaces that are essentially non- or more-than-representational, but imbued with an ethical quality that emerges in the spaces *between* objects, people and others.

Notes

1 The arguments advanced in this article have been further developed and published in the book *The Cosmopolitan Canopy: Race and Civility in Everyday Life* (Anderson, 2011).
2 See the introductory essay in a special issue on cosmopolitanism in *Theory, Culture and Society* (2002) 19 (1–2).

Bibliography

Ahmed, S. 2010. Happy objects, in M. Gregg and G.J. Seigworth (eds) *The affect theory reader* (pp. 29–51). Durham, NC: Duke University Press.
Anderson, E. 2004. The cosmopolitan canopy. *Annals of the American Academy of Political and Social Science*, 595: 14–31.
Anderson, E. 2011. *The cosmopolitan canopy: race and civility in everyday life*. New York: W.W. Norton & Co.
Bagnall, G. 2003. Performance and performativity at heritage sites. *Museum and Society*, 1(2): 87–103.
Beck, U. 2002. The cosmopolitan sociey and its enemies. *Theory, Culture and Society*, 19(1–2): 17–44.
Bondi, L. and Davidson, J. 2011. Lost in translation: a response to Steve Pile. *Transactions of the Institute of British Geographers*, 36: 595–8.
Chua, L. and Elliott, M. (eds) 2013. *Distributed objects: meaning and mattering after Alfred Gell*. Oxford and London: Berghahn Books.
Classen, C. and Howes, D. 2006. The museum as sensescape: Western sensibilities and Indigenous artifacts, in E. Edwards, C. Gosden and R.B. Phillips (eds) *Sensible objects: colonialism, museums and material culture* (pp. 199–222). Oxford: Berg.

Costello, L. 2013. Performative memory: form and content in the Jewish Museum Berlin. *Liminalities: A Journal of Performance Studies*, 9(4). http://liminalities.net/9-4/costello.pdf (accessed 9 January 2015).

Crouch, D. 2015. Affect, heritage, feeling, in E. Waterton and S. Watson (eds) *The Palgrave handbook to contemporary heritage research* (pp. 177–90). Basingstoke: Palgrave Macmillan.

Daugbjerg, M. 2009. Pacifying war heritage: patterns of cosmopolitan nationalism at a Danish battlefield site. *International Journal of Heritage Studies*, 15(5): 431–46.

Daugbjerg, M. and Fibiger, T. 2011. Introduction: heritage gone global. Investigating the production and problematics of globalised pasts. *History and Anthropology*, 22(2): 135–47.

Dawney, L. 2013. The interruption: investigating subjectivation and affect. *Environment and Planning D: Society and Space*, 31(4): 628–44.

DeLanda, M. 2006. *A new philosophy of society: assemblage theory and social complexity*. London and New York: Continuum.

Delanty, G. 2011. Cultural diversity, democracy and the prospects of cosmopolitanism: a theory of cultural encounters. *The British Journal of Sociology*, 62(4): 633–56.

Delanty, G. 2012. The idea of critical cosmopolitanism, in G. Delanty (ed) *Routledge handbook of cosmopolitan studies* (pp. 38–46). London: Routledge.

Deleuze, G. 1988. *Spinoza: practical philosophy* (R. Hurley, trans.). San Francisco: City Lights Books.

Deleuze, G. and Guattari, F. 1987. *A thousand plateaus: capitalism and schizophrenia* (B. Massumi, trans.). London and New York: Continuum.

Deleuze, G. and Guattari, F. 1994. *What is philosophy?* (G. Burchell and H. Tomlinson, trans.). London: Verso.

Dilthey, W. 1976. *Selected writings*. Cambridge: Cambridge University Press.

Diprose, R.J. 2011. Building and belonging amid the plight of dwelling. *Angelaki: Journal of the Theoretical Humanities*, 16(4): 59–72.

Dudley, S.H. 2009. Museum materialities: objects, sense and feeling, in S.H. Dudley (ed) *Museum materialities: objects, engagements, interpretations* (pp. 1–17). London: Routledge.

Edwards, E., Gosden, C. and Phillips, R.B. (eds) 2006. *Sensible objects: colonialism, museums and material culture*. Oxford: Berg.

Featherstone, M. 2002. Cosmopolis: an introduction. *Theory, Culture and Society*, 19(1–2): 1–16.

Gell, A. 1998. *Art and agency: an anthropological theory*. Oxford: Clarendon Press.

Gregg, M. and Seigworth, G.J. (eds) 2010. *The affect theory reader*. Durham, NC: Duke University Press.

Gregory, K. and Witcomb, A. 2007. Beyond nostalgia: the role of affect in generating historical understanding at heritage sites, in S.J. Knell, S. Macleod and S. Watson (eds) *Museum revolutions: how museums change and are changed* (pp. 263–75). London: Routledge.

Hakiwai, A. 2006. Māori Taonga – Māori identity in New Zealand, in B.T. Hoffman (ed) *Art and cultural heritage: law, policy and practice* (pp. 409–12). New York: Cambridge University Press.

Hakiwai, A. and Diamond, D. 2015. Plenary: the legacy of museum ethnography for indigenous people today – case studies from Aotearoa/New Zealand. *Museum and Society*, 13(1): 107–18.

Harding, J. and Pribram, D. 2002. The power of feeling: locating emotions in culture. *European Journal of Cultural Studies*, 5(4): 407–26.
Held, D. 2002. Culture and political community: national, global and cosmopolitan, in S. Vertovec and R. Cohen (eds) *Conceiving cosmopolitanism: theory, context, and practice* (pp. 48–60). Oxford: Oxford University Press.
Henare, A. 2007. Taonga Māori: encompassing rights and property in New Zealand, in A. Henare, M. Holbraad and S. Wastell (eds) *Thinking through things: theorising artefacts in ethnographic perspective* (pp. 47–67). London and New York: Routledge.
Henare, A., Holbraad, M. and Wastell, S. (eds) 2007. *Thinking through things: theorising artefacts in ethnographic perspective*. London and New York: Routledge.
Hughes-Freeland, F. 2004. Working images: epilogue, in S. Pink, L. Kürti and A.S. Afonso (eds) *Working images: visual research and representation in ethnography* (pp. 204–18). London and New York: Routledge.
IMM. 1998. *Opening exhibition catalogue*. Melbourne: Immigration Museum Melbourne.
Kendall, G., Woodward, I. and Skrbis, Z. 2009. *The sociology of cosmopolitanism: globalisation, identity, culture and government*. Basingstoke: Palgrave Macmillan.
Kirchberg, V. and Tröndle, M. 2012. Experiencing exhibitions: a review of studies on visitor experiences in museums. *Curator: The Museum Journal*, 55(4): 435–52.
Kirshenblatt-Gimblett, B. 1991. Objects of ethnography, in I. Karp and S.D. Lavine (eds) *Exhibiting cultures: the poetics and politics of museum display* (pp. 386–443). Washington and London: Smithsonian Institution Press.
Knell, S.J. (ed) 2007. *Museums in the material world*. London: Routledge.
Kreps, C. 2003. *Liberating culture: cross-cultural perspectives on museums, curation and heritage preservation*. London: Routledge.
Kreps, C. 2011. Non-Western models of musuems and curation in cross-cultural perspective, in S. Macdonald (ed) *A companion to museum studies* (pp. 457–72). Malden, MA: John Wiley and Sons.
Latham, K.F. 2007. The poetry of the museum: a holistic model of numinous museum experiences. *Museum Management and Curatorship*, 22(3): 247–63.
Latour, B. 2005. *Reassembling the social: an introduction to actor-network-theory*. Oxford: Oxford University Press.
Lorimer, H. 2005. Cultural geography: the busyness of being 'more than representational'. *Progress in Human Geography*, 29: 83–94.
Macdonald, S. 2002. *Behind the scenes at the science museum*. Oxford: Berg.
Macdonald, S. 2013. *Memorylands: heritage and identity in Europe today*. London: Routledge.
Mason, R. 2013. National museums, globalisation and postnationalism: imagining a cosmopolitan museology. *Museum Worlds: Advances in Research*, 1(1): 40–64.
McCarthy, C. 2007. *Exhibiting Māori: a history of the colonial cultures of display*. Wellington: Te Papa Press.
McCarthy, C. 2011. *Museums and Māori: heritage professionals, Indigenous collections, current practice*. Wellington: Te Papa Press.
Nash, C. 2000. Performativity in practice: some recent work in cultural geography. *Progress in Human Geography*, 24(4): 653–64.
Nava, M. 2006. Domestic cosmopolitanism and structures of feeling: the specificity of London, in N. Yuval-Davis, K. Kannabiran and U.M. Vieten (eds) *The situated politics of belonging* (pp. 42–53). London: Sage.
Pile, S. 2010. Emotions and affect in recent human geography. *Transactions of the Institute of British Geographers*, 31(1): 5–10.

Ricoeur, P. 1991. Life in quest of narrative, in David Wood (ed) *On Paul Ricoeur: narrative and interpretation* (pp. 20–33). London and New York: Routledge.
Sandell, R. 2007. *Museums, prejudice and the reframing of difference*. London and New York: Routledge.
Sassen, S. 2000. New frontiers facing urban sociology at the Millenium. *British Journal of Sociology*, 51(1): 143–59.
Schorch, P. 2013a. Contact zones, third spaces, and the act of interpretation. *Museum and Society*, 11(1): 68–81.
Schorch, P. 2013b. The hermeneutics of transpacific assemblages. *Alfred Deakin Research Institute Working Paper Series*, 2(41): 1–15.
Schorch, P. 2014a. The cosmohermeneutics of migration encounters at the Immigration Museum Melbourne. *Museum Worlds: Advances in Research*, 2(1): 81–98.
Schorch, P. 2014b. Cultural feelings and the making of meaning. *International Journal of Heritage Studies*, 20(1): 22–35.
Schorch, P. 2015a. Experiencing differences and negotiating prejudices at the Immigration Museum Melbourne. *International Journal of Heritage Studies*, 21(1): 46–64.
Schorch, P. 2015b. Museum encounters and narrative engagements, in A. Witcomb and K. Message (eds) *Museum theory: an expanded field* (pp. 437–57). Malden, MA and Oxford: Blackwell. [Volume 1 of International Handbooks of Museum Studies, series editors Sharon Macdonald and Helen Rees Leahy].
Schorch, P. 2015c. Assembling communities: curatorial practices, material cultures, and meanings, in B. Onciul, M. Stefano and S. Hawke (eds) *Engaging communities*. Suffolk: Boydell and Brewer.
Schorch, P. and Hakiwai, A. 2014. Mana Taonga and the public sphere: a dialogue between Indigenous practice and Western theory. *International Journal of Cultural Studies*, 17(2): 191–205.
Schorch, P., McCarthy, C. and Hakiwai, A. 2016. Globalizing Māori museology: reconceptualizing engagement, knowledge and virtuality through mana taonga. *Museum Anthropology*, 39(1): 48–69.
Schorch, P., Walton, J., Priest, N. and Paradies, Y. 2015. Encountering the 'other': interpreting student experiences of a multi-sensory museum exhibition. *Journal of Intercultural Studies*, 36(2): 220–39.
Smith, H. 2006. The Museum of New Zealand Te Papa Tongarewa, in C. Healy and A. Witcomb (eds) *South Pacific museums: experiments in culture* (pp. 10.11–10.13). Melbourne: Monash University ePress.
Smith, L. 2006. *Uses of heritage*. London: Routledge.
Smith, L. 2011. Affect and registers of engagement: navigating emotional responses to dissonant heritage, in L. Smith, G. Cubitt, R. Wilson and K. Fouseki (eds) *Representing enslavement and abolition in museums: ambiguous engagements* (pp. 260–303). London: Routledge.
Soren, B.J. 2009. Museum experiences that change visitors. *Museum Management and Curatorship*, 24(3): 233–51.
Staiff, R. 2014. *Re-imagining heritage interpretation*. Aldershot: Ashgate.
Thien, D. 2005. Affect or beyond feeling? A consideration of affect and emotion in geography. *Area*, 37(4): 450–4.
Thrift, N. 2008. *Non-representational theory: space, politics, affect*. London: Routledge.
Tolia-Kelly, D.P. 2006. Affect – an ethnocentric encounter? Exploring the 'universalist' imperative of emotional/affectual geographies. *Area*, 38: 213–17.

Uzzell, D. 1989. The hot interpretation of war and conflict, in Uzzell, D.L. (ed) *Heritage interpretation: the natural and built environment* (pp. 33–47). London: Belhaven Press.

Waterton, E. 2014. A more-than-representational understanding of heritage? The 'past' and the politics of affect. *Geography Compass*, 8(11): 823–33.

Waterton, E. and Dittmer, J. 2014. The museum as assemblage: bringing forth affect at the Australian War Memorial. *Museum Management and Curatorship*, 29(2): 122–39.

Waterton, E. and Watson, S. 2014. *The semiotics of heritage tourism*. Bristol: Channel View Publications.

Williams, R. 1977. *Marxism and literature*. Oxford: Oxford University Press.

Witcomb, A. 2009. Migration, social cohesion and cultural diversity: can museums move beyond pluralism? *Humanities Research*, 15(2): 49–66.

Witcomb, A. 2010. Using objects to remember the dead and affect the living: the case of a miniature model of Treblinka, in S. Dudley (ed) *Museum materialities: objects, engagements, interpretations* (pp. 39–52). London and New York: Routledge.

Witcomb, A. 2013. Understanding the role of affect in producing a critical pedagogy for history museums. *Museum Management and Curatorship*, 28(3): 255–71.

Woodward, I. and Skrbis, Z. 2012. Performing cosmopolitanism, in G. Delanty (ed) *Routledge handbook of cosmopolitan studies* (pp. 127–37). London: Routledge.

6 Constructing affective narratives in transatlantic slavery museums in the UK

Leanne Munroe

In recent years, there has been a 'growing sensitivity for complexity' (Schorch, 2014) in heritage and museums studies. The turn to complexity has prompted the development of new trajectories that examine the emotional and affective resonances inherent in the heritage project. For example, the study of the ephemeral and felt qualities of heritage has received increasing attention in studies that focus on visitor engagements. Such works have suggested that as we move about ancient and historic places, or engage with museum displays, our minds and bodies converge to elicit intensely vivid, meaningful experiences. For example, Witcomb (2013) has examined how affective display strategies can unsettle comfortable narratives and prompt critical thinking in visitors; Crang and Tolia-Kelly (2010) examine the affective registers of national identity for heritage visitors and chart how embodied responses, feeling and sentiment, rather than cognitive knowledge production, exclude certain people from the heritage experiences; and Schorch (2014) has used narrative interview methods to examine the simultaneously *felt* and *thought* visitor engagements within museums, arguing that our minds and bodies are concomitant partners in the process of meaning-making. In this chapter, I extend and add to this research by elucidating the ways in which affect and emotion suffuse processes of *narrative construction* (rather than visitor engagement) in the museum. The construction of museum narratives has received significant attention as a topic of enquiry from 'representational' studies of heritage, which seek to unpick the semiotic, hegemonic and discursive underpinnings behind these creations (see Hall, 1997; Harrison, 2010; Smith, 2006; Walsh, 1992). Such studies have been instrumental in analysing the inequalities which underlie the processes of narrative construction, yet, quite often, these studies fail to grasp the complex ways in which emotions, feelings and affects infuse this process from a number of directions. Narrative has often been portrayed as a cognitive process; a rational exercise in intellectual 'sense-making' often bound up in Foucauldian investigations into power/knowledge struggles. Yet recent enquiries into museum narrative suggest that it is a much more complex hermeneutic form than the above characterisation suggests (Macleod *et al*., 2012; Staiff, 2014). For example, Harrison states that heritage-making 'is not the inscription of meaning onto blank objects, places and practices … it is not primarily an intellectual endeavour, something that exists only in the human mind, but is one that emerges from dialogue, or practices of people and things' (2013, p. 217). Taking heed of Harrison, this investigation posits the idea that narrative-making is a process which is 'located in the world', crafted by emotionally situated agents and embedded in the corporeality of their lives. The representations curators make can never be completely emotionally neutral and may be fuelled by intense feelings connected to their socio-political positions or the material traces which surround them. As such, narrativisation is seen as a process that is intersected by a matrix of emotional and affective entanglements with both human and non-human elements: texts, people, discourses, objects, practices and spaces. Ultimately, this chapter suggests that stories about the past are not just purely political inventions but are a series of powerfully emotive, affective connections which attempt to hold us in place in the world, tell us who 'we' are and where 'we' come from, and, quite literally, 'mean' something.

Yet, the processes of narrativisation can never be wholly divorced from considerations of power and politics. As MacLeod *et al*. (2012, p. xxii) state, narrative is 'fundamentally a human construct, necessitating a process of exclusion and editing, with all the accompanying risks of bias and distortion'. This can lead to the perpetuation of exclusionary narratives of the past which are underpinned by intensely held attachments that feel 'emotionally authentic' to dominant groups in society. This chapter therefore seeks to understand how emotions and affects are inextricably connected to the discursive and the political during processes of narrativisation in the museum and grasp how narratives are forged from 'affective–discursive' networks (Wetherell, 2012). The case study examined in this chapter stems from my PhD research (2012–15), which aimed to conceptualise a more nuanced understanding of the processes of narrativisation regarding the history of transatlantic slavery in twelve British museums. Using a methodology that combined textual analysis, qualitative interviews, exhibition analysis and wider analysis of the social, material and cultural landscapes of the museums, it attempted to link museum narrative construction 'to the world, rather than separating it from it' (Witcomb, 2003, p. 10); to anchor the discursive in emotional, affective and material foundations; and to explore the subsequent nuances in narratives of slavery within British museums. The aim of this chapter, therefore, is to unspool and examine the many interrelated traces of emotions and affects that suffuse the discursive process of narrative construction in two slavery museums in the UK. This is achieved through examining museums which reside within very different affective–discursive milieus, and pulling out the many emotive and affective influences which work upon them. Ultimately, I suggest that narrative construction in the museum can be conceived of as a process of meaning-making that is embedded within 'affective practices' (Wetherell, 2012, p. 19).

Theoretical background: 'affective–discursive loops'

Before the case study can be analysed, some theoretical foundations need to be laid down which further unpack the relationship between affect, emotion and narrative construction. In the following section, I address the tensions that have arisen between some aspects of affect studies and the concept of narrative-making, and suggest that we need to reduce theoretical divisions between the affective and the discursive and see them as concomitant partners in the process of 'affective meaning-making' (Wetherell, 2012); I also suggest that we need to view narrative construction as constrained by affective practices (Wetherell, 2012, p. 53) that are 'situated' in the world (Bauman, 1986) and can be embroiled in hegemonic struggles. By doing this, we can recalibrate the relationship between the representational and the more-than-representational, and begin to explore the affective–discursive networks within which museum narrative construction is located.

Affect studies grew out of an ontological shift away from the dominant modes of thought that have characterised the social sciences, and indeed heritage studies, in the past few decades – namely, representational approaches which privilege 'discursive' analyses. It emerged from a concern that the 'discursive turn' was 'writing the body out of theory' (Hemmings, 2005, see also Tolia-Kelly, 2006) and sought to inject new life into the study of the material and physical affects of the world around us. As such, many scholars (e.g. Clough, 2007; Massumi, 2002; Thrift, 2009) turned towards analysing pre-/non-discursive engagements with the world. This strain of affect studies conceptualises the world around us as dynamic and flowing, full of 'potentials', 'forces' and 'intensities' that can work upon the body and elicit intensely felt experiences that transcend cognitive rationalisation. It attempts to understand the processes which lie underneath what are characterised as the 'thresholds of conscious contemplation' (McCormack, 2003, p. 488). In doing so, these understandings of affect often decentre or completely relegate analyses of discourse, and obscure historical and social contexts such as the impact gender, race, class and historical frames of reference can have upon how we interact with the world around us. This strain of affect studies therefore carves a gulf between the body, its cognitive processes and the historical–social realm. For example, Massumi (2002) draws a dividing line between the body and discourse, and privileges the track of the body. He states that affect is 'autonomous', meaning that physical stimuli provoke involuntary reactions from the body which are 'beyond' consciousness.

It is at this point that studies of narrative construction diverge from these conceptions of affect. The supposed division between the body and discourse provokes a separation between the subjective experiences of affect and emotion. It implies that affect stands apart from discussions of emotions, as affective experiences precede the mediation involved in translating these bodily 'intensities' into emotions, which are necessarily *cognitive* ruminations on these felt experiences: 'affect, unlike emotion or feeling, is something that has not yet been closed down, represented, labelled, communicated, shaped and structured ... the leading edge of the wave of any engagement with the world before human minds get to it' (Wetherell, 2012, p. 59). Affect, in this sense, implies a state of *being*, rather than *interpreted* emotions. If we apply this understanding of affect to narrative, it seems as though the two are incompatible. Narrative is fundamentally an *interpretative* form (Bruner, 1990; Ricoeur, 1991), whereas affect defies interpretation. Affect (in this particular conception) is precisely what narrative is not – unmediated, unstructured, unrepresentable, an ephemeral and fleeting imprint upon the body which cannot be tied down with language. Narrative, on the other hand, is a mediatory tool for translating bodily affects into coherent emotional stories.

Yet Wetherell suggests that the distinction between affect and emotion is untenable (2012, p. 19). She argues that it would be futile to attempt to pull apart the cognitive and bodily processes which spark both affective *and* emotional reactions, as they are inherently intertwined. Under the rubric of 'affective–discursive meaning-making', she discusses the ways in which 'spiralling affective-discursive loops [are] set in motion as initial affect is narrated, communicated, shared, intensified, dispersed, modified and sometimes re-awoken years later' (2012, p. 53). Burkitt (1997) similarly advocates a relational, social approach to emotions and affects which sees them as knotted together. He suggests that an emotion, such as anger, is always a response to a situation in the world: 'emotions rest on body responses that provoke, and turn into, feelings' (Burkitt, 1997, cited in Wetherell, 2012, p. 24). If we apply this line of thinking to museum narratives, we can see the ways in which they inescapably entwine the affective and the emotional, the bodily and the cognitive. For example, a museum narrative is always an interpretation; it can never, impossibly, capture and display the bodily pain *experienced* by an enslaved person who felt the lash of the whip on her back, or the sway of the slave ship as it crossed the ocean, or the rumble of aching hunger in her stomach. A curator therefore writes a narrative that translates these bodily affects into interpretations of those felt experiences through words, images and objects. The resultant narratives are subjective and situational mediations of the original bodily experiences, necessarily channelled through a present-day lens that is girded by a particular emotional regime. These representations can be designed to be highly emotive, so much so that they activate affective responses in visitors. We may recoil in disgust from drawings depicting the torture of the enslaved or feel sadness when we read an account of a child being sold (see Arnold-de Simine, 2013 for further discussion about mimetic and empathetic responses to viewing slavery installations). Unravelling these 'spiralling affective–discursive loops' will help us to understand the nature of the relationship between narrative and affect in the museum, which is always a complex mediation between the bodily and the cognitive. This chapter, therefore, adopts a 'grounded' model of affect which sees the construction of narratives as being a process of affective–discursive meaning-making in the museum. Museums can be seen as affective–discursive assemblages which oscillate between 'body-brain-narrative-feeling-response-context-history etc.' (Wetherell, 2012, p. 75).

Wetherell offers another discussion that enables us to examine the relationship between affect, emotion and narrative-making in the museum. She introduces the concept of 'affective practice' (2012, p. 22), which examines how routine ways of acting and thinking are bound together with certain emotional regimes that can constrain how we act, feel and think in the present and future. This offers an approach to affect and emotion which views them as located in social action. The concept of affective practice is useful for understanding how the past becomes associated with certain emotional repertoires that can change over time, depending on the socio-political or cultural mores of the day; for example, the emotional repertoires connected with the Holocaust can shape how we remember, commemorate, talk about and interpret this history in our museums. Stemming from this, Wetherell suggests that affective practices can act as a form of discipline and control, constraining or directing our emotional responses to certain events or histories. For trenchant adherents to 'pre-discursive' conceptions of affect, the consideration of power is not something they are concerned with, yet in order to study the relationships between narrative-making, affect and emotion, the consideration of power – or, as Thrift says, 'the politics of affect' – is crucial (Thrift, 2004). Through studying affective practices we can see that affects and emotions can be enlisted within society for various purposes and can be unevenly distributed (Crang and Tolia-Kelly, 2010). Affects can emotionally privilege one group while disadvantaging another. This view of affect is also discussed by Ahmed (2004), who introduces the concept of an 'affective economy' which assigns affective 'value' to some people at the expense of others, and the historian William Reddy (2001), who discusses an idea of historical 'emotional regimes'. Importantly, affective practices can have racialised (or gendered or classist, etc) modalities that foster exclusionary and unequal social relations (Hemmings, 2005). For example, Ahmed uses the example of Audre Lorde's account of a white woman's affective response to her blackness:

> When I look up the woman is still staring at me, her nose holes and eyes huge. And suddenly I realise that there is nothing crawling up the seat between us; it is me she doesn't want her coat to touch. The fur brushes past my face as she stands with a shudder and holds on to a strap in the speeding train … Something's going on here I do not understand, but I will never forget it. Her eyes. The flared nostrils. The hate.
> (Lorde, 1984, pp. 147–8, cited in Ahmed, 2000)

The white woman experiences an affect of disgust when she encounters Lorde's body, yet the white woman's affective experience cannot be disentangled from social and historical narratives and power relations which dictate that the woman *find* Lorde's body disgusting. What this demonstrates is that our affective engagements and emotional entanglements with the world are mediated socially, politically and culturally. Transposing the concept of affective practice into heritage and museums studies allows the researcher to see the emotional patterns, grooves and ruts which are overlaid onto the material and textual traces of the past that influence how they are interpreted. Power, then, is a crucial component of affect and emotion: it allows us to understand the uneven distribution of affective practices and see how affect and discourse are embedded within contested social situations and narratives.

'Relational' methods and approaches

A key concern for this chapter is to conceptualise narrative-making in the museum as a process that goes beyond the purely rational, intellectual or cognitive, and elucidate how it is intertwined with affective–discursive meaning-making and constrained by affective practices and emotional regimes. In order to understand the complexity of such affective and emotional modalities, we need to transcend textual analysis in isolation and view the processes of narrativisation as located 'in the world', connected to a network of social, material and political nodes that can exercise a range of affective and emotional influences. As such, 'relational' models of heritage provide useful frameworks for understanding the embodied and felt manifestations of the heritage experience that emerge out of the connections between people, texts, objects and places (e.g. dialogism, Bakhtin, 1981; Actor Network Theory, Latour, 2005; assemblage theory, see Waterton and Dittmer, 2014). Waterton and Watson (2013, p. 552) suggest that relational models 'stir [heritage] back in with being human and living, so that it emerges from the feelings of being, becoming and belonging in the flows and complexities that characterise life'. Relational studies of heritage are therefore sympathetic to the concept of affective–discursive meaning-making as they situate meaning-making within the interactions and negotiations between both human and non-human elements: narratives, discourses, museum practices, the materiality of the museum itself, objects, curators and other actors, places and the wider socio-political landscape of the museum.

Yet Waterton and Watson (2013) point out that one of the most striking things about relational theories is that 'theory' refers to 'what to study' rather than suggesting how we go about researching those networks. Harrison (2013) puts forward a relational ontology – a way of *thinking* about heritage – but there is no explicit methodology for examining these dialogical connections between people, objects, spaces and practices. Perhaps this is owing to the fact that this trajectory of research is in its relative infancy in heritage studies, and as yet we have not developed the appropriate tools to deal with the swirling and dynamic movements, flows and connections which characterise this way of thinking. Similarly, methodologies for studying the affective and emotive implications of heritage are thin on the ground. But there have been a number of studies which have proffered different tools of investigation. Waterton and Dittmer (2014; see also their study of the Australian War Museum as assemblage in this volume) use auto-ethnographic vignettes to explore the web of felt, embodied, emotive and cognitive influences that impinged upon them as they moved about the space (see also Light and Watson in this volume for a similar approach to the experience of castle visiting). Schorch (2014) suggests analysing biographical narratives of visitors who relay their thoughts and feelings about their experiences in the museum space to the researcher. Witcomb (2013) (also see Gregory and Witcomb, 2007) chooses exhibition analysis methods to examine the ways in which affective interpretative strategies can unsettle hegemonic narratives and encourage critical thinking.

This chapter adopts a case study approach, whereby the creation of the transatlantic slavery exhibitions in the Cowper and Newton Museum and the London Docklands Museum is analysed using a suite of methodologies that allow the many tangled affective, emotive and cognitive threads to be examined in a holistic, situated and relational manner. The flowing and dynamic traces of emotions and affect are captured through semi-structured interviews, immersive walks with the curators through the galleries, holistic exhibition analysis and analysis of the wider material and social landscape of the museum.

The emotional and affective traces quite often flowed most thickly between three primary nodes of the affective–discursive network of narrative construction: (1) agents; (2) objects and texts; and (3) the spatial. With regard to agents, the emotive and affective impingements upon curators and other museum stakeholders affected the ways in which the narratives of slavery were constructed. This includes the 'lived' experiences of their positions, the emotive discourses which influenced them and bound their speech and their relationships with other museum agents such as community groups. With regard to objects and texts, narrative construction was affected by the ways in which emotive and affective resonances settled on certain objects, and how language in text conveyed particular emotive timbres to different groups of people. And lastly, the spatiality and atmospheres of the museum space affected narrative creation through, for example, the location of the building, the content and arrangement of other displays and the intangible *felt atmosphere* of the place (such as a feeling of nostalgia, etc.). Taking all these together, it is possible to glimpse into the complex affective–discursive networks out of which museum narratives emerge in these two museums.

Background to case studies

The year 2007 was the bicentenary of the Act of Abolition in the UK, which ended the transatlantic slave trade within the British sphere of power. Prior to the bicentenary, the history of abolition and the slave trade was primarily refracted through a prism that foregrounded the work of white, male abolitionists such as MP William Wilberforce and largely obfuscated discussion of the enslaved, their experiences of enslavement, the role they played in abolition and the culpability of the British in establishing and perpetuating the trade in the first instance (Oldfield, 2007). The traces of this shameful period of history were effectively silenced within British collective memory. The bicentenary therefore provided a timely opportunity for reflection and enquiry about this past. A large number of exhibitions were redeveloped or created, and lectures, public debates and commemorative ceremonies took place in museums, which allowed these institutions to become public touchstones for a whole range of historical, cultural, social and political issues. Yet the commemorations and exhibitions were not without their critics. A vehement vein of criticism flowed from members of the African Caribbean community, who viewed the commemorations as a 'Wilberfest' or 'Wilberfarce' (Agbetu, 2011, p. 62) that crystallised the 'abolitionist discourse' established in the twentieth century around the memory of Wilberforce and the other abolitionist 'saints'. Agbetu argues that this celebration of Wilberforce reinforced the 'congratulatory' moment of abolition in British collective memory at the expense of other narratives that focused on the enslaved. The debates in public forums (such as newspapers and internet comment pages) about the commemorations became electrically charged with emotive language: expressions of shame, blame, grief, outrage and anger were commonplace, as were expressions of pride and moral righteousness (Waterton and Wilson, 2009). Smith's (2011) research shows that feelings about the displays, whether positive or negative, ran high as visitors walked around the museum and engaged with the exhibits. These emotional responses, Smith demonstrates, were not always linked directly to the content on display but were tied together with a knot of socio-cultural issues such as national identity and race, inflamed further by heated political questions of whether there should be apologies and reparations. The bicentenary was therefore marked by palpable anxiety, clustered around a tangle of emotive and affective issues, expressed most evidently through the highly charged language used to discuss it in public forums.

The museum responses to the bicentenary have been examined largely from a 'representational' viewpoint, which focused on the content of the displays, the tone in which the narratives were couched, the displays' relationships to socio-political issues, the importance of imagery and language in silencing certain voices and other concerns such as moral questions over whether to display images of brutality or address racism (Benjamin, 2012; Cubitt, 2012; Hamilton *et al.*, 2012; Smith *et al.*, 2010; Smith *et al.*, 2011; Waterton, 2010; Waterton *et al.*, 2010). While Smith (2011) examined the emotional responses of visitors to these exhibitions, little work was done to understand the how the creation of the narratives emerged out of thick affective–discursive networks. The following case studies attempt to unravel the complex affective–discursive networks which created slavery exhibitions at two museums at this time.

Case study one: the Cowper and Newton Museum

Located in the north of rural Buckinghamshire, on the banks of the River Ouse, Olney is a small town that is picture-postcard perfect, filled with houses buried underneath crawling ivy, quaint inns and an impressive fourteenth-century Gothic church. On its historic marketplace stands a grand Georgian double-fronted house, which used to be home to the eighteenth-century writer and abolitionist, William Cowper. The house is now the Cowper and Newton Museum, an independent historic house museum which charts the lives and works of William Cowper and the other man referenced in its name, the Reverend John Newton. John Newton stakes his name in history as one of the most famous abolitionists; an ex-slave trader turned evangelical campaigner who renounced his former life in favour of the church. He wrote the famous hymn, *Amazing Grace*, as well as penning an influential abolitionist tract, *Thoughts on the African Slave Trade*, and published his diaries which detailed his life as a slave trader, his moment of conversion to Christianity and his subsequent renunciation of the slave trade. Newton lived in Olney for sixteen years from 1767, as the curate of St Peter's and St Paul's Church. His vicarage was located directly behind Cowper's house and the two men became close friends, lending their support to the Parliamentary campaign for abolition led by William Wilberforce MP through writing songs, letters and poems.

Cowper's house was turned into a 'memorial museum' for the two men in 1900. From the outset, the museum cultivated an atmosphere that was nostalgic, evocative and had a deep sense of reverence, encouraging veneration for the 'relics' of the men. An article in the *Northampton Herald* from 1907 describes the museum as a 'shrine' with its 'wealth of relics of the two great [men]'. Another article stated that Olney was flooded with visitors from far and wide, 'like devoted pilgrims at a cherished shrine'. The museum displays combined naturalistic 'historic house' elements with more explicitly exhibitionary techniques. It was around this time (1906) that the Wilberforce House Museum also opened and fostered similarly reverential engagements through portraying Wilberforce as a 'statesman-saint' and a 'hero'. Wilberforce House Museum fostered a cult of personality around Wilberforce, displaying a life-size wax effigy, filling his 'birth room' with portraits of the man and summoning the language of superlatives during the dedication ceremony which opened the museum in 1906. This ceremony described Wilberforce as 'a victorious leader whose fame has gone to the four corners of the earth'. The type of breathless ardour which surrounded the memory of Wilberforce was echoed in Olney during the 1907 commemorations of Newton's death (which incidentally coincided with the centenary of abolition). During the addresses at the church, Newton was described as 'a Divine' and a man of 'heroic faith and foresight' and was credited with 'immortalizing' the town of Olney (*Northampton Herald*, 26 April 1907). This emotive discourse has echoes of the language used in what Oldfield has termed a 'culture of abolition' in the early twentieth century. In this 'abolitionist' discourse, moralistic language was used to celebrate the 'triumph' of abolition and describe the 'heroes' of the movement in saint-like terms (2007, p. 2). This emotional, affective 'shrine-like' atmosphere at the Cowper and Newton Museum continued to influence the ways in which the museum presented itself during the twentieth century.

As the bicentenary of abolition approached in 2007, the museum received two grants to redevelop the displays on Newton's role in the slave trade and the abolitionist movement. Cubitt (2009) has suggested that in 2007 many museums at the local level struggled with how to craft a narrative of the slave trade that both 'particularized' and 'generalized' this history, highlighting and exceptionalising the locality yet also integrating the local story into the national narrative framework. While undoubtedly discursive struggles between the local and the 'other', the particular and the general, proved challenging to the Cowper and Newton Museum, the process of narrativisation was also constrained by a set of intensely held emotional resonances and affective practices. Significant factors in determining the timbre of the narrative were the persistent nostalgic, 'shrine-like' atmosphere of the house itself, the nature of the objects and texts in the collection and the positions of the staff as figures that were deeply embedded within the reality of the town's civic and historic identity.

The Cowper and Newton Museum has a very small staff: one paid house manager and twelve voluntary trustees who are well-connected and active members of the local community (at the time of research some trustees sat on the local council; others were involved with the local tourism organisation; others in church activities; a few with the local charity; others with the historic Olney Lace Circle; and one was the chair of the local Masonic Lodge). The museum was therefore deeply embedded in the civic identity of the town through the actions and positions of its staff. The trustees, as prominent public figures who were 'gatekeepers' of Olney's quaint, 'olde-worlde' image (especially to tourists), therefore had vested interests in the cultivation of a nostalgic tone, content and overall design of the museum that aggrandised Cowper and Newton's legacies. But their involvement was deeper than that: the trustees had intense, emotional ties to the museum, fostered from the role that the museum played in their everyday lives. For example, one of the trustees had been on the board for more than forty years and lived behind the museum, taking on the role of taming the vast and beautiful gardens. She worked in the gardens every day and had renovated them completely, working tirelessly to transform them into a replica of their state in the eighteenth century. She, her husband and a group of elderly friends used the gardens to grow a range of historic fruits and vegetables such as quinces, which they distributed among the trustees, volunteers and visitors. They also installed a beehive with the intention of selling 'museum honey' to friends, townspeople and visitors. Other trustees (some of whom had been on the board for more than twenty years) gave their time and creativity generously, making up for the lack of funding from local and national government with endless voluntary hours and personal contributions. They took care not only of the 'business' side of the museum, but also of the exhibitions, the collections, the building itself, the volunteers and the education/events policies. The museum was therefore very much a dynamic, 'living' and *emotional* project for the people involved with it – a tight-knit 'interpretive community' (in Witcomb's words: 2003, pp. 79–101) who gave up their unpaid time to ensure that the museum continued to open. And, as such, the emotional attachment to the displays, the objects and the building itself was palpable.

As part of the grant to redevelop the slavery displays in 2007, the trustees worked with an external curator to write the new narrative of the slave trade, abolition and Newton's role within it. The curator had numerous meetings with the trustees but no external consultations (such as with African Caribbean groups or academics) took place – the trustees and volunteers had creative control over the direction the curator took. The curator's lived position as an 'external' agent, someone who was not 'a part' of the museum itself, an 'outsider', meant that he had to ensure that the narrative he created would meet the expectations and hopes of the emotionally invested community of people who were letting him into 'their' museum. He had to ensure that the narrative would *feel* as though it belonged in this historic museum and would coincide with the nostalgic/celebratory tone of the rest of the museum, and indeed of Olney's proud civic identity. Far from being a mere intellectual exercise in crafting a historic narrative, the curator had to work within the realities of the small museum: juggling interactions with staff and trustees who acted as intellectual and emotional gatekeepers to the town's past, the limitations of the collections, the atmosphere and tone of the house itself and also the location and position of the museum as a town 'institution' which was a central pivot in Olney's civic pride. Such everyday realities of museum practice and institutional production have been highlighted by Witcomb (2003) as crucial for understanding the complexity of museum work in determining the meanings and resonances of the past.

It became clear to the consultant that the trustees felt the focus of the slavery exhibition should not stray too far from the perspectives of the two 'great men'. The location and history of the house as being 'the actual place' where the men lived and worked, as well as the nature of the rest of the museum displays (primarily naturalistic 'historic house' displays) and the collections, influenced the content and tone of the new exhibition. Gregory and Witcomb (2007) have discussed the intensely affective and emotional resonances that historic house museums elicit. They suggest that historic house museums excite 'pleasure, delight and wonder ... when we perceive the imprints of past lives that are somehow embodied within the house' (2007, p. 265). An intangible atmosphere is created through negotiation of the absences and presences of the people who used to live there, invoking a particular type of embodied reaction – wonder, a sense of 'haunting', a reverence for things that the owners might have once touched. This is bundled together with the valuing of 'authentic' objects and experiences, as the past and the present somehow become collapsed in the museum space. Within the Cowper and Newton Museum, this haunting atmosphere is palpable. The dark wood, the creaking floors, the closed curtains, the woody smell of furniture polish and the dim lighting create an experience which is visceral and provokes a *felt* sense that this is an important, authentic remnant of a past age, where the presences of the two great men inhabit the spaces of the house. In some senses, the 'shrine-like' atmosphere of the early twentieth century still lingered, embodied in the display of intimate personal effects such as Cowper's nightcap and locks of his hair.

The decision was therefore made to address the slave trade and slavery through the eyes of John Newton's experiences as a slave trader, so that the narrative would *feel* as though it belonged in the house. Display boards were created to address such topics as 'John Newton, slave trader', 'John Newton, a changed man', 'John Newton, abolitionist' and 'John Newton, Hymn-writer'. The narrative used excerpts from Newton's diaries, sermons and hymns to illustrate the text, as well as poems written by Cowper such as 'Pity for the Poor Africans' and 'The Negro's Complaint'. The texts of Cowper and Newton were used almost exclusively and greatly influenced the tone of the narrative which was celebratory. Ahern (2013) states that the literary style of eighteenth-century abolitionists was overtly 'sentimentalist' and emotional, full of the language of pity, sadness and moral righteousness. It was a style that was deliberately emotive in order to encourage sympathy for the abolitionist cause by highlighting the terrible degradation of the enslaved Africans (as 'passive' victims) and the heroic, moral deeds of the abolitionists. Newton's texts are similarly peppered with emotional language. His hymn *Amazing Grace*, which charts his conversion from slave trader to abolitionist, has taken on a strong emotional timbre in the present day: positive, celebratory and primarily associated with the American civil rights movement, set to rousing music which stirs the emotions. Waterton and Dittmer (2014) have highlighted how the use of sound can conjure affective atmospheres within museums, drawing the visitor into a sensual experience that envelops them in an embodied sense of history. A speaker was installed in the gallery so that once a button is pressed, the words of *Amazing Grace* echo about the room, read in a rich, deep and resonant (male) voice, which imbues the museum atmosphere with an almost spiritual sense of historical significance – that *this* is the place where the world's most famous hymn was written.

The collecting policies of the Cowper and Newton Museum focused on acquiring the texts and ephemera of the 'two great men'; the curator therefore had very little 'slavery' material to deal with when crafting the narrative, and was subsequently limited to displaying and utilising the abolitionist texts which were suffused with the emotional/sentimentalist language of the era. Other items displayed were 'relics' of Newton such as his cleric bands, his bible, his chair, his church pew and a copy of the stained glass window from Olney Church which shows Newton with a slave kneeling at his feet. The religious nature of these objects only further coloured them with a patina of reverence. The Cowper and Newton Museum *did* try to counter this by including a cross-cut model of a slave ship, an outline of the amount of space an enslaved person would have on a ship and the display of replica yokes and manacles, but these elements remained somewhat buried beneath the memorial atmosphere and the display of abolitionist/religious ephemera. The nature of the collection and the affective/emotional resonances of those items, as well as the *atmosphere* of the house itself, therefore greatly affected the process of constructing the narrative.

Case study two: London, Sugar and Slavery Gallery

The London, Sugar and Slavery Gallery is located in the Museum of London Docklands, near Canary Wharf. The museum is a social history museum which commemorates the history and legacies of the ports and waterways of London. The museum houses galleries which discuss the ancient history of the Thames, trade and empire, the docklands during wartime, the lives of sailors and how the river contributes to life in the city today. The London, Sugar and Slavery Gallery is the newest permanent exhibition in the museum. The gallery opened in November 2007 and received a Heritage Lottery Fund grant as part of the approximately £14 million made available for the bicentennial commemorations of abolition. The focus of the gallery is local, examining how London benefited from the wealth brought by its involvement in the slave trade, as well as examining how London interacted with different peoples and places located within the 'triangular trade'. The building itself carried an affective pull on the construction of the museum narrative. The building was described by the curatorial team as being one of the most important buildings in London involved with the slave trade, as it was an old shipping warehouse which housed transhipments of sugar from the plantations. The history and feel of the building itself therefore directed the narrative to focus on the production of sugar, and trace the wealth of this product. The museum space itself, as a warehouse, is airy, square and devoid of overtly nostalgic/affective resonances – a blank canvas for creating social history displays, primarily using text panels and standalone displays.

As a social history museum, as opposed to a historic house museum, the aims, tone and display of the gallery differed greatly from those of the Cowper and Newton Museum. Just as the Cowper and Newton Museum's narrative was influenced by the nostalgic and memorial atmosphere of their building, the emotional and sentimental nature of their collections and the embodied position of its agents as gatekeepers of Olney's past, the London, Sugar and Slavery Gallery was subject to the interactions of its own unique affective–discursive network. This affective–discursive network offers a very different set of relations and interactions than the Cowper and Newton case study. If the Cowper and Newton Museum example is characterised by emotive and affective resonances that are nostalgic, comfortable and pleasurable, then the London, Sugar and Slavery example is marked by ones which unsettle these affects and emotions.

From the outset, the narrative-making process became embroiled in a deep, complex negotiation of emotional and affective interplays. This was partially owing to the relationships between the various different agents who had a hand in crafting the narrative. As a nationally funded, large museum in the capital city, there were a significant number of internal and external stakeholders who sought to become involved in the narrative-making process. The museum assigned a number of permanent curators to the task, as well as hiring an external guest curator with in-depth knowledge of black British history and a number of academic advisors from universities and people from African Caribbean organisations, and also arranged a public consultative group. The narrative-making process was therefore highly collaborative and was greatly influenced by the lived experiences, positions, thoughts and feelings of those involved, as well as by the physical/cognitive interactions between the different agents at the monthly meetings within the museum. This process produced some highly emotional exchanges which were derived from the position each agent was perceived to inhabit. For example, one male curator felt as though his lived experience as a white, middle-class curator would make it inappropriate for him to discuss some issues, especially concerning the racist legacies of slavery. He stated: 'we realised that to get the story right, to get the narrative right, we couldn't tell the story on our [the curators'] own.' He went on to state: 'I knew that *I* couldn't talk about the inherited experiences of people from the Caribbean or West Africa or racism in London in the modern day. I can talk about that from a white middle-class perspective, I can't talk about that from a black perspective.'[1] This quote highlights a number of deeply complex issues about emotional 'ownership' over history narratives. In this instance, emotions, affects and politics cannot be disentangled. African Caribbean people have long been denied a voice with regard to studying the legacies and histories of enslavement, which has contributed to feelings of hurt, marginalisation and anger (Agbetu, 2011). For African Caribbean people, as recipients of racist legacies of enslavement, the right to finally be able to 'speak' about their ordeals carries tremendous affective and emotional power. As such, the inclusion of the African Caribbean group in the narrative-making process nudged the story in a direction which was more pedagogical, political, emotional and 'hot' (Uzzell and Ballantyne, 1998) than the usual uncritical, comfortable narratives of abolition.

This process of negotiating differing affects, emotions and politics bled through every stage of the narrative-making process. The process of narrative-making in the gallery consisted of cognitive *and* embodied engagements with other agents, enacted through 'long, protracted meetings' which could invariably stray into tense and angry confrontations with other members of the consultative group. One curator described such a meeting when signing off the panel text:

> We didn't quite understand why the discussion wasn't flowing – you know, the discussion was just [gestures with hands and face as if to say 'tense']. And finally one of the consultative group turned round to say 'Look! We don't like the language! When we read this, what we're reading is the voice of the curator. It's not the voice of the African Caribbean community'.

The curator goes on to explain that his experience as an 'academic curator' in a national institution had induced him to write a narrative in a more 'neutral' tone than the African Caribbean group (as stakeholders with strong emotional attachments to the history of slavery and a lived experience of the racist legacies of it) were comfortable with. Undoubtedly the emotional response of the community group was tied together with deeper issues concerned with the prevalence of the obfuscating 'abolitionist discourse' that circulated at a national level during the bicentenary. This discourse was perceived by the African Caribbean community to be inadequate with regard to discussing the brutalities of enslavement and the resilience and resistance of enslaved Africans. A number of high-profile protests (such as Toyin Agbetu's disruption of the commemoration service at Westminster Abbey) attested to the real physical and emotional anger and hurt that this discourse caused. It also highlights that not all narratives carry the same affective 'weight' or 'resonances' for all groups of people. While the history of slavery was viewed as highly important as a topic of history for the curators, the community group connected with the narrative of enslavement on a more personal, and hence emotional, level. The narrative-making process was therefore intersected by a wide range of emotional discourses, socio-political entanglements and affective realities which were brought into contact during the monthly meetings. With only two weeks until the deadline for the text, the panels were completely rewritten, harnessing a voice which eschewed neutrality in favour of a more didactic, 'hot' (Uzzell and Ballantyne, 1998) interpretation, which made use of emotive language and critical viewpoints. One aspect of this rewrite was ensuring that the emotive and offensive implications of the language used in the panels were properly contextualised. As some of the terms used in eighteenth-century texts are highly offensive to modern-day African Caribbean groups (for example, the uncritical use of the term 'slave' as opposed to 'enslaved African'), a panel which explains 'terminology' was erected at the start of the gallery in order to explore the emotive issue of language.

The Museum of London Docklands had a lack of objects with regard to slavery. A significant number of items had to be purchased especially for the gallery, including a plantation archive from St Kitts and Nevis consisting of mostly two-dimensional texts and books owned by the Mills brothers. The issue of slavery objects and images has received attention from researchers (Araujo, 2012; Guyatt, 2000; Trodd, 2013; Webster, 2009; Wood, 2000) who have explored the semiotic underpinnings and unintended reinforcement of hegemonic viewpoints through the uncritical and de-contextualised display of these artefacts. Yet these studies, in examining what the objects may represent, have overlooked the emotional and affective weight that some objects carry. Objects have the capacity to affect the text and tone of the museum, turning something which, at first glance, appeared fairly neutral into something which conveys a vastly different meaning. In the London, Sugar and Slavery Gallery, a number of items had the capacity to affect the tone and trajectory of the narrative. In particular, the consultative group highlighted the affective weight of images and texts of brutality, punishment and torture of enslaved people. The curator stated:

> One of the issues that we were made aware of is that some of these items have different significances for the African Caribbean community. If you put a slave collar up, for example, it is seen in a different way [by them]. They made us aware that these relics had a very different emotional impact [for them].

The African Caribbean group realised the affective capacity of the instruments of torture and worried that the message of the narrative (of resistance and bravery) would be undercut by the display of shackles, chains and manacles. The display of the shackles could also elicit a number of purely bodily reactions in visitors such as fascination, revulsion or horror, which, when unaccompanied by cognitive ruminations on the objects, could reinforce the status of the enslaved as dehumanised, helpless victims. The decision, then, was taken to limit images and artefacts concerned with brutality and, instead, construct an oral, immersive narrative designed to create both bodily and cognitive engagement. This took the form of a sound and light show in the gallery every twenty minutes. The show is an arresting and unexpected occurrence for visitors, as the lights dim and a loud, sinister voice floods the exhibition space. The visitors cannot help but pay attention as the lights are too dim to view the exhibitions and the voice too loud to ignore. The voice tells visitors 'you are to be stripped of your name, you are to be taken from your loved ones, you are to be brutalised', etc. The show lasts for a number of minutes and is a powerful affective exhibitionary strategy. The show provokes a type of 'empathetic unsettlement' (La Capra, 2001 p. 78), to 'make people think that it can happen to anyone, regardless of race'. It is both a *felt* and *thought* engagement. The creation of this immersive narrative strategy therefore emerged out of a clutch of negative affective and emotional responses to images of brutality from the consultative group.

Conclusion

In this chapter, I have highlighted the swirling interrelations of affects, emotions and discourses that are partners in the process of narrative-making in the museum. I have sought to show that narratives are not purely intellectual creations, but are hitched to an array of *lived, felt, thought* and *experienced* interplays between the mind and the body. Using a relational model of heritage, I have demonstrated the complex interaction between agents, objects and texts and spatialities which are imbued with affective and emotive resonances in the museum. Theoretically, this investigation has been supported by a recalibration of the relationship between representational and more-than-representational theories, which allows issues of the sociality of affect to be taken into consideration. Using Wetherell's conceptions of 'affective–discursive meaning-making', one can examine the 'spiralling affective–discursive loops' which translate between the brain and the body in concrete social situations. By viewing museum narrative-making as a type of 'affective practice', narrative-making is extracted from the realm of the purely cognitive and transposed into situated, corporeal worlds which are built through both discursive and affective regimes.

My first case study illustrated how the slavery narrative at the Cowper and Newton Museum emerged out of the intense emotional and affective connection the staff had to the museum, the nostalgic atmosphere of the house itself and the complex emotive/affective nature of the texts and objects in the collection. The second examined how the narrative at the London, Sugar and Slavery Gallery was built out of highly emotional negotiations between curators and community groups, differing conceptions of the affective 'weight' of certain objects and texts and differing 'lived' experiences which dictated the strength of the emotional connection with slavery and the right to 'speak' about it. These situated and lived processes of narrative construction also coincided with a national discourse about the role of slavery and abolition in British history. This national discourse was entangled with the emotive language of national identity and pride, as well as anger and marginalisation for alienated groups. The complexities of these interactions highlight the difficult, but rewarding, task ahead for museum studies as a discipline. Museologists are only just beginning to clear the tangled thickets which grow tightly around the theory of affect and, as such, we need to sift through the many variants of this theory in order to develop an application which is suitable for our discipline. I have suggested that considering museums as places entangled in affective practices and places that are engaged in affective–discursive meaning-making may allow us to combine the vital strains of representational and more-than-representational work which make heritage and museums studies such a dynamic discipline.

Note

1 My own interviews here are supplemented by transcripts of interviews conducted by the 1807 Commemorated research project at the website: www.history.ac.uk/1807commemorated/interviews/wareham.html, interview carried out by Cubitt (n.d.), accessed 29 October 2014. Also see Wilson (2011) for more information about the curatorial experiences during the bicentenary.

Bibliography

Agbetu, T. 2011. Restoring the Pan-African perspective. In L. Smith, G. Cubitt, K. Fouseki and R. Wilson (eds) *Representing enslavement and abolition in museums: ambiguous engagements* (pp. 61–74). London: Routledge.

Ahern, S. (ed) 2013. *Affect and abolition in the Anglo-Atlantic, 1770–1830*. Farnham and Burlington: Ashgate.

Ahmed, S. 2000. *Strange encounters*. London: Routledge.

Ahmed, S. 2004. *The cultural politics of emotion*. New York: Routledge.

Araujo, L.A. (ed) 2012. *Politics of memory: making slavery visible in the public space*. London: Routledge.

Arnold-de Simine, S. 2013. *Mediating memory in the museum: trauma, empathy, nostalgia*. Houndsmill: Palgrave Macmillan.

Bakhtin, M. 1981. *The dialogic imagination*. C. Emerson and M. Holquist (trans. and eds.). Austin: University of Texas Press.

Bauman, R. 1986. *Story, performance and event: contextual studies of oral narrative*. Cambridge: Cambridge University Press.

Benjamin, R. 2012. Museums and sensitive histories: the International Slavery Museum, in L.A. Araujo (ed) *Politics of memory: making slavery visible in the public space* (pp. 178–96). London: Routledge.

Bruner, J. 1990. *Acts of meaning*. Cambridge: Harvard University Press.

Burkitt, I. 1997. Social relationships and emotions. *Sociology*, 31(1): 37–55.

Clough, P. 2007. *The affective turn: theorizing the social*. Durham, NC and London: Duke University Press.

Crang, M. and Tolia-Kelly, D.P. 2010. Nation, race and affect: senses and sensibilities at national heritage sites. *Environment and Planning A*, 42: 2315–31.

Cubitt, G. n.d. 1807 commemorated website: interviews: Museum of Docklands London. www.history.ac.uk/1807commemorated/interviews/wareham.html (accessed 29 October 2014).

Cubitt, G. 2009. Bringing it home: making local meaning in 2007 bicentenary exhibitions. *Slavery and Abolition*, 30(2): 259–75.

Cubitt, G. 2012. Museums and slavery in Britain: the bicentenary of 1807, in L.A. Araujo (ed) *Politics of memory: making slavery visible in the public space* (pp. 159–77). London: Routledge.

Gregory, K. and Witcomb, A. 2007. Beyond nostalgia: the role of affect in generating historical understanding at heritage sites, in S. Watson, S. MacLeod and S. Knell (eds) *Museum revolutions: how museums change and are changed* (pp. 263–75). London and New York: Routledge.

Guyatt, M. 2000. The Wedgwood slave medallion: values in eighteenth-century design. *Journal of Design History*, 13(2): 93–105.

Hall, S. 1997. *Representation: cultural representations and signifying practices*. London: Sage Publications.

Hamilton, D., Hodgson, K. and Quirk, J. (eds) 2012. *Slavery, memory and identity: national representations and global legacies*. London: Pickering and Chatto.

Harrison, R. 2010. *Understanding the politics of heritage*. Manchester: Manchester University Press.

Harrison, R. 2013. *Heritage: critical approaches*. London and New York: Routledge.

Hemmings, C. 2005. Invoking affect: cultural theory and the ontological turn. *Cultural Studies*, 19: 548–67.

La Capra, D. 2001. *Writing history, writing trauma*. Baltimore and London: Johns Hopkins University Press.
Latour, B. 2005. *Reassembling the social: an introduction to actor-network theory*. Oxford: Oxford University Press.
Macleod, S., Hourston Hanks, L. and Hale, J. (eds) 2012. *Museum making: narratives, architectures, exhibitions*. London: Routledge.
Massumi, B. 2002. *Parables of the virtual: movement, affect, sensation*. Durham, NC: Duke University Press.
McCormack, D. 2003. An event of geographical ethics in space after affect. *Transactions of the Institute of British Geographers*, 28(4): 488–507.
Oldfield, J. 2007. *Chords of freedom: commemoration, ritual, and British transatlantic slavery*. Manchester: Manchester University Press.
Reddy, W. 2001. *The navigation of feelings: a framework for the history of emotions*. Cambridge: Cambridge University Press.
Ricoeur, P. 1991. Life in quest of narrative, in D. Woods (ed) *On Paul Ricoeur: narrative and interpretation* (pp. 20–33). London and New York: Routledge.
Schorch, P. 2014. Cultural feelings and the making of meaning. *International Journal of Heritage Studies*, 20(1): 22–35.
Smith, L. 2006. *Uses of heritage*. London and New York: Routledge.
Smith, L. 2011. Affect and registers of engagement: navigating emotional responses to dissonant heritages, in L. Smith, G. Cubitt, K. Fouseki and R. Wilson (eds) *Representing enslavement and abolition in museums: ambiguous engagements* (pp. 260–303). London and New York: Routledge.
Smith, L., Cubitt, G. and Waterton, E. (eds) 2010 Special volume: Museums and the bicentenary of the abolition of the slave trade. *Museum and Society* 8(3): 122–7.
Smith, L., Cubitt, G., Fouseki, K. and Wilson, R. (eds) 2011. *Representing enslavement and abolition in museums: ambiguous engagements*. London: Routledge.
Staiff, R. 2014. *Reimagining heritage interpretation: enchanting the past-future*. Farnham: Ashgate.
Thrift, N. 2004. Intensities of feeling: towards a spatial politics of affect. *Geografiska Annaler: Series B, Human Geography*, 86(1): 57–78.
Thrift, N. 2009. Understanding the affective spaces of political performance, in L. Bondi, L. Cameron, J. Davidson and M. Smith (eds) *Emotion, place and culture* (pp. 79–96). Aldershot: Ashgate.
Tolia-Kelly, D.P. 2006. Affect – an ethnocentric encounter? Exploring the 'universalist' imperative of emotional/affectual geographies. *Area*, 38: 213–17.
Trodd, Z. 2013. Am I still not man and a brother? Protest memory in contemporary anti-slavery visual culture. *Slavery and Abolition*, 34(2): 338–52.
Uzzell, D.L. and Ballantyne, R. 1998. Heritage that hurts: interpretation in a post-modern world, in D.L. Uzzell and R. Ballantyne (eds) *Contemporary issues in heritage and environmental interpretation: problems and prospects* (pp. 502–13). London: The Stationery Office.
Walsh, K. 1992. *The representation of the past: museums and heritage in the post-modern world*. London: Routledge.
Waterton, E. 2010. Humiliated silence: multiculturalism, blame and the trope of 'moving on'. *Museum and Society*, 8(3): 128–57.
Waterton, E. and Dittmer, J. 2014. The museum as assemblage: bringing forth affect at the Australian War Memorial. *Museum Management and Curatorship*, 29(2): 122–39.

Waterton, E. and Watson, S. 2013. Framing theory: towards a critical imagination in heritage studies. *International Journal of Heritage Studies*, 19(6): 546–61.

Waterton, E. and Wilson, R. 2009. Talking the talk: policy, popular and media responses to the bicentenary of the abolition of the slave trade using the 'abolition discourse'. *Discourse Society*, 20(30): 381–99.

Waterton, E., Smith, L., Wilson, R. and Fouseki, K. 2010. Forgetting to heal: remembering the abolition act of 1807. *The European Journal of English Studies*, 14(1): 23–36.

Webster, J. 2009. The unredeemed object: displaying abolitionist artefacts in 2007. *Slavery and Abolition*, 30(2): 311–25.

Wetherell, M. 2012. *Affect and emotion: a new social science understanding*. London, New York, New Delhi and Singapore: Sage.

Wilson, R. 2011. The curatorial complex: marking the bicentenary of the abolition of the slave trade, in L. Smith, G. Cubitt, K. Fouseki and R. Wilson (eds) *Representing enslavement and abolition in museums: ambiguous engagements* (pp. 131–46). London: Routledge.

Witcomb, A. 2003. *Re-imaging the museum: beyond the mausoleum*. London and New York: Routledge.

Witcomb, A. 2013. Understanding the role of affect in producing a critical pedagogy for history museums. *Museum Management and Curatorship*, 28(3): 255–71.

Wood, M. 2000. *Blind memory: visual representations of slavery in England and America, 1780–1865*. Manchester: Manchester University Press.

Part II
Places

7 Overlooking affect?
Vertigo as geo-sensitive industrial heritage at Malakoff Diggins, California

Gareth Hoskins

This chapter explores the affective intensities that emerge at and below Chute Hill campground overlook, a viewpoint across Malakoff Diggins in Nevada County California that has, since 1965, been designated as a State Historic Park. The cliffs and canyons excavated by hydraulic gold miners in the 1870s and 1880s are an important part of the industrial and environmental history of California and are commemorated as such. I examine how, as heritage, these ruins are depoliticized through their incorporation within a recreational pursuit of scenery. It was scenery, primarily, that drove the US preservation movement in the late nineteenth century and led to the production of viewing spaces that physically orientated and aesthetically trained individuals to appropriate nature as an object of display. This ocular-centric encounter is problematic, especially when extended to industrial landscapes, because it resolves past environmental ruin into pleasing picturesque. I use vertigo as a kind of embodied unsettling of the picturesque that helps us to actively politicize the industrial past and use it to foster engagement with contemporary social and environmental issues. Vertigo can be understood within an emerging suite of geo-philosophical approaches (Deleuze and Guattari, 1988; Clark, 2011; De Landa, 1995, 1997; Grosz, 2008; Serres, 1995) as a means of mapping an alternative earthly affectivity at Malakoff Diggins; one that better attends to the troubling on-going material legacies and persistences of extraction so often ignored by the focus of industrial heritage on societal progress and technological advance. In this 'earthly re-mapping' I am prompted by Whatmore and subsequent work (for example Last, 2013; Yusoff, 2013) addressing the critique that 'Geographies, like histories, [have] become stories of exclusively human creativity and invention played out over and through an inert bedrock of matter and objects made up of everything else' (Whatmore, 2003 p. 165). We need a way of engaging with the world that does not obscure these conceits of human exceptionalism.

Industrial heritage sites tend to be organized to affirm a 'conquest over nature' narrative that reinforces an absolute distinction between nature and culture (Latour, 1993). Malakoff Diggins provides an additional narrative about wise-minded restraint. This rampage-and-restraint couplet reassures contemporary fears of ecological catastrophe by asserting 'now we know better' and does little to address the uneven distribution of wealth and harm that industrial development precipitates. If industrialization is at the root of our contemporary environmental crisis then why does it remain an object of such historical veneration? In this chapter I explain how heritage creates an interpretive fix that confines industry to history and a scenographic fix that renders despoliation picturesque. I trouble these fixes by employing vertigo as an affective relation with the geological conditions of our own existence.

In strict medical terms, vertigo is a dysfunction of the vestibular system of the inner ear that creates the perception of rotational movement and linear acceleration and leads to an experience of confusion over height and distance. I employ vertigo here as an embodied disposition; a way of encountering the world in which the uneasy sense of the ground coming up to meet us helps to disturb Enlightenment distinctions between subject and object and the associated fantasy of our elevated mastery over nature which overlooks and viewpoints are implicitly designed to rehearse. The swaying associated with vertigo can come, according to Brandt *et al.* (1980), from an inter-sensory mismatch between the eye and the visual cue it searches for because the distance between the two is too large to properly establish position and orientation. The head moves imperceptibly to attempt focus and the body moves with it.

Vertigo shows us that seeing is an embodied and distributive practice, an affective relation between forces and materials that organize 'above, below, and alongside the subject' (Protevi, 2009, p. 3). It can work as a form of a radical post-anthropocentrism (Braidotti, 2013); a way to redistribute agency (Bennett, 2009); a way of helping us think about how we extend into the world and how the world extends into us. Vertigo flags up and fleshes out parallels between the park and the mine as emblematic places of human domination. Attending to these affects at heritage sites of extraction makes it hard to deny the all-too-often obfuscated connections between our contemporary wealth and convenience and the mining required to maintain it.

Over the past few years increasing attention has been paid to the emotional and affective relations at play at heritage sites and heritage-related institutions (Gregory and Witcomb, 2007; Picard and Robinson, 2012; Schorch, 2014; Waterton, 2014). Investigations into techniques of role adoption (Bagnall, 2003; Hoskins, 2012) and the vicarious experience of trauma in places of past violence and tragedy (Edkins, 2003; Landsberg 2004; Till, 2005) have helped to enrich our understanding of the politics associated with embodied visitor engagement. For instance, Crang and Tolia-Kelly's (2010) critique of the default white sensibility within the Lake District National Park and British Museum, and more recently Modlin *et al.*'s (2011) account of affective inequality at a Louisiana Plantation house museum, both outline how affective experience and emotional orientation can work to reproduce established social hierarchies even when the stated aim is precisely the opposite.

Other research engages affect and the non-representational to level critique in an environmental direction. DeSilvey (2013, p. 47) writes of the circuitries and residual memories of copper mining invoking 'the elements of experience not reducible to the narrative structures that heritage requires'; Morris (2013) describes the sea haar on the east coast of Scotland as a transient heritage atmosphere ungoverned

Overlooking affect? 137

by direct human influence; and Hill (2013) narrates a walk through the Forest of Dean to tap into the non-representational qualities of that place's mining history. There remains room, however, to more thoroughly employ these approaches in ways that challenge industrial heritage as a legitimation of progress and development. Affect theory's attunement to materiality and sensed relations within and between bodies, loosely defined, is particularly useful in countering the promotion of grand histories of industrial development as necessary, inevitable, unifying moments in the advance of humanity. Affect steers us toward the inhuman, non-human and extra-human arrangements and intersections of a planet indifferent to the role we have romantically assigned it as cradle of life. As Protevi notes about Hurricane Katrina in his book *Political Affect*,

> You have to understand geomorphology, meteorology, biology, economics, politics, and history. You have to understand how they come together to form ... the bodies politic of the region. You have to understand what those bodies could do, what they could withstand, and how they intersected in the event of the storm.
>
> (2009, pp. 163–4)

Similarly, if industrial heritage is to properly engage with the most pressing environmental questions of our time it needs to acknowledge that the social is also material, elemental and geological. Instead, industrial heritage renders disparate machines, buildings and landscapes illustrative of national progress, a perpetual distancing from nature, and the environment as little more than a passive foil in a story of conquest.

The new transdisciplinary wave of 'geo-humanities' (Dear *et al.*, 2011; Dixon *et al.*, 2013; Hawkins *et al.*, 2015) is exposing flaws in this modernist fairytale and critical heritage studies is well placed to contribute to the discussion. Critical heritage can bring together four kinds of engagement with the affective turn in social theory. The first is a broad and growing interest in the non-representational that extends beyond immediate precognitive experience to incorporate historical context, habit and memory (Chadha, 2006; Griffin and Evans, 2008; Jones, 2011). The second is work on collective feelings and public emotions and how they are bound up with securing social hierarchy (Ahmed, 2004; Berlant, 2008). The third is emerging research on atmospheres exploring how indistinct structures of feeling seem to hold in particular places (Anderson, 2009; Böhme, 1993; Edensor, 2012; Stewart, 2011). Finally, the fourth directs us to the realm of the subterranean through geo-philosophy (Deleuze and Guattari, 1994, Woodard, 2013), neomaterialsm (Bennett, 2009; De Landa, 1997), geontology (Povinelli, 2014) and notions of geopower (Grosz, 2008; Yusoff *et al.*, 2012; Clark, 2011), helping us relate materiality and affect through multiplicities of force, capacity, flux and process.

All four kinds of engagement with affect offer insight into the ways in which heritage operates and can be understood. In this example an affective lens is valuable for promoting contemporary environmental awareness because it can upset the conceit of human exceptionalism reaffirmed in the staging of industrial

138 Gareth Hoskins

heritage generally, and overlooks and viewpoints in particular. Indeed, affect, with its more inclusive non-human sensitivities, challenges the central impulse of heritage to conceive of decay negatively as a 'loss' or 'problem' for us to solve, rather than a stance adopted for a range of self-interested reasons. We need to examine those self-interested reasons more carefully, and an affective heritage of the industrial past can help us to do that.

I move now to outline precisely how Malakoff Diggins, once the largest hydraulic gold-mining operation in the world and an infamous space of brutal environmental transformation (Brechin, 1999; Dasmann, 1998), was stripped of its radical potential by incorporation into a scenic picturesque made available for appreciation on an elevated platform. The discussion then drops down to the surface of the pit with a walk along the Diggins Loop Trail, where I explore the Diggins as a self-organizing machinic ecology with agencies, intensities and capacities that force its high walls to move, erode, slough, collapse, accrue and accumulate. I show how the surface of the pit gives us a better sense of the various ways in which the geologic is active in fabricating social events, and how this approach gets us closer to a notion of geologic subjectivity (Yusoff, 2013) (see Figure 7.1).

Overlooks as moral high ground

The rim overlook next to Chute Hill campground at Malakoff Diggins State Historic Park is in a grassy clearing encroached by cedar, interior live oak and manzanita bushes. It faces east in front of an excavated canyon that is 600 feet deep and runs a mile and a half along the San Juan Ridge of California's Sierra Nevada. The vast scale of the hydraulic mining operations in the area during the

Figure 7.1 Panoramic strip of view from Chute Hill Campground Overlook.

1870s and 1880s attracted journalists, photographers and geomorphologists from all over the world wanting to witness, and then write about, the awesome and gloriously destructive power of modern engineering (Merchant, 1998; Worster, 1982). Hydraulic miners washed around one hundred million cubic tons of sediment into the Sacramento Basin, so much that it raised the level of the San Francisco Bay by up to six feet. Such enormous quantities of debris caused flooding of housing, transportation networks and farmland which prompted the formation of an anti-debris association in the 1870s and led ultimately to a battle now famous in the canons of environmental legislative history: Woodruff vs. North Bloomfield Gravel Mining Company. The so-called Sawyer Decision of the 1884 Supreme Court that resulted prohibited the dumping of tailings and removed the profitability of hydraulic mining as a large-scale means of extraction (Kelley, 1959). Redundant as a going concern and with the surrounding towns abandoned and dismantled, Malakoff Diggins slowly evolved into a landmark. It was established as a State Historic Park in 1965 and was listed in the National Register of Historic Places in 1973.

Standing at the overlook prompts a sense of bewilderment, because the extent of the Diggins confounds conventional distinctions between natural and 'man-made'. On one level we look out over a romantic wilderness that conforms to the famed nineteenth-century American Sublime paintings of nearby Yosemite Valley by Albert Beirstadt and Thomas Moran. But these now opposing valley sides at Malakoff Diggins are an artefact of human excavation – a coordination of muscular energy, technology and topography. The overlook therefore sets up a troubling substitution of nature for humanity which park agencies are usually so careful to avoid. It is a substitution commonly referred to elsewhere as the industrial, technological or toxic sublime (Peeples, 2011; Solnit, 2007).

The sublime is famously described by Edmund Burke in his book *A Philosophical Enquiry into the Origin of our Ideas of the Sublime and the Beautiful* as 'a sort of delightful horror, a sort of tranquillity tinged with terror' (1998, p. 114) and has developed into a loosely agreed upon nature aesthetic with particular representational significance and symbolic meaning. For Burke in the eighteenth century, however, it was a straightforward result of limitations in our own bin-ocular equipment. His description parallels the aforementioned research on vertigo:

> For if but one point is observed at once, the eye must traverse the vast space of such bodies with great quickness, and consequently the fine nerves and muscles destined to the motion of that part must be very much strained; and their great sensibility must make them highly affected by this straining.
> (1736, Note 1, Part 2 Section 7 – 'Why Visual Objects of Great Dimensions are Sublime')

The sublime, in contrast to current theorizations of affect as transpersonal, has been most notably marshalled to establish a distinction between subject and object (Kant, 1764). For Stormer, the sublime exposes us to the limits of the self; its landscapes 'act as witness to the possibility of the unrepresentable' (2004, p. 416) and the subject as a result is forced to recognize his own vulnerability in relation to the magnitude of the external world. A sublime aesthetic contributed to the international renown of Malakoff Diggins in the 1870s specifically via photographer Carleton E. Watkins (Hult-Lewis, 2011), whose pictures of denuded cliffs and dark threatening holes would, in a contemporary context, be registered as an environmental critique. But the Diggins as historical ruins are politically neutralized, enjoyed as curiously obscure and also, more importantly, as calm, comforting, pleasing, painterly, logical and ordered (Ackerman, 2003).

Malakoff Diggins' aesthetic transformation to scenic resource began in the 1950s, a few years before the park was designated, as this account from the Mineral Information Service (a California State monthly news release published by the Department of Natural Resources) makes clear:

> Today the old pits are scenic features of exceptional interest to the tourist; some of them – such as Malakoff at North Bloomfield near Nevada City ... are easily accessible by modern automobile; others are off in the back country. Any one of them is well worth seeing, if only to gain an idea of the tremendous effectiveness of water under pressure as a mining tool.
> (Mineral Information Service, 1953)

A few years later the Mineral Information Service explicitly listed Malakoff Diggins under a subheading of 'Scenic Resources' in a glossy promotional publication. Large format photographs accompany a description of the site that reads:

> The enormous pit has, with the passing of the years become one of the state's precious scenic resources – a miniature Bryce Canyon set in a dusky evergreen forest, with the little old town of North Bloomfield drowsing quietly in filtered sunlight near its eastern edge. It may be difficult to grasp, in the quiet of today, the tremendous fury of activity with which men inadvertently created this lovely spot. The miners are gone, and the giants no longer inhabit this part of the earth: but they have left their mark, carving in a few short years beauty that nature measures in the long eons of geologic time.
> (Egenhoff, 1965, p. 47)

In these and other examples, hydraulic mining is cast as a kind of inadvertent archaeology revealing nature's 'timeless' beauty through a benign uncovering of the surface. Such landscapes, as Cosgrove (2008) notes in his work on the affective role of vision in American environmentalism, are a collaborative effort; as much to do with aesthetic preference shaped by gender divisions, the broader social context of an urbanising population and the development of a new transport infrastructure as the increasing capacity of camera and lens technologies to capture the scene. The following description of the diggings by a resident in the local press is one example:

> In the autumn the water clears somewhat, and is then of a jade color or turquoise blue. Then the carved banks of multi-coloured crenelated spires, like

battlements of ancient castles, blend with the jade-coloured water to make a paradise of hues that only color photography could do justice in picture.

(Kallenberger, 1974)

Indeed, affection for the town's past glory means that hydraulic mining is rendered as a form of landscape gardening, as if technology and nature have cooperated and share the same scenic ambition:

> during this time natural erosion continued to chisel the hillsides into great fluted columns and palisades. Minerals in many layers of exposed earth painted the cliffs all the shades of red and orange, pink and violet that are famous in Bryce Canyon, which Malakoff resembles in miniature. And it could be that the most striking feature of coloration is the vivid contrast of the bluffs with the deep green conifer forest that surrounds the mined surfaces at this thirty-three-hundred foot elevation.
>
> (Truscott, 1974, p. 1)

These representational frameworks privilege vision as the principal means of knowing the world and position hydraulic mining as environmentally benign, even environmentally beneficial. Moreover, the eroding hills and recovering vegetation conveniently illustrate nature's resilience enough to allow for the veneration of extraction technology. See for instance, and finally, this description from a visitor writing in the travel and lifestyle magazine *Westways*:

> At the campsite, I walked down to the rough wooden bench to view what could be seen there of the great Malakoff diggings. Pine trees are beginning to spring up in the deep hole in the ground and there is even some vegetation here and there over the scarred, gouged face of the mutilated mountain. If I didn't know anything of Malakoff's history I would think that the cliff was formed by the eroding forces of nature and that the hole in the ground was a canyon cut by some long ago great river.
>
> (Boynoff, 1977, n.p.)

An important part of the development of Malakoff Diggins' visual economy is the platform from which the view occurs (see Figure 7.2). It is crucial to emphasize that viewing, to use the terms of Jacobs *et al.* (2011) in their examination of the high-rise window, is an 'embodied, materialised and practiced event' 'that happens' (p. 256). In the context of an expanding North American park movement, overlooks are an enabling technology reinforcing a sense of mastery over nature. Along with scenic parkways, overlooks became increasingly popular throughout the twentieth century as a growing middle class gained access to automobile travel and escaped from the city on weekend vacations to mountain retreats (Smith, 1984, p. 227). Overlooks were designed into the expanding national and state-park system and the national scenic parkway movement (Carr, 1999; Davis, 2008). In the 1930s, for instance, the New Deal-instigated Civilian Conservation Corps built

Figure 7.2 Chute Hill campground overlook sign.

thousands of embankments, parking spaces, viewpoints and overlook shelters in state parks across North America, as well as providing substantial labour for forest improvement and 'vista cutting' activities (Carr, 1999, p. 268). In this sense, overlooks are mediating technologies that connect psychology, physiology and aesthetics. They tend to be directional rather than 360-degree panoramas, complimenting the biological arrangement of our two forward-facing eyes (common to predators) that together excel at locating an object, fixing it in a gaze and making it graspable. Overlooks make the environment graspable as an external comprehensible thing that is singular, coherent and contained and disguised as providing 'pure' access to the landscape.

In his now classic instructional text, *Interpreting our Heritage* (1957), Freeman Tilden gave guidance to rangers on how best to prepare a park's viewpoints for visitors. Tilden noted:

> We sometimes call overlooks in the National Park areas of 'ohs and ahs' from the fact that these exclamations are the spontaneous manner in which the visitor expresses his wonderstruck feeling. Thus in an interpretive sign you are not wise to describe any definite object as beautiful; besides being impertinent by infringing on the visitor's taste, you are imposing between him and the scene.
>
> (1957, p. 85)

As a supplement to John Muir and Henry David Thoreau, influential advocates of 'the wild' in the 1860s, Tilden played a significant role in engineering a wilderness gaze that emotionally orientated subjects toward nature. Cosgrove (2003, p. 257) tells us that 'images do not merely represent a powerful reality[,] they are powerful agents in shaping that reality'. In his extensive instructions on scenery, however, Tilden tacitly ignores the ideological underpinnings of his own scopic regime:

> It is axiomatic that natural beauty, as perceived by the organs of sense, needs no interpretation: it interprets itself. Here the interpreter acts only as a scout and a guide. He leads his groups to the most alluring scenes he has discovered, and is silent. Would you varnish the orchid? He refrains even from using the word beauty.
>
> (1957, p. 110)

The assumption of a neutral universal point of view was common to many members of the social elite who made up the North American preservation movement in the late nineteenth and early twentieth centuries. New western historian Patricia Nelson Limerick explains the implications of this conceit:

> Human beings act as if they are only vehicles and amplifiers for transmitting nature's messages; nature truly does seem to them to decree what human beings can, should, and must do in particular places. But this assumption that nature carries political and economic messages and mandates is finally a failure of self-consciousness, a failure of a recognition that humans implant and install the meaning in these messages and mandates.
>
> (2001, p. 176)

At an overlook we are enveloped by the panorama and seduced by the physical elevation to such an extent that it becomes easy to forget that these are carefully engineered ideological platforms situated, designed and maintained with planting schemes and interpretive furniture to appropriate nature as an object of display. The first 'Master Plan' for Malakoff Diggins State Historic Park compiled in 1967, for example, demonstrates how visual rather than ecological imperatives often took precedent in management routines:

> This natural environment will be kept inviolate except where it becomes necessary to remove selected trees in order to enhance the scene, and/or achieve historical authenticity and/or where trails are required to bring visitor to overlooks and historical mining equipment and signs used in depicting the area and its beauty ... Care will be exercised to provide scenic buffers between historical and natural environ and to insure that vehicular traffic will not intrude upon historic elements. ... Because the principal objective here is historical in nature, it is not important that the vegetation be managed to achieve an early return to a natural or virgin condition. A partially

modified appearance in the vegetation is more in keeping with the history and emphasis of this unit.

(1967, pp. 4–5)

These guidelines show how visual and material concerns are bound together with practices in 'ecologies of the visual' (Rose and Tolia-Kelly, 2012, p. 4). Sometimes these ecologies are in tension. The sunlight we detect as humans and mentally organize as 'a view', for instance, has a different physical relation with the surrounding trees that promotes leaf growth obscuring that view. For the park service, things, plants and vehicles 'intrude', compromise 'authenticity' and demand certain kinds of mitigation for the landscape to convey a sense of the past.

In a similar way, the more overt industrial landscapes of toxic wastelands, mining spoils and huge holes in the ground can reinforce tired old divisions between nature and culture by playing on incongruity (Storm, 2014) and demonstrating the horror of human–nature infiltration (Davis, 1993; Kuletz, 1998; Misratch and Misratch, 1990). Overlooks and viewpoints, whether their objects are wild nature or designated industrial ruin, invariably limit possibilities for political engagement by encapsulating the landscape as a resource. But overlooks can also cause vertigo: a spinning, moving, swaying unsteadiness that breaks out a sweat and prompts feelings of anxiety and disorientation. Between 2 and 5 per cent of the population suffer from the height-related condition more specifically referred to as acrophobia, where vertigo, the symptom profile, can be triggered by looking down from a high place or looking up to a high spot from the ground. Vertigo, while certainly unpleasant and debilitating, can also be thought of as more broadly productive; as an instability that challenges human exceptionalism and disturbs the sense of our elevated mastery over nature. If an elevated gaze implies a 'head' for heights, then vertigo is the embodied rejection of cognitive bias. Vertigo encourages us to move down from the overlook and closer to the ground where haptic, kinaesthetic, ambulatory encounters have more purchase. While no less culturally and technologically mediated (see MacNaghten and Urry, 1998; Michael, 2000), these more-than-visual modes of encounter are more able to avoid the superiority so often rehearsed at overlooks and viewpoints.

Down in the Diggins

Nigel Clark (2011), Katherine Yusoff (2013, inspired by the philosophy of Deleuze and Guattari (1994)), Michel Serres (1995) and others have asked that we attend to the geologic dimension of subjectivity and think about how we extend into the mineral world. Their post-anthropocentric position develops from a critique of the Enlightenment-rooted elevation of humans above nature which leaves us so ill equipped to deal with pressing environmental concerns. For De Landa, a key contributor in this discussion, the *modern* form of reasoning reflects an 'organic chauvinism' (1997). In his essay 'The Geology of Morals' (1995), a nod to Nietzsche's 1887 book *On the Genealogy of Morals*, De Landa outlines a neo-materialist interpretation that undermines common-sense distinctions between living and inert, bringing the worlds of geology, biology and sociology together as combinations of matter and energy. He explains:

... our individual bodies and minds are mere coagulations or decelerations in the flows of biomass, genes, memes and norms...our languages are also momentary slowing-downs or thickenings in a flow of norms that give rise to multiple different structures.

(1995, p. 258)

Grosz makes a similar point when she writes '[life] is a temporary detour of the forces of the earth through the forces of the body, making them an endless openness' (2012, p. 975). The geo-centric position has profound intellectual implications for our analysis of the social world, the way we define our object of enquiry and the temporal scales at which our theories are conceived and understood to operate. In her book *The Posthuman* (2013), Braidotti states: 'The issue of geocentric perspectives and the change of location of humans from mere biological to geological agents calls for recompositions of both subjectivity and community' (2013, p. 83). The political dimensions of this new 'geo' sensitivity have been explored on a number of occasions, albeit briefly. Massey (2006, p. 35) discusses how the notion of rocks as immigrants can work to undermine essentialist claims to place; Saldana (2013) uses it to challenge the hypocrisy of western humanism and push for an anti-capitalist geo-communism 'that resists all vitalism and holds that the physical universe is absolutely devoid of prior purpose' (4.1); while Yusoff (2013, p. 780) explores fossils 'in a narrative arc of human becoming' to question current environmental ethics of the Anthropocene.

My objective here is to demonstrate how geo-sensitivity might inform an industrial heritage not geared toward demonstrating human progress and national technological advance but attuned instead to our intimate integrations and becomings with the rocks, soils and minerals that are otherwise defined as *our* 'resources'. I try to achieve this by flagging the affective links, contacts, shared moments of sensing and common faculties of feeling that come with moving through and along a trail.

To access the Diggins Loop Trail, you take a left off the main road about a mile west of town and pick up the path that takes you clockwise around the lake and along the bottom of the Diggins' high walls. The trail runs nearly three miles over a part-boardwalk, part-dirt-and-scrub surface. It is marked by occasional square posts topped with yellow paint. In the late 1870s, descent into the Diggins was often the highlight of a visitor's experience. Journalists writing about the rapidly industrializing far west for audiences in the east depicted Malakoff Diggins as a 'battlefield of antediluvian giants and monsters' (see Kolodny, 1984, p. 5) and a 'wild barren amphitheatre' (Reyer, 1886, p. 371).

From the elevated perspective of the overlook, time stands still, fixed by distance. At the bottom of the pit, however, geomorphological agency is much more readily apprehended. Up close the cliff walls can be recognized as gravels on the move. Red layers, indicating the presence of iron oxide, take on their colour only when exposed to the air. Blue gravel, much lower down, around 130 feet from the bottom, was the richest section of the mountain and is today barely visible. Observations of nature's progressive healing, mellowing, softening recovery and regrowth so often used in tandem with the pictorial view are less easy to sustain on the trail. All around is evidence that Malakoff Diggins has emergent properties of its own; that it is a generative assemblage not following any purposeful trajectory (see Figure 7.3).

Movement, run-off, rock falls, landslides, slips and slumps have brought large second-growth ponderosa pine toppling over the edge. As the perimeter erodes, Malakoff Diggins extends steadily outwards. In the same movement, other establishing saplings are covered with the infill, strangled by the sudden inundation with roots pushed beyond reach of the new water level. More incremental accumulations are evident in the trail markers that are swallowed up at the rate of 30 centimetres a year. In some spots, particularly in the north-west corner, markers barely protrude above the ground. By 1979 the perimeter of the pit was estimated to have expanded by as much as 300 feet. This reduced its depth substantially. The Diggins in this way refutes its purpose as a historical resource, much to the frustration of researchers: 'To date, the amount of infilling and revegetation of the Malakoff Pit Basin is significant enough to have compromised the "historic" quality and visual integrity of the pit basin' (Lindstrom, 1990, p. 6). Indeed, archaeological investigations reported that the constant accumulation of sediment from mass wasting has left the pit void of surface remains. But that notion of compromised history works only if history's remit is exclusively human. The Diggins has been resistant to coordinated human endeavour for a long time, as this report illustrates:

Figure 7.3 Marker posts on the Diggins Loop Trail.

Sometime in the early 1930s a great slide occurred in the Malakoff ravine. The bank of the digging on the west rim of the ravine that was left, broke away and slowly moved thousands of tons of earth, trees, and brush into the pit. The slide stopped only after encountering a higher ridge of bedrock on the south rim of the ancient river.

(Jackson, 1967, p. 126)

Holding to a geo-centric perspective along the Diggins Loop Trail, it is easy to see these gravel walls as bodies that sense things and cultivate affects. They have an inclination to move, an affective capacity to fall down, to collapse and re-organize in collaboration with planetary forces of gravity, chemical bonds of adhesion and meteorological promotions of frost, wind, rain and fire. Heritage is blind to the productive aspects of ruin, decay, loss and wasting that attention to affect can help to reveal. Instead of keeping those materials and capacities at bay, a geo-sensitive approach can engender humility, give us a sense of our own momentary presence on the planet and force us to reckon with the unanticipated legacies of our interventions.

Much of the heritage scholarship concerning affect explores its apprehension by a sovereign human subject where affect is registered as a pre-personal, non-conscious experience of intensity (Massumi, 2002; Thrift, 2004). From a geo-centric perspective, the body in question need not be a human body at all. The human is not a necessary component in affective relationships. The false premise that it is, as Clark reminds us, is a legacy of the Kantian settlement between subject and object which 'gives rise to the idea that thinking beings are restricted to accessing only the correlation between their thought and the outside world, and not what things get up to "in themselves"' (2011, p. 85). McCormack's affective materialism of remotely sensing bodies during an 1897 balloon expedition to the North Pole assures us again that the human is not required: 'the sensing bodies in question here do not need to be human; they can also be anything that produces an affect in the world – an idea, a gust of wind, a thing' (2010, p. 644). With industrial heritage, specifically, it is valid, indeed necessary, therefore to examine the vectors, forces and relations between things over which we have little or no control.

The intimate human–non-human collaborations set up through various industrial practices confront us, for instance, with our own fleshy vulnerabilities to the elevated levels of arsenic, nickel, zinc, barium and mercury still found in the Humbug Creek Watershed (California Natural Resources Agency, 2014). While gold has eclipsed all others as the key metallic element in the heritage story, it was mercury, mined in the nearby Coastal Ranges, that enabled the technical integration of forestry, water management and human labour in the hydraulicking process. Millions of pounds of the neurotoxin in its elemental form, quicksilver, was scatted over the banks of the creeks and allowed to work down with the gravel into sluices before being lost to streams and rivers. This mercury continues to move down the Sacramento Delta with every storm event and will continue doing so for the next 10,000 years (Alpers *et al.*, 2008). There remains a huge amount of highly contaminated unconsolidated sediment that waits ready to be remobilized onto the fields and orchards of the Central Valley. Mercury provides a very literal, if not unexceptional, example of the on-going geological infiltration of life that began, as De Landa points out, about five hundred million years ago, when 'some of the conglomerations of fleshy matter-energy that made up life underwent a sudden mineralization' (1997, p. 24). Mercury and corporeality are not compatible. Through processes of alluvial turbidity and biomagnification, mercury's presence alters the interior geographies of the body, causing damage to the nervous system and the functioning of major organs. It impairs early cognitive development and creates disturbances in sensation, lack of coordination and muscle weakness. These effects become part of a broader planetary dynamics linked to climate change as the episodic redistribution of mercury from Malakoff Diggins is 'coupled to the changing frequency and duration of extreme floods' (Singer *et al.*, 2013, p. 18346).

The Diggins Loop Trail can impart a more earthly and emergent apprehension of the industrial past that predates and outlives us. Moving along the trail can challenge the conquest-of-nature fixations that accompany overlooks and elevated platforms more generally. From the bottom of the pit we can be much more aware of our own sometimes precarious position within a set of forces and agglomerations that are autonomous and indifferent to our presence. In 1967 the parks department interviewed a long-time resident of North Bloomfield, the settlement adjacent to Malakoff Diggins. Charlie Gaus worked at the mine as a piper and recounted how the indifference of the Diggins, and our failure often to comprehend it, could have tragic consequences:

> And if a big chunk of clay come down or big rock come on hitting that it would make a big splash and throw rocks for 400 or 500 feet maybe. That was what you had to look out for. We had several men killed and hurt that way. I remember accidents that happened while I was there. There was one man, he was piking, he got hit with a rock and lived but a short time. It was during the night and a heavy storm and you couldn't see good. The floodlights didn't show up on the bank good. It was foggy. And then at another time a cave come down and caught three of them – killed one of them. It threw him right out of his boots. Must of throws him – the wind or something – from the cave. And the other fellow it crippled up, and he didn't live very long, and one got his arm broke. At another time a cave came down on a fellow and killed him but he could have avoided that if he'd run. But he just stood there. He let the cave come up to him and it caught him.
>
> (Lindstorm, 1990, p. 117)

Industrial heritage too can root us to the spot with pleasing spectacles so alien and apparently distant from our contemporary everyday lives. As commemoration of technological achievement or testament to wise-minded adaptation, industrial heritage can create a kind of paralysis that freezes us of a politics of the present and from an imperative to act.

Conclusion

Places marked out as heritage provide us with a huge opportunity to engage with contemporary social and environmental justice. Yet, framed exclusively as icons of technological triumph, they function only as tools of legitimation. This chapter has explored the affective dimensions of one kind of industrial heritage in California. I have outlined how overlooks can reinforce long-held divisions between nature and culture and subject and object so crucial in depoliticizing the industrial past and keeping it apart from contemporary social and environmental debate.

As a visual technology, overlooks cultivate a reassuring sense of the world as stable, unchanging, eminently graspable and available for comprehension. This is a conceit. There is no underlying order available for us to access or disseminate, no matter how elevated our perspective. Indeed, we need to be aware that the ground we label heritage is indifferent to its naming and might work against our well-intentioned desire to 'save' it. Bruce Braun, in his advocacy of a non-essential environmental politics, gets close to the position toward which I have been trying to steer us:

> To be faithful to the 'terrible truth' of an amodern ontology, then, would be to follow Deleuze and Guattari and face without flinching the reality that ethics and politics must be understood in terms of force and affect rather than truth and essence, that eco- politics must be about creation, even experimentation, about 'eco- art' rather than preservation.
>
> (2006, p. 206)

A geo-sensitive approach to the industrial past means rejecting bipolar interpretations of nature as either pristine or comprehensively colonized. It means moving beyond a conventional heritage practice that acquires authority by distinguishing natural from artificial, authentic from inauthentic. It means questioning the viability of heritage itself as an enterprise structured around the human-centric logics of life linked to notions of inheritance, lineage, entitlement, endangerment and vulnerability.

Affect can help us map our intimate infiltration with the world. If the overlook impresses with elevation and a distanced sense of superiority, affect can engage vertigo as a disruptive force and productive anxiety. By flagging the links, contacts and shared moments of sensing between bodies human and inhuman, organic and inorganic, affect enrols us more directly into an environmental politics that confronts the destructive legacies of extraction and our continued reliance upon it.

Bibliography

Ackerman, J.S. 2003. The photographic picturesque. *Artibus et Historiae*, 24(48): 73–94.
Ahmed, S. 2004. Collective feelings: or, the impressions left by others. *Theory, Culture & Society*, 21(2): 25–42.

Alpers, C., Eagles-Smith, C., Foe, C., Klasing, S., Marvin-DiPasquale, M., Slotton, D. and Winham-Myers, L. 2008. *Mercury conceptual model*. Sacramento, CA: Delta Regional Ecosystem Restoration Implementation Plan.

Anderson, B. 2009. Affective atmospheres. *Emotion, Space and Society*, 2(2): 77–81.

Bagnall, G. 2003. Performance and performativity at heritage sites. *Museum and Society*, 1(2): 87–103.

Bennett, J. 2009. *Vibrant matter: A political ecology of things*. Durham, NC: Duke University Press.

Berlant, L. 2008. Thinking about feeling historical. *Emotion, Space and Society*, 1(1): 4–9.

Böhme, G. 1993. Atmosphere as the fundamental concept of a new aesthetics. *Thesis Eleven*, 36(1): 113–26.

Boynoff, S. 1977. A landmark decision. *Westways Magazine*, September (Searls Historical Library Reference H 127).

Braidotti, R. 2013. *The posthuman*. Cambridge: Polity Press.

Brandt, T., Arnold, F., Bles, W. and Kapteyn, T.S. 1980. The mechanism of physiological height vertigo: I. Theoretical approach and psychophysics. *Acta oto-laryngologica*, 89(3–6): 513–23.

Braun, B. 2006. Towards a new earth and a new humanity: nature, ontology, politics, in N. Castree and D. Gregory (eds) *David Harvey: A critical reader* (pp. 191–222). Oxford: Blackwell.

Brechin, G. 1999. *Imperial San Francisco: urban power, earthly ruin*. Berkeley: University of California Press.

Burke, E. 1998 [1736]. *A philosophical enquiry into the sublime and beautiful*. London: Penguin.

California Department of Parks and Recreation. 1967. *Master plan narrative, Malakoff Diggins State Historic Park*. Sacramento: Interpretive Services Division.

California Natural Resources Agency. 2014. *Humbug Creek watershed assessment and management plan*. Sacramento: California Natural Resources Agency.

Carr, E. 1999. *Wilderness by design: landscape architecture and the National Park Service*. Lincoln: University of Nebraska Press.

Chadha, A. 2006. Ambivalent heritage between affect and ideology in a colonial cemetery. *Journal of Material Culture*, 11(3): 339–63.

Clark, N. 2011. *Inhuman nature: sociable life on a dynamic planet*. London: Sage.

Cosgrove, D. 2003. Landscape and the European sense of sight–eyeing nature, in K. Anderson, M. Domosh, S. Pile and N. Thrift (eds) *Handbook of cultural geography* (pp. 249–68). London: Sage.

Cosgrove, D. 2008. Images and imagination in 20th-century environmentalism: from the Sierras to the Poles. *Environment and Planning A*, 40(8): 1862–80.

Crang, M. and Tolia-Kelly, D.P. 2010. Nation, race and affect: senses and sensibilities at National Heritage sites. *Environment and Planning A*, 42(10): 2315–31.

Dasmann, R.F. 1998. Environmental changes before and after the Gold Rush. *California History*, 77(4): 105–22.

Davis, M. 1993. The dead west: ecocide in Marlboro country, *New Left Review*, 1/200: 49–73.

Davis, T. 2008. The rise and decline of the American parkway, in C. Mauch and T. Zeller (eds) *The world beyond the windshield: roads and landscapes in the United States and Europe* (pp. 35–58). Athens: Ohio University Press.

De Landa, M. 1995. The geology of morals: A neo-materialist interpretation. *Virtual Futures, 95* www.t0.or.at/delanda/geology.htm (accessed 16 February 2016).

De Landa, M. 1997. *A thousand years of nonlinear history*. New York: Zone Books.

De Landa, M. 1998. The machinic phylum. *TechnoMorphica*. www.egs.edu/faculty/manuel-de-landa/articles/the-machinic-phylum/ (accessed 19 June 2014).

Dear, M., Ketchum, J., Luria, S. and Richardson, D. (eds) 2011. *GeoHumanities: art, history, text at the edge of place*. London: Routledge.

Deleuze, G. and Guattari, F. 1988. *A thousand plateaus: capitalism and schizophrenia*. London: Bloomsbury.

Deleuze, G. and Guattari, F. 1994. *What is Philosophy?* London: Verso.

DeSilvey, C. 2013. Copper places: affective circuitrie, in O. Jones and J. Garde-Hansen (eds) *Geography and memory: explorations in identity, place and becoming* (pp. 45–57). London: Routledge.

Dixon, D.P., Hawkins, H. and Straughan, E.R. 2013. Wonder-full geomorphology: sublime aesthetics and the place of art. *Progress in Physical Geography*, 37(2): 227–47.

Edensor, T. 2012. Illuminated atmospheres: anticipating and reproducing the flow of affective experience in Blackpool. *Environment and Planning D*, 30(6): 1103–22.

Edkins, J. 2003. *Trauma and the memory of politics*. Cambridge: Cambridge University Press.

Egenhoff, E.L. 1965. *Scenic resources of California: a page from history*. Sacramento: Mineral Information Service.

Gaus, C. 1967. Oral history of the Malakoff Diggins operations, interview in Lindstrom, S. 1990, *A historic sites archaeological survey of the main hydraulic pit basin, Malakoff Diggins State Historic Park*. North Bloomfield, Nevada County California, California, Department of Parks and Recreation, Resource Protection Division, Sacramento.

Gregory, K. and Witcomb, A. 2007. Beyond nostalgia: the role of affect in generating historical understanding at heritage sites, in S. Watson, S. MacLeod and S. Knell (eds) *Museum revolutions: how museums change and are changed* (pp. 263–75). London: Routledge.

Griffin, C. and Evans, A. 2008. Historical geographies of embodied practice and performance. *Historical Geography*, 35: 5–162.

Grosz, E.A. 2008. *Chaos, territory, art: Deleuze and the framing of the earth*. Durham, NC: Duke University Press.

Hawkins, H., Marston, S.A., Ingram, M. and Straughan, E. 2015. The art of socioecological transformation. *Annals of the Association of American Geographers*, 105(2): 331–41.

Hill, L. 2013. Archaeologies and geographies of the post-industrial past: landscape, memory and the spectral. *Cultural Geographies*, 20(3): 379–96.

Hoskins, G. 2012. On arrival: memory and temporality at Ellis Island, New York. *Environment and Planning D: Society and Space*, 30(6): 1011–1027.

Hult-Lewis, C.A. 2011. *The mining photographs of Carleton Watkins, 1858–1891, and the origins of corporate photography*. PhD Thesis, Boston University.

Jackson, W.T. 1967. *Report on the Malakoff Mine, the North Bloomfield Mining District, and the Town of North Boomfield, for Division of Beaches and Parks*. Sacramento, CA: Department of Parks and Recreation.

Jacobs, J.M., Cairns, S. and Strebel, I. 2011. Materialising vision: performing a high-rise view, in S. Daniels, D. DeLyser, J.N. Entrikin and D. Richardson (eds) *Envisioning landscapes, making worlds: geography and the humanities* (pp. 256–68). London: Routledge.

Jones, O. 2011. Geography, memory and non-representational geographies. *Geography Compass*, 5(12): 875–85.
Kallenberger, W.W. 1974. Noble scar, *Grass Valley Union Newspaper* (Searls Historical Library Reference H62).
Kant, I. 1960 [1764]. *Observations on the feeling of the beautiful and sublime*. Berkeley: University of California Press.
Kelley, R.L. 1959. *Gold vs. grain: the hydraulic mining controversy in California's Sacramento Valley: a chapter in the decline of the concept of laissez faire*. Glendale, CA: Arthur H. Clark Company.
Kolodny, A. 1984. *The land before her: fantasy and experience of the American frontiers, 1630–1860*. Chapel Hill, NC: University of North Carolina Press.
Kuletz, V. 1998. *The tainted desert: environmental ruin in the American West*. London: Routledge.
Landsberg, A. 2004. *Prosthetic memory: The transformation of American remembrance in the age of mass culture*. New York: Columbia University Press.
Last, A. 2013. Negotiating the inhuman: Bakhtin, materiality and the instrumentalization of climate change. *Theory Culture and Society*, 30(2): 60–83.
Latour, B. 1993. *We have never been modern*. Cambridge: Harvard University Press.
Limerick, P.N. 2001. *Something in the soil: legacies and reckonings in the new west*. London: W.W. Norton & Co.
Lindstrom, S. 1990. *A historic sites archaeological survey of the main hydraulic pit basin Malakoff Diggins State Historical Park North Bloomfield*, Nevada County California, California Department of Parks and Recreation, Sacramento.
MacNaughton, P. and Urry, J. 1998. *Contested natures*. London: Sage.
Massey, D. 2006. Landscape as a provocation: reflections on moving mountains. *Journal of Material Culture*, 11(1–2): 33–48.
Massumi, B. 2002. *Parables for the virtual: movement, affect, sensation*. Durham, NC: Duke University Press.
McCormack, D.P. 2010. Remotely sensing affective afterlives: the spectral geographies of material remains. *Annals of the Association of American Geographers*, 100(3): 640–54.
Merchant, C. 1998. *Green versus gold: sources in California's environmental history*. Washington DC: Island Press.
Michael, M. 2000. These boots are made for walking…: mundane technology, the body and human–environment relations. *Body and Society*, 6(3–4): 107–26.
Misrach, R. and Misrach, M.W. (1990). *Bravo 20: the bombing of the American west*. Baltimore, MD: Johns Hopkins University Press.
Modlin, E.A., Alderman, D.H. and Gentry, G.W. 2011. Tour guides as creators of empathy: the role of affective inequality in marginalizing the enslaved at plantation house museums. *Tourist Studies*, 11(1): 3–19.
Peeples, J. 2011. Toxic sublime: imaging contaminated landscapes. *Environmental Communication*, 5(4): 373–92.
Picard, D. and Robinson, M. (eds) 2012. *Emotion in motion: tourism, affect and transformation*. Farnham: Ashgate.
Povinelli, E. 2014. Geontologies of the otherwise, fieldsights – theorizing the contemporary, *Cultural Anthropology Online*. http://culanth.org/fieldsights/465-geontologies-of-the-otherwise (accessed 23 October 2014).
Protevi, J. 2009. *Political affect: connecting the social and the somatic*. Minneapolis: University of Minnesota Press.

Reyer, E. 1886. Californische Skizzen: 1, Die hydraulischen Goldwaschen, *Deutsche Rundschau Berling* XLIX (1996) pp. 371–97, translated and printed as E. Reyer and S.K. Padover, Documents: placer mining, *California Pacific Historical Review*, 4 1935: 398.

Rose, G. and Tolia-Kelly, D.P. (eds) 2012. *Visuality/materiality: images, objects and practices*. Farnham: Ashgate.

Saldana, A. 2013. Some principles of geocommunism. www.geocritique.org/arunsaldanha-some-principles-of-geocommunism/ (accessed 5 March 2015).

Schorch, P. 2014. Cultural feelings and the making of meaning. *International Journal of Heritage Studies*, 20(1): 22–35.

Serres, M. 1995. *The natural contract*. Ann Arbor: University of Michigan Press.

Singer, M.B., Aalto, R., James, L.A., Kilham, N.E., Higson, J.L. and Ghoshal, S. 2013. Enduring legacy of a toxic fan via episodic redistribution of California gold mining debris. *Proceedings of the National Academy of Sciences*, 110(46): 18436–41.

Smith, N. 1984. *Uneven development: nature, capital, and the production of space*. Athens: University of Georgia Press.

Solnit, R. 2007. *Storming the gates of paradise: landscapes for politics*. Berkeley: University of California Press.

Stewart, K. 2011. Atmospheric attunements. *Environment and Planning D*, 29(3): 445–53.

Storm, A. 2014. *Post-industrial landscape scars*. Basingstoke: Palgrave Macmillan.

Stormer, N. 2004. Addressing the sublime: space, mass representation, and the unpresentable. *Critical Studies in Media Communication*, 21(3): 212–40.

Thrift, N. 2004. Intensities of feeling. *Geografiska Annaler: Series B, Human Geography*, 86(1): 57–78.

Tilden, F. 1957. *Interpreting our heritage: principles and practices for visitor services in parks, museums, and historic places*. Chapel Hill: University of North Carolina Press.

Till, K.E. 2005. *The new Berlin: memory, politics, place*. Minneapolis: University of Minnesota Press.

Truscott, J. 1974. 'Malakoff Diggins' magazine article title unlisted (H74 Searls Ref) p. 1.

Waterton, E. 2014. A more-than-representational understanding of heritage? The 'past' and the politics of affect, *Geography Compass*, 8(11): 823–33.

Whatmore, S. 2003. Introduction: more than human geographies, in K. Anderson, M. Domosh, S. Pile and N. Thrift (eds) *Handbook of cultural geography* (pp. 165–7). London: Sage.

Woodard, B. 2013. *On an underground earth: towards a new geophilosophy*. Brooklyn, NY: Punctum Books.

Worster, D. 1982. Hydraulic society in California: an ecological interpretation. *Agricultural History*, 56: 503–15.

Yusoff, K. 2013. Geologic life: prehistory, climate, futures in the Anthropocene. *Environment and Planning D*, 31(5): 779–95.

Yusoff, K., Grosz, E., Clark, N., Saldanha, A. and Nash, C. 2012. Geopower: a panel on Elizabeth Grosz's 'Chaos, Territory, Art: Deleuze and the Framing of the Earth'. *Environment and Planning D*, 30(6): 971–88.

8 The castle imagined

Emotion and affect in the experience of ruins

Duncan Light and Steve Watson

In this chapter we focus on the experience of visiting one type of heritage place: the ruined medieval castle. While architectural styles vary, castles are common throughout Europe, where they have become emblematic not only of the Middle Ages but of ruins in general, and the way they are represented and experienced as cultural objects. We begin with two scene-setting vignettes based on our own experiences:

> **Steve** – I grew up in a landscape with castles: Northern England, close to the border with Scotland, is full of ruinous and restored fortifications. My first memories of them, from the 1960s, are of glimpses from my parents' car – in the Yorkshire Dales, travelling over to the Lake District or on day trips to Northumberland. Even their names seemed to evoke feelings of something stirring and powerful: Richmond, Middleham, Castle Bolton, Warkworth, Dunstanburgh, Bamburgh – strong, sturdy names that seemed to reach back into an unknowable but romantic past. From the passing car, distance lent enchantment and a sense of mystery. My yearning to explore was, if anything, fuelled by the limited prospects of doing so as a child whose parents had better things to do.
>
> When opportunities did appear, they were in the course of family outings when, after surreptitiously consulting maps and guidebooks, the castle had to be somehow worked into the itinerary. I have memories of each visit: the rich mixture of excitement and discovery, the melancholy of rain-swept stone, the physical presence and scale, the rough-textured and fragmentary senescence. Even the weather seemed complicit in the experience and created what felt like a mood, a disposition of some kind that weighed on me, affected me. The iron safety railings, installed by the Ministry of Public Building and Works in the 1920s, and the ubiquitous cast-iron plaques drawing attention to obscure architectural features and warning of the danger of slippery surfaces only added to the 'atmosphere'. It has long occurred to me that castles, either up close and personal or as distant prospects, are there to be felt as much as seen.

The castle imagined 155

Duncan – It's 1987 and I'm at Carew Castle in Pembrokeshire on a wet day in late spring. I'm in the first year of my PhD, which is about how visitors make use of interpretive facilities at ancient monuments. I'm on a reconnaissance mission to identify sites where I can collect data later that summer. The department where I'm based has a very quantitative ethos so it's assumed that my main method of data collection will be questionnaires – and lots of them. As I walk round this castle I become aware that I'm not paying any attention to the interpretive facilities. It also dawns on me that there are lots of things about this building that are more interesting that the interpretation. I *like* this castle. There's lots to do and see here – it's full of intriguing rooms, long dark corridors and treacherous staircases. I'm enjoying exploring it, finding new rooms, coming across dead ends and retracing my steps, climbing up the towers. I'm relishing this building – and the sense of excitement and challenge that it offers me. It becomes clear that the experience of visiting this place is a complex entanglement of the cognitive and the emotional – and interpretation often has very little to do with it.

Over the following weeks I start to think about how I could 'capture' the excitement of the 'visit experience' at castles. My supervisors are sceptical. The only theoretical perspective that seems to offer any promise is humanistic geography (this was in the days before the cultural turn). I spend about a month ploughing through the phenomenology literature but I can't understand most of it, still less see how I could apply it to the heritage experience. Reluctantly I return to the safe option. Urged on by my supervisors (and their warnings about the importance of a representative sample), I undertake more than 1,500 questionnaires at seven sites over two summers. I master SPSS and do thousands of Chi-Square tests to explore how the characteristics of visitors are related to the ways that they interact with interpretive media. I submit a thesis and successfully defend it.

And yet … I'm well aware that my research has completely overlooked a whole 'other' dimension of the experience of being in castles and other ruins. Moreover, this experience probably has a major influence on the ways that visitors interact (or don't) with interpretive facilities. In other words, there's a lot that happens when visiting a castle that my research didn't even come close to capturing. Nowadays I would recognise this as being the emotional and affective dimension of the visit experience.

Our experience of castles is thus found and received, represented through a variety of cultural expressions but also encountered, experienced and constituted *in situ*. There is no doubt that as tourists we are both culturally equipped to visit castles and at the same time open to their affordances as objects and physical spaces (see Figure 8.1). In the latter guise they address our sensory capacities and evoke the kinds of responses that are recorded in these vignettes. But in the end what makes them interesting is the interplay between what we know and what we feel, what we expect and what we encounter: castles, as an experience, and 'castleness' as an

156 *Duncan Light and Steve Watson*

Figure 8.1 Landscape with castles: Dunstanburgh, Northumberland.

evocation of that experience are the products of these subtle reciprocities. In this chapter we look at the cultural and experiential framework that 'organises' castle visiting: its cognitive content and the accumulation of affects and emotion that is registered in these heritage encounters.

Castles, for us, are affective–discursive assemblages of the sort proposed by Wetherell (2012, pp. 53, 76), where representations form the core of affective practices. Edensor (2011) argues that the material properties and capacities of buildings as assemblages shape the ways in which they affect (and are affected by) other entities and agencies (both human and non-human). In addition, building assemblages are not fixed and stable but instead are open and dynamic, fluid and in a process of emergence, often in unpredictable ways (Waterton and Dittmer, 2014). This is especially the case for castles which have been used and appropriated in diverse ways at different times, something that has shaped not only their materialities but also the ways in which they are represented, experienced and imagined.

The castle assembled

Castles and fortifications are major heritage objects in Europe and Asia, and as such have become significant tourism resources which are – despite their lack of contemporary functionality – readily recognised for what they are. In many countries they receive statutory protection of some kind, depending on their age, condition and historical or architectural significance, and their size and commanding position in the landscape or in urban settings endows them with a visibility that contributes much to their contemporary role as tourist sites. Architectural and historical interpretations support this role, providing a cultural dimension that is of increasing importance in a tourism industry for which heritage has become a valuable commodity (Light, 2015). How then have castles, which might otherwise be seen as the redundant capital of an earlier age, come to play such an important role in the cultural production of heritage, and how do we experience them as such?

Other buildings that have long outlived their original function have found their way into contemporary life through adaptations and new uses. Power stations and warehouses have become art galleries and redundant churches have become, among other things, homes, community centres, offices and even bars and restaurants. Castles have also experienced such transformations, but they present another dimension that has extended their existence as objects of tourism: they are one of the most striking signifiers of the European medieval period and its historical and mythical associations, something expressed clearly in various 'revivals' and the continued culture of *medievalism*, in the centuries following the renaissance (Workman, 1985).

Unlike other medieval buildings, however, the survival of castles as real estate, preserved ruins, film sets (real and purpose-built) and tourist attractions has endowed them with a unique physical and cultural presence in contemporary culture. In England, English Heritage is responsible for the care and presentation of a portfolio of 400 buildings, monuments and sites, of which sixty-six are castles. The attraction value of this estate can be measured in visitor numbers – more than 10 million per year – of which the majority are to castles and ecclesiastical sites (English Heritage, 2015). Latterly the organisation is expected to become self-financing from a mixture of visitor admission revenue and donations; a 'visitor economy', in other words, that trades on a naturalised sense of what *the* heritage is and what it represents as an authorised discourse about the national past (Smith, 2006; Waterton, 2010). But to understand castles as an assemblage is to understand the various ways in which these buildings have been perceived over the centuries since their defensive role ceased and their cultural trajectory began, appearing and reappearing at various times as signifiers of the medieval and the pre-modern, and symbols of the mystery and imagination of another time (Watson, 2001).

In fact, the castle as a military structure had a relatively short lifespan, and its decline in the face of gunpowder, the consequential need for a new architecture of fortification and relative political stability (in the UK at least) made its defensive role more or less redundant. After this time castles were either remodelled for domestic use, as at Belsay and Chipchase in Northumberland (UK), or were abandoned altogether and fell into ruin. Sometimes, as at Castle Howard and Harewood House in Yorkshire, and Hardwick in Derbyshire (UK), they were abandoned for more comfortable and fashionable accommodations built close by, the abandoned structure being either dismantled or left to decay. Whatever its specific circumstances, any castle standing today is likely to have devoted only a fraction of its lifespan to the needs of defence. The cultural significance of castles, therefore, must rely on some other value, and to understand that we must follow their socio-historical trajectory and the story of their assemblage as cultural objects from the end of their period as functional buildings.

The wholesale destruction of castles in England received its most dramatic impetus in the activities of the Parliamentary forces under Cromwell after the Civil War. At the end of hostilities, many of the castles that had played a part in the war were deliberately demolished or 'slighted'. Conventional opinion is that this was to save the cost of garrisoning them and to prevent their further use by any remaining Royalist forces, although more recent interpretations stress the social and economic context of their destruction, and not least the financial gain from salvaging materials such as glass, metal and timber, especially where cheap labour could be employed to do the work (Rakoczy, 2008). Either way, the demolition was so extensive that many castles, save for a few that were restored, remained ruinous until their material value was transformed into cultural capital as objects of antiquarian and ultimately touristic interest – a role that was confirmed when so many of them eventually came under the guardianship of successive government agencies responsible for their preservation, conservation and presentation to the public (Thompson, 1981). This, together with the equally fortuitous activities of Henry VIII in destroying the nation's abbeys and monasteries, has provided English Heritage and others (such as the National Trust) with a portfolio of ruined splendour that now lies at the heart of the heritage tourism industry. As ruins, rather than simply as old buildings, castles offer a particular kind of engagement and experience, although not all castles are ruins, and not all ruins are castles. So we also need to consider what it is about the castle *as a ruin* that adds to the assemblage experience we have already discussed.

A well-known theoretical basis for the aestheticisation of ruins is the eponymous essay by Georg Simmel (1958), for whom ruins were interesting and appealing because they represented the opposing powers of culture and nature. Here also was a source of fascination with decay and decadence, things that showed the look of age and were losing the battle against nature's inevitable onslaught:

> This unique balance – between mechanical, inert matter which passively resists pressure, and informing spirituality which pushes upward – breaks, however, the instant a building crumbles. For this means nothing else than that merely natural forces begin to become master over the work of man: the balance between nature and spirit, which the building manifested, shifts in favor of nature. This shift becomes a cosmic tragedy which, so we feel, makes every ruin an object infused with our nostalgia; for now the decay appears as nature's revenge for the spirit's having violated it by making a form in its own image.
> (p. 379)

But ruins are not only evocative because of their apparent age. They also contain the sense of a lost future, of what might have been, and in their decay and irredeemable brokenness we might detect in what we see and touch and feel the sensuality of loss beyond hope, in a 'shock of vanishing materiality' and a 'visceral experience of the irreversibility of time', as Svetlana Boym (2011, n.p.) has described it. And yet over time these structures have mellowed into their landscapes until at last they address the senses aesthetically (Ginsberg, 2004). Rose Macaulay (1953) is perhaps the most well-known proponent of the 'ruin-gaze', with an emotional thread that she was happy to separate from the more intellectual engagements offered by archaeology and antiquarianism, in something of a stream of consciousness:

> When did it consciously begin, this delight in decayed or wrecked buildings? Very early, it seems. Since down the ages men have meditated before ruins, rhapsodized before them, mourned pleasurably over their ruination, it is interesting to speculate on the various strands in this complex enjoyment, on how much of it is admiration for the ruin as it was in its prime – *quanta Roma fuit, ipsa ruina docet* – how much aesthetic pleasure in its present appearance – *plus belle que la beauté est la ruine de la beauté* – how much is association, historical or literary, what part is played by morbid pleasure in decay, by righteous pleasure in retribution (for so often it is the proud and the bad who have fallen), by mystical pleasure in the destruction of all things mortal and the eternity of God (a common reaction, in the Middle Ages), by egotistic satisfaction in surviving (where now art thou? here still am I) by masochistic joy in a common destruction – *L'homme va méditer sur les ruines des empires, il oublie qu'il est lui-même une ruine encore plus chancelante et qu'il sera tombé avant ces debris* – and by a dozen other entwined threads of pleasurable and melancholy emotion, of which the main strand is, one imagines, the romantic and conscious swimming down the hurrying river of time, whose mysterious reaches, stretching limitlessly behind, glimmer suddenly into view with these wracks washed on to the silted shores.
> (1953, pp. xv–xvi)

The ruin-gaze thus engendered is Macaulay's project, and she provides at least the basis for an account of the *experience* of the ruins with which she has engaged. Ginsberg (2004, pp. 315–34) offers an even more experiential aesthetic and contrasts the romantic and classical theories as sources for such engagement. For the romantic, it is the sense of ruination in itself that frames engagement, reflecting the passage above from Macaulay: it speaks mainly of irrevocable loss and the lessons for our own mortality. What Ginsberg calls classical theory, on the other hand, sees the ruin as a more cognitive experience, a resource from which an original might be reconstructed and understood. Here, the past is recoverable as an artefact and a source of knowledge and understanding, whereas for the romantic it is the mystery that is the thing. For the classicist it is construction or reconstruction that provokes engagement; for the romantic it is destruction and all that is implied by that melancholic process. Yet both are emotive in their provocations, one prompting curiosity and the other sadness, and both, as Ginsberg makes clear, employ the imagination as a vector for their respective meanings (2004, p. 325). The issue of imagination seems to be key to the understanding of these engagements. For Ginsberg, not only does it unify (to an extent) the romantic and the classicist (a seemingly impossible task); it also provides a locus for the embodied engagements that provoke our interest in castles as heritage objects. To the imagination, therefore, we will return.

If we go on to make the inevitable connection between the dualities of romanticism and classicism and those of affect and cognition, we find in the castle a physical and cultural space that will easily accommodate both. In doing so, castles register feelings of both an affective and expressive sort, and moments of imaginative intensity that have not been lost on writers and artists who have contributed their representations to what is known and felt about castles. Turner's colour studies of Norham Castle and the major oil painting that he based on these are perhaps emblematic of this.

It is the gap between the demise of the castle as a fortification and its reappearance as a ruin in an aestheticised landscape, an orderly 'arcadia redesigned' (Schama, 1996, p. 530), that endowed it with a unique quality of antiquity, the physical sublimation of a 'time before' that was soon reflected in cultural production. In this sense the castle enters an iconography of landscape (Daniels and Cosgrove, 1988), part of the symbolic imagery that constitutes a recognisable and empirically knowable past, perhaps best expressed in the cultural axis of the 'rural-historic' (Watson, 2013).

Early contributors to this iconography were the illustrators Nathaniel and Samuel Buck, who had begun to publish engravings of the most notable buildings in the English landscape in the early eighteenth century. The prints are somewhat naïve in style, with faltering perspectives and excessive formality, yet they do, for the first time, indicate the venerability of castles and, perhaps more importantly, the families who owned them. Gilpin, another early tourist of the medieval, published his *Observations Relating Chiefly to Picturesque Beauty* in 1786. For Gilpin, the picturesque was expressed in the scopic regime of contemporary painting and the furnishing of the English landscape, with a good many decorative ruins (real ones and 'follies') supporting these aesthetic principles (Brett, 1996). Important to Gilpin's concept of the picturesque were the qualities of roughness and asymmetry and the effects of these on creating variety and effects of light, shade and contrast, the essential characteristics of the romantic ruin:

> But among all the objects of art, the picturesque eye is perhaps most inquisitive after the elegant relics of ancient architecture; the ruined tower, the gothic arch, the remains of castles, and abbeys. These are the richest legacies of art. They are consecrated by time; and almost deserve the veneration we pay to the works of nature itself.
>
> (Gilpin, 1794, p. 46)

From this period can be derived the first thematic meanings that have come to be significant in the contemporary experience of visitors to castles, and which are thus contributory to the assemblage of what is received culturally and what is understood *a priori* and taken to the castle to connect with embodied experience: notions of power and grandeur, the look of age, monumentality, the built semiotics of the medieval and its cultural assemblage.

In the eighteenth century, the 'k' in 'gothick' seemed to reflect an embellished and more fanciful form of medievalism, but while the mock ruins and Strawberry Hill fripperies of the mid-eighteenth century provided a backdrop for picturesque excursions, the interest in its aesthetics came to provide the castle with a new impetus as a cultural object. As the gothic sensibility gathered pace in the latter half of the century, Horace Walpole's novel *Castle Otranto*, and the spate of novels that followed in the gothic genre, presented the medieval castle as a mysterious, menacing place full of half-light and moonbeams, a place of malaise and of horror, replete with affective and emotional affordances. The castle is never fully achieved descriptively – we see it only in fragments – and this adds to its oppressive atmosphere and a sense of brooding apprehension (see Potter, 2005). As such, it controls and encompasses the dramas that unfold within it and imbues them with the sense of horror and terror necessary to stir the reader's imagination.

The medievalism of the nineteenth century saw a movement from the picturesque aesthetic to the romantic and sublime, where the perception of the untamed natural environment evoked wonder and awe as well as fear and apprehension. The tendency for castles to be perched for defensive purposes on top of precipices reinforced such feelings. Edmund Burke's 1757 *Philosophical Enquiry into the Origin of Our Ideas of the Sublime and Beautiful* contrasted beauty, as something aesthetically pleasing, with the sublime, as something powerful and dangerous, a delightful horror (Schama, 1996, p. 447). Mountains in particular – the Alps visited by Grand Tourists and the Lake District by more domestic souls – offered much to the sublime sensibility, where great block-like structures reflected Burke's doctrine that 'irregular symmetry was to be shown in dark and massive forms' (Schama, 1996, p. 447) and where a well-placed and well-ruined castle could easily and perfectly complement such a scene.

At the same time, a keener interest in the historicity of the medieval was engendered by a desire to locate the new industrial society within a framework of the national past that neutralised its uncertainties and offered re-assurance in a sense of permanence and depth, while reinforcing nationhood and supporting the social and economic relations of emergent capitalism (see Figure 8.2). At its most brutal, this movement led to the 'great hulking castles' that Robert Smirke designed for Lords Lowther and Somers (Lowther and Eastnor castles respectively); physical manifestations of the siege mentality of the aristocracy in the wake of the French Revolution (Mandler, 1997, p. 14).

Lowther, now a spectacular ruin, 'quotes' and almost parodies the medieval with its exaggerated battlements, scale and monumentality, and while as a whole it bears little resemblance to an authentic medieval castle, its purpose is clear, which is to impress in a way that goes beyond the mannerisms of the earlier 'gothick' revivalists. Nearby, the Citadel at Carlisle was built as the county's law courts, and as such they dominate the centre of the city in a way that its real Norman castle simply does not. The medievalist main hall is thus a setting within which the power of the law, derived largely from medieval precedents and the Common Law, is exercised.

Figure 8.2 The castle (re-)assembled, Bamburgh, UK: Norman with later restoration.

As exercises in the semiotics of power, these buildings might present quotations of medieval detail, but they do so in a way that is exaggerated and idealised. The turrets are a little higher, the design a little more fanciful than a faithful rendition or copy would be. And yet in their enhanced visuality they eventually begin to stand in for the real thing, an impulse that found its fullest expression in Ludwig II's late nineteenth-century Neuschwanstein Castle in Bavaria which, significantly, required the demolition of an existing medieval castle ruin before its construction could begin. It is perhaps also significant that Neuschwanstein began its life as a tourist attraction almost immediately upon the death of its owner in 1886, its affordances as a place of wonder all too obvious in its wowing visuality and its role as inspiration for Walt Disney's *Sleeping Beauty* castle. The castle newly built must reflect and re-constitute medieval splendour, a restorative and peculiarly Anglo-Saxon version of classicist order, just as the castle ruined evoked romantic awe.

In literature, the castle becomes the stage set of the great medieval and chivalric drama. From his earliest works to his last, such as *Castle Dangerous* published in 1831, Walter Scott used real places as the settings for his novels, some of them already famous, others quite obscure, but to which reference was readily made by early tourists. For example, Walter White (1858), in describing Teesdale, quotes extensively from Scott's poem, *Rokeby*, and especially its references to the castles of Raby and Ravensworth and Mortham Tower. While it is difficult to estimate the effects of Scott's medieval epics on the development of tourism in the nineteenth century, they undoubtedly contributed something to the assemblage that became the castle as an object of cultural heritage and, as such, a semiotic landscape within which this assemblage is combined with the immediacy of experience.

Later, John Ruskin and William Morris (the latter responsible for establishing the Society for the Protection of Ancient Buildings in 1877) were even more influential in the gathering pace of preservation (Hunter, 1996). All of this culminated in an acceptance that the state had a central role in preservation, and the 1882 *Ancient Monuments Protection Act* enabled forty-three monuments to gain statutory protection over the following twenty years and, through that, practical measures to preserve them. Morris, much influenced by Ruskin, was disturbed by the gothic revival of the nineteenth century, which had seen a great deal of 'restoration' at the hands of architects such as Anthony Salvin, Augustus Pugin, George Edmund Street and George Gilbert Scott. For Morris, the emphasis lay in preservation rather than restoration, with minimal impact on the original. This 'preserve as found' ethic is largely responsible for the unaltered state of ruins in England, although it might also be criticised for a later tendency towards a certain primness in their presentation by the succession of statutory agencies that were responsible for them (Thompson, 1981).

But even these new institutional measures could not undo the centuries-old assemblage that was now the castle, and the connection between gothic medievalism and later touristic experiences that depended to a large extent on the feelings these associations evoked in the body and mind of the visitor. For example, Riley, an early motorised tourist, made the following observation on Castle Bolton in Yorkshire:

> That Dungeon has a horrible fascination for me … The floor is wet, and in one corner there is a ring in the rock to which prisoners were chained. They tell a gruesome story of how the bones of a human arm were found in that ring when the dungeon was opened out; but one's own imagination can supply the dismal pictures without the help of facts.
>
> (Riley, 1934, p. 96)

Dungeons provoke particularly strong imaginative and emotional responses, and we often find ourselves wondering what it might have been to be a prisoner thrown into one. Castles are often haunted by ghosts (Edensor, 2011) and dungeons are the places where we are most conscious of hauntings and of the past disturbing the present. Here we might experience the affective thrill of the spooky, the spectral, the mysterious and the absent (Holloway and Kneale, 2008). In fact, some castles do have their own ghost stories (often enthusiastically recounted in the place's promotional or interpretive materials). Knowing this enchants the building in all sorts of unexpected ways (Holloway, 2010), producing an affective atmosphere (Anderson, 2009) rich with tension and excitement.

The castle assembled in contemporary culture is transformed, again, as a place of leisure and recreation: a place to play, an imaginary of all we know about castles recast as the backdrop to a family picnic, a day out, and at the same time offering a connection with a past that is assembled in moments of engagement with stone fragments, audio-guides, narrow spiral staircases, damp under-crofts and arrow slits. And the assemblage works in the way in which visitors bring to that materiality a medieval imagination that has been fed by the image of the castle as it has developed over the centuries, bristling with turrets and machicolations, the model of Camelot in all those Hollywood films.

The castle as a contemporary heritage object is thus an assemblage of its historical and cultural trajectory, often reproduced now as the context for a medieval otherness associated with proto-mythologies and the fantasy genre in movies and other media. 'Real' castles, either preserved as found or restored, are very much the material of the heritage tourism industry. As attractions they present and represent the cultural assemblages of the medieval and the medievalist tradition that emerged from it. In a reverse projection, they materialise, in their stony reality, the cultural assemblage that they now represent in the eyes of the visitor, a cultural assemblage that took on a life of its own when the castle as an idea detached from the castle as a building.

Castles, then, provide a richness of experience and intensities that illustrate well the affective nature of heritage encounters, not least because as objects and as spaces they are an assemblage that has changed, uniquely, the ways in which they are engaged. When castles reappeared after their dormancy in the sixteenth and seventeenth centuries, they were shorn of their militaristic functions (which were in any case redundant) and invested with the aesthetics of the picturesque, and later the sublime and the romantic. In this guise, they gained their capacities as places of intensity and affect, where emotion could be mobilised and expressed in both representational practices and the direct experience of engagement with their materiality, setting and the sensual affordances that they possessed. A 'feeling of knowing' (Hart, 1965) is what people bring with them as contemporary castle visitors – the genealogies of what is known and experienced of the castle as a cultural assemblage and what in turn this contributes to the reproduction of that assemblage in further experience and representation. The castle is now transformed again, through heritage practices and tourism, acting as a place of recreation and of play as well as of quiet contemplation, reflection and exploration. It is the experience of the contemporary visitor to which we now turn.

The castle experienced

In this section we focus on the experiences of visiting ruined castles and seek to explore some of the emotional and affective dimensions of the encounter with such places. The starting point is our personal interest in such buildings, as enthusiastic visitors to castles, and not just with a professional interest in heritage and tourism. We have both been *affected* by them and, indeed, it was our discussions around this that prompted the conference paper that led to our writing this chapter. The vignettes at the beginning of the chapter express something of this feeling, but now we would like to unpick this a little more and draw on our experience of castle visiting to explore the castle both as a cultural assemblage and as a direct experience. That said, our experiences are quite different: Steve's experience is mostly in Northern England, with forays into Spain and Greece, while most of Duncan's castle visiting has been in Wales. Between us, we have visited hundreds of ruined castles, sometimes as researchers and sometimes for the pure pleasure of it. However, since we live and work at opposite ends of the country there are, in fact, very few that we have both visited.

In the following account, we use a form of autoethnography to critically reflect on our own experiences of visiting castles. Autoethnography is 'an approach to research and writing that seeks to describe and systematically analyse ... personal experience ... in order to understand cultural experience' (Ellis *et al.*, 2011 p.1). It aims for a synthesis of autobiography with cultural critique (Grant *et al.*, 2013). As such, it allows researchers to use and foreground their own lived experiences to better understand some aspect of the contemporary world (Morgan and Pritchard, 2006). By definition, autoethnography is a subjective process and it makes no claim to produce universal or totalising accounts of a social phenomenon. Instead, it aspires to 'creatively written, detailed, local and evocative first person accounts of the relationship between personal autobiography and culture' (Grant *et al.*, 2013, p. 2). Autoethnography is relatively uncommon within tourism studies (although see Morgan and Pritchard, 2006; Noy, 2007a, 2007b; Scarles, 2010; Waterton and Dittmer, 2014) but it seems entirely appropriate to the present topic, where we are exploring a complex confluence of potentialities: the representational (through the assemblage), the personal–historical, the pre-personal–affective, the emotionally expressive and the cognitive. If it is in the ragged intersections of these registers, in between the representational and non-representational, or the embodied and expressive, that we hope to explore the experience of castle visiting, then something akin to a personal account seems methodologically apposite. And there are precedents, in the work of Denis Byrne (2013) and Russell Staiff (2012, 2014), both of whom employ more expressive modes of writing to explore the somatic and sensory aspects of experience of places and objects and the ways in which those objects themselves become charged with affective and emotional potential. Research-theoretical precursors of such approaches have also begun to appear (Büscher and Urry, 2009; Clough, 2009) and to filter into heritage studies (Crouch, 2015; Waterton and Watson, 2015). This amounts to an extended repertoire of research methods as well as an openness to innovation and a more agnostic approach to any particular research paradigm.

But perhaps our enthusiasm masks a sensitivity to criticism for our use of this 'unscientific' method. Well, a lack of science in our undertakings bothers us very little and, in a way that we never anticipated, our methods reflected the aesthetic commentary offered by Ginsberg (2004) in his own engagement with ruins:

> The ruin comes into its own by jumping into our space. In gardening, sculpture and architecture, we have to take into account the space between work and visitor. In these arts, the work may press upon the space of the visitor, but not every moment need be a springing forth, as is the case in the ruin. The ruin comes at once, without warning, because its aesthetic unity is taking place just at that moment in our space.
>
> (2004, p. 164)

So here is a subjective experience presented as an aesthetic one, but a subjective experience that is available collectively and brought together under a recognisable signifier – the ruin – and a collectively received cultural assemblage – the medieval castle. We are not making aesthetic judgements about castles, at least not in any systematic architectural–historic way, but we appear to be using the same frames of reference and the same registers that are available for aesthetic judgement as we make our explorations of the castle experience. For us, what is important is the way these frames facilitate an exciting and potentially productive application of new approaches to our field.

The analysis that follows is thus a composite of our individual experiences and reflections. Following ethnographic tradition, we use the first person to narrate our experiences. However this does not mean that our experiences of visiting ruins are identical (far from it!), so the use of 'we' may, in fact, refer to something which only one of us has encountered. We are also aware that these subjective accounts are just one form of representation, and that they are as problematic as any other representation (Waterton and Watson, 2014). We are equally aware of the difficulties in attempting to represent subjective 'inner' life (cf. Grant *et al.*, 2013), particularly since some aspects of individual experience – such as affect – seem to defy representation (Pile, 2010).

In a number of ways, ruined castles are a quite distinctive form of heritage place. By definition they are incomplete: rooms do not have complete walls and are often without roofs; doorways and windows are often damaged; and entire floors may be missing from towers. In addition, most of the site is open to the elements (unlike many heritage sites which are 'indoors'). Ruined castles are also among the least regulated and commodified of heritage places. Most have a small staff (whose main role is to sell tickets and souvenirs, normally at an entrance booth which may be some distance from the castle itself). This means that the custodians and other staff are not usually present within the main structure to monitor visitor behaviour (as in some museums or galleries) or direct the way in which they move around (as in country houses). Neither (unlike other types of heritage attraction) is there much use of volunteers at ruins. Indeed, some ruined castles have no permanent staff at all and are open to the public at all times. This all means that the encounter with a ruin is much less proscribed than those at other types of heritage sites (such as museums or country houses). There is limited signposting and usually little attempt to direct the order in which different parts are experienced. However, there are usually some parts (those that are unsafe) which are off-limits to visitors. Overall, ruined castles are sites of relative freedom which invite spontaneous and improvised performances among their visitors (Edensor, 2001). These performances are, in turn, associated with particular types of 'feeling states generated in place' (Duff, 2010, p. 885).

The encounter with a ruined castle is an open-ended one: such places present their visitors with multiple possibilities and opportunities, and we have already indicated our pleasure in these engagements. There is an expectation, therefore, as we approach the entrance, that this will be an enjoyable experience. They are places which invite flirtatious encounters that offer the continual prospect of the unexpected (Crouch, 2005, 2012). Often we find ourselves in a central courtyard (sometimes a quite substantial space). There are usually multiple exits: doorways to walk through (which often lead to further courtyards, wards or baileys), dark rooms to enter and unfamiliar spiral staircases to challenge our coordination and

The castle imagined 167

stamina. There are multiple claims for our attention. At this stage, everything is unknown and open for discovery. The whole experience of visiting a ruin (particularly for the first time) is structured by a vivid sense of exploration, provoking emotional registers of excitement, curiosity and enjoyment of what awaits to be discovered. The larger the castle (and the more there is to explore), the richer the experience (see Figure 8.3).

Unlike many heritage attractions where the visitor's movement through the site is carefully determined, exploring a ruin is an improvised and disorderly process. Since there is rarely a defined trail to follow, we have to devise our own routes through the building. We may enter a doorway or climb a staircase only to find it leads nowhere. On the other hand, we may find ourselves following passageways, corridors or stairways for some distance into the heart of the structure. We pass doorways, rooms or staircases and make a mental note to return to them later. We frequently come to a dead end and have to retrace our steps. We get fleeting and unexpected views through the windows (which are often very small) of the rest of the building and of other visitors. Exploring an unknown building in this way is exciting! We relish making our way through ruins, not knowing what lies ahead, what's in the next room or what's around the next corner. The interiors of castles are usually poorly lit and so exploring them reminds us of playing hide and seek, and for one of us it brings back childhood memories of playing 'murder in the dark'. This is fun! It is an illustration of what Edensor (2012) calls the playful consumption of space. On the other hand, exploring a ruin can be confusing and

Figure 8.3 The castle experienced: the remains of a staircase at Bolton Castle, UK.

disorientating (which can even lead to frustration). It is not difficult to get lost inside the structure, or to be unable to find our way back to something we noted earlier as worth looking at.

The very unpredictability of the encounter illustrates how the experience of heritage is dynamic and emergent (see Crouch, 2010; Staiff et al., 2013; Waterton and Watson, 2014) rather than pre-determined. Ruined castles gradually reveal themselves to their visitors (cf. Ginsberg, 2004) – who are also active co-participants in the creation of their visit experiences and their associated meanings. Experience and meaning here are fluidly connected. Meaning flows from previous experience and the current engagement, and together they combine with what is known from the cultural assemblage of the castle. 'Cold and creepy' thus recalls the gothic, while turrets and battlements recall something of Hollywood movies, Robin Hood and knights of old. Uneven, slippery surfaces pose a threat and the signage tells us to beware. Modern walls and fences tame dangerous heights and stand between us and certain death. This is horror contained by a railing, scary but enjoyable, and recalling Edmund Burke's concept of the sublime and the dual emotional registers of fear and attraction.

Height is a particular characteristic, along with the feeling of ascent. Most castles have towers and indeed we might be disappointed if they do not. The whole experience of climbing a tower is framed by anticipation, eager expectation and the uncertainty of not knowing what is at the top. Sometimes the climb ends at a metal grille just below the summit – which remains tantalisingly out of reach. If visitors are allowed access to the very top of the tower then arriving there produces a sudden sensation of satisfaction and accomplishment. The controlled danger of height has already been mentioned, but other emotions can also come into play. The view from a tower (whether over surrounding countryside, or down on to the rest of the structure) can be enough to provoke a sense of awe, wonder and delight (see Robinson, 2012). We've noticed how often the first thing that we (and other visitors) say when we reach the top is 'wow', and some managers of the more commercial attractions will trade on this feature. Oxford Castle, with its heightened viewing platform, is typical, and the guide almost invites you to mumble a gasp before pointing out the various sights. And as we look down from the castle's summit we may also enjoy a sense of exclusivity ('I'm the king of the castle and you're a dirty rascal!'), and for a time can feel masters of all we survey. If the climb is a particularly difficult one then we may also feel a sense of quiet superiority over those we can see on the ground who do not have the stomach for the challenge (see Figure 8.4).

However, exploring a ruin can also be associated with more negative emotions. The most common are disappointment and frustration. Indeed, disappointment has long been recognised as one of the most common emotions associated with the tourism experience (Robinson, 2012; Rojek, 1997; Tucker, 2009). Ruins have long been associated with solitude and the romantic gaze (Ginsberg, 2004; Urry, 1990), and when we visit castles we might secretly hope that we will be the only ones there. Occasionally we are lucky; more often than not, though, there will be plenty of other visitors around, much to our disappointment. But more mundane

The castle imagined 169

Figure 8.4 Dizzy heights at Bolton Castle, UK.

things can cause disappointment for the castle enthusiast. When we explore a castle we frequently find parts of it that are not open to visitors (with a bar or grille to prevent access). We are afforded tantalising glimpses of other parts of the ruin – intriguing passageways leading off to who-knows-where, or sometimes whole floors of towers that are seemingly intact – that are waiting to be explored but which are inaccessible. These often make us feel irritated, annoyed or even cheated. Sometimes we feel resentment about such unnecessary restrictions; we feel 'controlled' when we want to feel free to roam at will.

Some emotional responses can be more negative still: on occasions we find ourselves feeling disgust – itself one of the six 'basic' emotions (Ekman, 1992). Other visitors frequently drop litter in some of the most remote parts of the castle. A particularly common practice is throwing beer cans into dungeons or cellars. We ask ourselves irritably: 'why do people do this?' and 'why don't the site's managers clear it up'? Other experiences may be more unpleasant. For some reason there seems to be a part of every castle (often a remote, damp tower room) which smells strongly of urine. To judge from the smell, a lot of people have relieved themselves there. It can get worse. Occasionally we have come across piles of excrement, sometimes in the most inaccessible parts of the building. We recoil, suddenly and acutely aware that local youth have little difficulty breaking into the castle once the visitors have all gone home. Once we start to look around we can see plenty of evidence that they claim it as their own illicit social space – not only the excrement but also the empty beer cans, discarded joints and cigarette ends. We may feel distaste or disgust at such transgressive behaviour, but such responses also expose our normative assumptions about who castles are for and what is the 'correct' way to behave in them. While we may like to think of ourselves as viscerally engaged, we realise that we have internalised, cognitively, the principles of visitor management.

Castles also present their own opportunities for transgression and disobedience – this time among their visitors! We mentioned earlier our frustration when we find a part of the ruin that is closed off to visitors, especially when we cannot understand why it is closed. But there is minimal surveillance within many ruins and, if we are visiting at a quiet time, we are not subject to the disciplinary gaze of other visitors (see Edensor, 2007). Hence it is not difficult to squeeze under bars or ropes to visit those elusive parts of the building that are off-limits. And sometimes only basic climbing skills are required to get into those rooms and towers that are not open to the public. Such 'resistant performances' (Edensor, 2001, p. 76) illustrate, in a heritage setting, the ways through which ordinary people can quietly elude and subvert an unseen authority (de Certeau, 1984). This is selfish and sometimes foolhardy behaviour, but it produces an affective thrill and a sense of accomplishment and satisfaction at seeing what we were not supposed to see. It also produces a sense of exclusivity and distinction (at being more daring than the average visitor and getting to enjoy something that they will not).

Up to now we have focused on (some of) the emotional dimensions of the encounter with a ruined castle, but we are well aware that 'emotion is located within the body' (Crouch, 2005, p. 30). The emotion–body relationship is a complex one. In all sorts of ways, our emotional state (how we 'feel') is reflected in how we comport our bodies. For example, we would appear very differently to an observer if we were either elated or frightened. Similarly, our embodied experiences of place can affect us – and shape our emotional states – in complex ways which are often difficult to articulate. We now turn to consider such embodied dimensions of the encounter and the ways in which this is intertwined with emotional and affective responses to ruined castles.

Ruins are open to the elements, meaning that the bodily experience is very different from other types of 'indoor' heritage attractions where the environment is more controlled and regulated. In fine weather, visiting a ruin can be a pleasant experience (although even in bright sunshine the inside of a castle can be surprisingly cool and often damp). In poor weather (rain or strong wind) ruins are chilly and clammy, which makes them uninviting and occasionally unnerving places. There is an immediate sensation of bodily discomfort. This is accentuated by the particular (not always pleasant) smells associated with such places: damp, mould, decay, wet vegetation and, in some places, urine. The whole matter of illumination also affects bodies and emotions. Edensor (2012, p. 1106) has examined how engagement with light (and the absence of light) is a 'deeply embodied experience' (see also Waterton and Dittmer, 2014). In the winter sun long shadows are cast and the texture of weathered stone is accentuated in the sharp relief light. The insides of ruins are often surprisingly dark and gloomy (even when there is bright sunshine outside) and, apart from at the larger and more popular ruins, there is rarely any artificial lighting. This modulates the way and the pace that we move through the building: we have often found ourselves stumbling along in semi-darkness, arms in front of us, ever conscious of potential hazards. The absence of light recalls the gothic component of the castle assembled. It is unsettling and this, in turn, shapes our emotional and affective responses to the building (Edensor, 2013). Our eyes have to constantly adapt to the changing light: one minute we are picking our way along in the gloom, the next we are temporarily dazzled from a shaft of bright life that comes through a window or arrow slit.

Exploring a ruin can be a physically demanding activity and nowhere is this more apparent than when climbing a tower. This can involve quite serious physical exertion: we soon become out of breath, our knees ache, our hearts pound and we become acutely conscious that we're not as young as we used to be. Spiral staircases can be difficult to climb due to the need to stay on the widest part of the steps and it can be awkward and uncomfortable to pass other visitors who are coming down. They also leave us giddy and disorientated, and we quickly lose all sense of direction. When we arrive at the top there is the sensation of accomplishment that we described earlier, but this also mingles with a range of other bodily experience: our eyes adjusting to the sudden brightness, breathlessness from the ascent and the wind in our faces. As castle enthusiasts we usually make sure that we explore every tower, so that this experience is repeated a number of times (although each tower is slightly different from the last). To miss something would be a travesty of castle visiting.

The physical fabric of ruins also adds another dimension to the visit: apprehension. Controlled danger has already been mentioned in connection with a sublime aesthetic, but there are real dangers that need to be assessed and avoided, particularly in unmanaged sites. Surfaces underfoot are rough and uneven and the material is frequently weathered and friable. Corridors and stairs can be slippery, even in fine weather. Stones may be missing, or the edges of steps may be so eroded that it is easy to trip and fall. All this can make visiting a ruin a risky, even dangerous experience (particularly in the most dimly lit parts of the building). On our visits we have both been acutely aware that we could fall and seriously injure ourselves. This sets the mind racing. How long would it be until somebody finds me? Would anybody find me? We might lie there in a pool of blood until closing time. We can illustrate this with a recent experience (recorded by Duncan soon after the visit):

> One tower sticks in my mind: there was some sort of 'climb these stairs at your own risk' sign at the bottom. I decided to climb them anyway. The steps were seriously uneven, slippery, bordering on treacherous. It had rained the previous day so that the steps were also damp. There was a hand rope to hold on to (on the right-hand side) but I wasn't convinced that it would hold me if I slipped. I picked my way to the top, conscious that if I fell and injured myself then at best that would ruin my holiday … and at worst … who knows? But I got to the top and was rewarded by having the tower to myself. I was at the highest point of the building, commanding great views over everybody else

172 *Duncan Light and Steve Watson*

in the courtyard below. Nobody else disturbed me; nobody else was courageous or foolhardy enough to make it this far. I felt elated and supreme – I had conquered this building. But getting down was even harder. It seemed to take twice as long because I was twice as careful. The rope was now on my left-hand side where my grip was less strong. I picked my way down, one step at a time, bringing both feet onto a step and ensuring that I was stable before tentatively moving down to the next step. My heart was pounding, my mouth was dry and adrenaline was coursing through my veins. In some way I had a very heightened sense of my own existence. But when I got to the bottom I knew it had been worth it.

For Steve, the danger is also tempered with the excitement and atmosphere to be found in fragmentary ruins in isolated places. The castles of Siana and Monolithos on the island of Rhodes (see Figure 8.5), and La Estrella Castle in Malaga Province, Spain, have each combined the excitement of discovery and exploration with a sense of solitude that seems to weigh in the air:

At La Estrella I stumbled over rubble and broken stones that seemed to materialise my own uncertainties about being there. I felt an immense solitude. Should I be there at all? At Sania I was with two colleagues who went off exploring while I felt rooted to the spot, unsettled by the sheer drops all around, disturbed rather than awed by the view over the surrounding countryside. I lost sight of my colleagues and felt giddy, sharing the moment with an

Figure 8.5 Dangerous places, Monolithos, Rhodes.

inquisitive mountain goat. Monolithos was exciting; a jagged castle perched precipitously on a small table of rock. But Estrella was an enchanted place, perched on its windswept grassy crag, heavy with atmosphere, as if something clung to it and then to me. I knew it was a Moorish fortress reconquered in 1326 and the sense of that abandonment seems never to have escaped it. I felt its emptiness; a fragment of what once had been and an atmosphere that seemed to link the stones and the wind with a presence of some kind, the presence of absence.

Atmospheres are also active in the intersections of affect and emotion and, as Anderson (2009) so clearly puts it, atmosphere unsettles the distinction between the two, and does not fit neatly into either:

> [Atmospheres] mix together narrative and signifying elements and non-narrative and asignyfying elements. And they are impersonal in that they belong to collective situations and yet can be felt as intensely personal. On this account atmospheres are spatially discharged affective qualities that are autonomous from the bodies that they emerge from, enable and perish with.
>
> (2009, p. 79)

Through (and in) atmospheres, ineffable affective sensations mingle with feelings that are expressible to produce a 'mix', as Anderson (2009) put it in his description of atmosphere, or a blurring of affect and emotion, as Edensor (2012) put it in his. However, atmospheres do not exist in a vacuum: they are frequently anticipated, subject to their consistencies and recurrences and shaped by prior experiences (Edensor 2012). This, in turn, points to the kind of cultural antecedents expressed in the castle as a historically formed assemblage.

The castle imagined

In conclusion, we might think about the ways in which the castle assembled and the castle experienced combine expressively and cognitively. For this we have invoked the idea of imagination. The imagination is an essential (if frequently overlooked) aspect of the tourism experience (Hennig, 2002; Lean *et al.*, 2014; Salazar and Graburn, 2014) and tourists can flit between the real and imagined world with practised ease (Robinson, 2012). Visiting a castle can stimulate and trigger a wide range of imaginative processes such as dreamwork, reverie and mind-voyaging (cf. Rojek, 1997). These 'inner' mental processes can, in turn, induce their own emotional responses (Picard, 2012) and embodied sensations. The moment of engagement is therefore a dialogue between what the visitor experiences during the material encounter with a castle and what they bring with them as the cultural assemblage discussed earlier. The imaginative reworking of what we already know (or think we know) about castles is thus central to this dialogue (see Figure 8.6).

Some ruins stimulate us to engage in imaginative time travel and imagine ourselves being there at some point in the past. We might imagine (just for a moment)

Figure 8.6 The castle imagined: a moment of reverie at Warkworth, UK.

that we are in a Tudor Court, or part of the audience for a medieval jousting display. But we do not need interpretation. We bring our own assemblage, formed from previous experience or the last 'historical'/fantasy/horror movie we watched, or the last History Channel documentary, or a lifetime of similar accretions; and we combine them *in situ* with our embodied and emotional responses. What we take away from this exists in the imaginative realm of the half-remembered, reconstituted as meaning when returned to at some future date, perhaps even a subsequent castle visit.

This is perhaps the most important aspect of castles as places of affect and emotion, because the imagination becomes the melting pot of experience and assemblage. Sometimes the assemblage is employed interpretively in order to enhance the affective and emotional affordances of the space. Can we imagine *what it was really like*? And can this imaginative reconstruction, provoked by the interplay of assemblage and interpretation with direct experience *in situ*, produce its own affects and emotions? For David Crouch (2010, 2012, 2015), the experience of heritage and its meanings are constituted and reconstituted in such moments of engagement, and following Crouch we expect to find the castle imagined and re-imagined in such moments and in what follows, as memory supersedes experience.

This chapter has been an exploration of the feelings prompted by the experience of a particular heritage object, the ruined medieval castle. We do not pretend to have arrived, through this exploration, at a theory of heritage affect or an emotional geography of heritage. But we are clear, having examined the castle assembled as a cultural artefact and the castle experienced through our own life-long engagement with them, that there are some important conclusions that might be drawn that are of interest in wider debates in *affecting* heritage. We conclude, for example, that in the more-than-representational experience of castle visiting the representational is an assemblage that contains not only discursive and narrative elements but also echoes of embodied engagement, affective registers and emotional expression. The aesthetic of the sublime, gothic sentiment and the emergence of the modern tourist are all implicated in this. The assemblage, in turn, is constituted and reconstituted in moments of encounter and engagement, and in those moments we experience atmosphere as an autonomous force, even though it forms in the coalescence of pre-personal affect and our own subjective emotional registers. These feelings come to rest on the way home; they settle in layers drawn from many sources, brought together in a dream, a memory, a moment, and in the castle imagined.

Bibliography

Anderson, B. 2009. Affective atmospheres. *Emotion, Space and Society*, 2(2): 77–81.
Boym, S. 2011. Ruinophilia: appreciation of ruins. http://monumenttotransformation.org/atlas-of-transformation/html/r/ruinophilia/ruinophilia-appreciation-of-ruins-svetlana-boym.html (accessed 23 October 2014).
Brett, D. 1996. *The construction of heritage*. Cork: Cork University Press.

Burke, E. 2015 [1757]. *A philosophical enquiry into the sublime and the beautiful*. Oxford: Oxford University Press.

Büscher, M. and Urry, J. 2009. Mobile methods and the empirical. *European Journal of Social Theory*, 12(1): 99–116.

Byrne, D. 2013. Love and loss in the 1960s. *International Journal of Heritage Studies*, 19(6): 596–609.

Clough, P.T. 2009. The new empiricism, affect and sociological method. *European Journal of Social Theory*, 12(1): 43–61.

Crouch, D. 2005. Flirting with space: tourism geographies as sensuous/expressive practice, in C. Cartier and A.A. Lew (eds) *Seductions of place: geographical perspectives on globalization and touristed landscapes* (pp. 23–36). London: Routledge.

Crouch, D. 2010. The perpetual performance and emergence of heritage, in E. Waterton and S. Watson (eds) *Culture, heritage and representation: perspectives on visuality and the past* (pp. 57–71). Farnham: Ashgate.

Crouch, D. 2012. Meaning, encounter and performativity: threads and moments of space-times in doing tourism, in L. Smith, E. Waterton and S. Watson (eds) *The cultural moment in tourism* (pp. 19–37). London: Routledge.

Crouch, D. 2015. Affect, heritage, feeling, in E. Waterton and S. Watson (eds) *The Palgrave handbook of contemporary heritage research* (pp. 177–90). Basingstoke: Palgrave Macmillan.

Daniels, S. and Cosgrove, D. 1988. *The iconography of landscape*. Cambridge: Cambridge University Press.

De Certeau, M. 1984. *The practice of everyday life*. Berkeley: University of California Press.

Duff, C. 2010. On the role of affect and practice in the production of place. *Environment and Planning D: Society and Space*, 28(5): 881–95.

Edensor, T. 2001. Performing tourism, staging tourism: (re)producing tourist space and practice. *Tourist Studies*, 1(1): 59–81.

Edensor, T. 2007. Mundane mobilities, performances and spaces of tourism. *Social and Cultural Geography*, 8(2): 199–215.

Edensor, T. 2011. Entangled agencies, material networks and repair in a building assemblage: the mutable stone of St Ann's Church, Manchester. *Transactions of the Institute of British Geographers*, 36(2): 238–52.

Edensor, T. 2012. Affective atmospheres: Anticipating and reproducing the flow of affective experience in Blackpool. *Environment and Planning D: Society and Space*, 30(6): 1103–22.

Edensor, T. 2013. Reconnecting with darkness: gloomy landscapes, lightless places. *Social and Cultural Geography*, 14(4): 446–65.

Ekman, P. 1992. An argument for basic emotions. *Cognition and Emotion*, 6(3–4): 169–200.

Ellis, C., Adams, T.E. and Bochner, A.P. 2011. Autoethnography: An overview. *Forum Qualitative Sozialforschung/Forum: Qualitative Social Research*, 12(1), art 10. www.qualitative-research.net/index.php/fqs/article/view/1589/3095 (accessed 29 September 2014).

English Heritage. 2015. About us. www.english-heritage.org.uk/about-us/ (accessed 20 October 2014).

Gilpin, W. 1794. *On picturesque beauty, three essays on picturesque beauty, on picturesque travel and on sketching landscape, to which is added a poem on landscape painting*. London: R. Blamire.

Ginsberg, R. 2004. *The aesthetics of ruins*. Amsterdam: Rodopi.

Grant, A., Short, N.P. and Turner, L. 2013. Introduction: storying life and lives, in N.P. Short, L. Turner and A. Grant (eds) *Contemporary British autoethnography* (pp. 1–16). Rotterdam: Sense Publishers.
Hart, J.T. 1965. Memory and the feeling-of-knowing experience. *Journal of Educational Psychology*, 56(4): 208–16.
Hennig, C. 2002. Tourism: enacting modern myths. In G.M.S. Dann (ed) *The tourist as a metaphor of the social world* (pp. 1–19). Wallingford: CABI.
Holloway, J. 2010. Legend-tripping in spooky spaces: ghost tourism and infrastructures of enchantment. *Environment and Planning D: Society and Space*, 28(4): 618–37.
Holloway, J. and Kneale, J. 2008. Locating haunting: a ghost-hunter's guide. *Cultural Geographies*, 15(3): 297–312.
Hunter, M. 1996. *Preserving the past: the rise of heritage in modern Britain*. Stroud: Sutton Publishers.
Lean, G., Waterton, E. and Staiff, R. 2014. Reimagining travel and imagination, in G. Lean, R. Staiff and E. Waterton (eds) *Travel and imagination* (pp. 10–22). Farnham: Ashgate.
Light, D. 2015. Heritage and tourism, in E. Waterton and S. Watson (eds) *The Palgrave handbook of contemporary heritage research* (pp. 144–58). Basingstoke: Palgrave Macmillan.
Macaulay, R. 1984 [1953]. *The pleasure of ruins*. London: Thames and Hudson.
Mandler, P. 1997. *The fall and rise of the stately home*. London: Yale University Press.
Morgan, N. and Pritchard, A. 2006. On souvenirs and metonymy: narratives of memory, metaphor and materiality. *Tourist Studies*, 5(1): 29–53.
Noy, C. 2007a. The poetics of tourist experience: An autoethnography of a family trip to Eilat, *Journal of Tourism and Cultural Change*, 5(3): 141–56.
Noy, C. 2007b. The language(s) of the tourist experience: an autoethnography of the poetic tourist, in A. Ateljevic, A. Pritchard and N. Morgan (eds) *The critical turn in tourism studies: Innovative research methods* (pp. 349–70). Oxford: Elsevier.
Picard, D. 2012. Tourism, awe and inner journeys, in D. Picard and M. Robinson (eds) *Emotion in motion: tourism, affect and transformation* (pp. 1–19). Farnham: Ashgate.
Pile, S. 2010. Emotions and affect in recent human geography. *Transactions of the Institute of British Geographers*, 35(1): 5–20.
Potter, F.J. 2005. *The history of gothic publishing, 1800–1835, exhuming the trade*. Basingstoke: Palgrave Macmillan.
Rakoczy, L. 2008. Out of the ashes: destruction, reuse and profiteering in the English Civil war, in L. Rakoczy (ed) *The archaeology of destruction* (pp. 261–86). Newcastle-upon-Tyne: Cambridge Scholars Publishing.
Riley, W. 1934. *The Yorkshire Pennines of the North West: the open roads of the Yorkshire highlands*. London: Herbert Jenkins.
Robinson, R. 2012. The emotional tourist, in D. Picard and M. Robinson (eds) *Emotion in motion: tourism, affect and transformation* (pp. 21–47). Farnham: Ashgate.
Rojek, C. 1997. Indexing, dragging and the social construction of tourist sites, in C. Rojek and J. Urry (eds) *Touring cultures: transformations of travel and theory* (pp. 52–74). London: Routledge.
Salazar, N.B. and Graburn, N.H.H. 2014. Introduction: towards an anthropology of tourism imaginaries, in N.B. Salazar and N.H.H Graburn (eds) *Tourism imaginaries: anthropological approaches* (pp. 1–28). Oxford: Berghahn.
Scarles, C. 2010. Where words fail, visuals ignite: opportunities for visual autoethnography in tourism research. *Annals of Tourism Research*, 37(4): 905–26.

Schama, S. 1996. *Landscape and memory*. New York: Vintage.

Simmel, G. 1958. The ruin. *The Hudson Review*, 11(3): 371–85.

Smith, L. 2006. *Uses of heritage*. London: Routledge.

Staiff, R. 2012. The somatic and the aesthetic: embodied heritage tourism experiences of Luang Prabang, Laos, in L. Smith, E. Waterton and S. Watson (eds) *The cultural moment in tourism* (pp. 38–55). London: Routledge.

Staiff, R. 2013. Swords, sandals and togas: the cinematic imaginary and the tourist experiences of Roman heritage sites, in R. Staiff, R. Bushell and S. Watson (eds) *Heritage and tourism: place, encounter, engagement* (pp. 85–102). London: Routledge.

Staiff, R. 2014. *Re-imagining heritage interpretation: Enchanting the past-future*. Farnham: Ashgate.

Staiff, R., Watson, S. and Bushell, R. 2013. Introduction – place, encounter, engagement, in R. Staiff, R. Bushell and S. Watson (eds) *Heritage and tourism: place, encounter, engagement* (pp. 1–23). London: Routledge.

Thompson, M.W. 1981. *Ruins: their preservation and display*. London: British Museum Publications.

Tucker, H. 2009. Recognizing emotion and its postcolonial potentialities: discomfort and shame in a tourism encounter in Turkey. *Tourism Geographies*, 11(4): 444–61.

Urry, J. 1990. *The tourist gaze: leisure and travel in contemporary societies*. London: Sage.

Waterton, E. 2010. *Politics, policy and the discourses of heritage in Britain*. Basingstoke: Palgrave Macmillan.

Waterton, E. and Dittmer, J. 2014. The museum as assemblage: bringing forth affect at the Australian War Memorial. *Museum Management and Curatorship*, 29(2): 122–39.

Waterton, E. and Watson, S. 2014. *The semiotics of heritage tourism*. Bristol: Channel View.

Waterton, E. and Watson, S. 2015. Methods in motion: affecting heritage research, in B.T. Knudsen and C. Stage (eds) *Affective methodologies* (pp. 97–118). Basingstoke: Palgrave Macmillan.

Watson, S. 2001. Touring the medieval: tourism, heritage and medievalism in Northumbria. *Studies in Medievalism*, 11: 239–61.

Watson, S. 2013. Country matters: the rural-historic as an authorised heritage discourse in England, in R. Staiff, R. Bushell and S. Watson (eds) *Heritage and tourism: place, encounter, engagement* (pp. 107–26). London: Routledge.

Wetherell, M. 2012. *Affect and emotion: a new social science understanding*. London: Sage.

White, W. 1858. *A month in Yorkshire*. London: Chapman and Hall.

Workman, L. 1985. Medievalism, in N.L. Lacey (ed) *The Arthurian encyclopedia* (pp. 387–91). New York: Garland.

9 From Menie to Montego Bay

Documenting, representing and mobilising emotion in coastal heritage landscapes

Susan P. Mains

Introduction: this land is your land

> I have never seen such an unspoiled and dramatic sea side landscape and the location makes it perfect for our development
> (Donald Trump, 2014, 'Greetings from Donald J. Trump', *Trump International Golf Links*, www.trumpgolfscotland.com)

A vast expanse of sea stretches out as far as the eye can see. Seagulls gust in and out of view while the striking scene is framed by the hazy borders of waving ambery-green grass. The scene ahead conjures up images of beguiling tourism commercials and seems almost too dramatic to be real. A slight turn to the right reveals that this landscape is not quite so serene and remote as it might at first appear: a gaggle of people jostle around a central figure; some hold microphones aloft, while others jot down commentaries. A series of suited, sunglassed and ear-pieced figures stand guard in elevated positions above the scene, gazing into an unidentifiable distance. The crowd's central character suddenly walks a few paces, stretching his arm ahead of him in an all-encompassing arc – his hair flowing in a manner paralleling that of the dune grass behind him. The crowd attempts to re-group, re-establishing prime positions. Even further away, but not so far to be completely out of view, a sombre figure watches the activities: a local resident smoking defiantly, silently documenting the re-visioned landscape that has become the narrative centrepiece of the story currently being announced to the captive press invitees – a story which declares proposed coastal golf developments as a critical survival strategy for rural communities.

While painting a picture of a specific physical location and social activities (i.e. a press conference held celebrating a ground-breaking ceremony at the golf course site at Trump International Golf Links, Aberdeenshire, Scotland), the scene described above also embodies a narrative that is personal, global and paradigmatic. Situated in an area that has been hailed by many as a valuable ecological and aesthetic site, the Trump International Links golf course has become a nodal point in debates around heritage and tourism. It is a story of landscape change that embraces and masks conflicting emotions surrounding the role of tourism and media organisations, contentious concepts of heritage and shifting perspectives of coastal 'development'. This chapter explores the importance of affective geographies in contested representations of coastal landscapes such as this, and how these mobilise conflicting notions of heritage, development and sovereignty.

In the discussion to follow, I argue that media representations of place can provide a controlling and catalytic role in approaches taken towards specific sites, particularly in the context of potential tourism development. Specifically, I draw on two case study documentary films, *Jamaica for Sale* (Figueroa and McCauley, 2008) and *You've Been Trumped* (Baxter, 2011) – films set in Jamaica's North Coast and northeast Scotland, respectively – to illustrate the ways in which depictions of landscapes that are often associated with leisure and relaxation also contain and inspire narratives of loss, intimidation, joy, resignation, indignation and anger. These spaces and their related representations demonstrate the dynamic and affective nature of heritage landscapes, as well as demonstrating their wider relevance for more nuanced understandings of the politics of tourism, decision-making processes and the negotiation of place.

To undertake this exploration, the chapter is divided into five sections. The first section sets out the conceptual framing of the discussion to follow, in particular the theoretical connections (and implications) between heritage, destination, emotion and political geographies. The second section provides a brief overview of the two films to be highlighted as a central part of the analysis, illustrating key narrative and contextual features. The third section explores emotional tensions between media depictions of heritage, development and violence, while the fourth section analyses representations of coastal heritage and beaches, communication and sovereignty. The final concluding section provides critical reflection on what the examples discussed offer for future understandings of affective (and inclusive) heritage.

Theorising heritage: emotion, politics and (banal) destinations

Heritage is an inherently spatial and temporal concept. It is also rooted in the notion of meaningful ties to specific landscapes and events – a connection to place that has emotional resonance (Mains, 2014). Places that are designated as having a significant heritage value often include locations where particular events have taken place (a battle, protest, popular gatherings), places where well-known figures once lived (the houses of famous authors or inventors) or tours that seek to recreate the story of a past landscape (themed walks commemorating past trading activities or recreated village sites). Heritage is not, however, only associated with human-made artefacts and practices; it may also be used to recognise and protect environmental settings that are regarded as having unique features (for example, endangered species or a unique habitat system). While recognition is given to the scientific (and aesthetic) value of physical landscapes, therefore, it is worth noting that they are also afforded this value and protection through their ability to induce emotive responses in their human audiences (for example, a sense of awe at a sweeping landscape or feelings of happiness while walking through a wild beach setting). These heritage sites and their accompanying affective realms coalesce and collide to produce complex social and material 'destinations'. Although we tend to think of destinations as physical locations that mark a specific arrival or fixed time spent in a particular place, as we more closely interrogate and open up the concept of heritage, the notion of destination also loosens to encompass the expectations, representations and emotional states that are intertwined with concepts of travel, proximity and a sense of achievement, connection and/or entitlement.

Destinations are places that are often described (in tourism promotions, business investment literature, civic boosterism campaigns, for example) as places which we should aspire to visit. They are places that are promoted as being 'extra' ordinary, and thus desirable locations, perhaps because they enable us to recuperate and reinvigorate ourselves away from the everyday. But if we turn this notion of destination as a place that is somehow unexpected and unusual on its head, and explore instead the notion of the destination as something that actually heightens the everyday and encourages us to be more reflective and appreciative of it, then we suddenly have a somewhat different understanding of our relationships with tourism and place. Through this new understanding, destinations also include places with which we are already familiar – for example a local park, a nearby beach or a waterfront café – and through an increased sense of awareness, or appreciation, we may be able to experience them in new ways and see the value of their particular contributions to our quality of life. This new sense of place, and the degrees of value that we give it, may also be forged through the specific rubric of heritage.

Heritage tourism, in particular, offers an intriguing context in which to explore how our experiences and expectations intersect with our desires and emotional connections to specific sites before, during and after our physical experience of them. Reflecting on their study examining the different responses made by visitors to Israeli heritage destinations, Poria *et al.* (2003, p. 247) note: 'It is suggested that those who perceive a site as a part of their personal heritage are the basis of the phenomenon called heritage tourism, and they are distinguished from others by their behaviour.' Based on these conclusions, the study by Poria *et al.* suggests that heritage or a heritage destination is not a 'fixed' idea and that the same place may be experienced in many different ways by people visiting at the same time, partly due to their awareness of how such a place resonates with their own beliefs and emotional connections.

While the matter may not be as straightforward as the categorising of visitors into 'heritage tourists' or simply 'tourists', the study does point out the importance of understanding visitors' personal context and the cultural milieu within which tourism takes place. For example, a site related to a particularly well-known historical event may be seen as an interesting place to visit, but may not necessarily evoke the same kind of emotional response as a location known to have had an impact for a related social group or distant ancestors, which is seen as being more personally connected to the visitor. The heritage site, and the view of specific places as a particular kind of destination, is, therefore, a relative concept. One person's heritage site may be another person's car park, and that same car park may be transformed in the eyes of a visitor when they start to view it through a differently informed lens. A critical issue emerges, however, when one perspective – or version of history – is given more status than another, often leading to specific investments/removal of funding for a particular landscape or increased/lack of protection for sites, depending on which perspective is given greater credence or aligned with more powerful allies. As Timothy and Boyd (2006, p. 3) note: 'Dissonant views of history and the perceived superiority of various heritages have been at the root of many conflicts.'

The examples discussed in the sections below build on this notion that heritage, and the idea of creating and visiting destinations, is a relative concept and one that may shift over time and vary for different social groups visiting, or living in, the same site. In addition, the chapter interrogates the emotive and political dimensions of heritage and heritage tourism destinations. The analysis builds on Timothy and Boyd's (2006, p. 3) observation – 'Heritage is a complex and highly political phenomenon. There are few social elements and types of tourism that are more hotly contested at so many levels' – to suggest that heritage tourism is as much about political geographies as it is about leisure and recreation.

Conversations between tourism studies and political geography have emerged relatively recently (see, for example, Harrison and Hitchcock, 2005; Johnson, 1999; King and Flynn, 2014). Similar to the process of conceptualising how we think about heritage outlined above, when we think of political geographies we tend to think of 'big' formal events and institutions, such as elections, political parties and government structures, but these are only part of the picture. The 'everydayness' of political and politicised identities and places – for example, hotels as places of work (and conflict), places of rest, the home, roads – means that they are ubiquitous and are interwoven with tourist spaces. As Sara Fregonese (2012) highlights in the context of Beirut's urban redevelopment and 'discrepant cosmopolitanism', hospitality and hostility co-exist and are necessarily bound up in socio-spatial inequalities. Although the idea of cosmopolitanism suggests (a possibly elite sense of) openness and mobility, Fregonese's (2012, p. 317) interrogation of the term within a contested city landscape (through, for example, the co-existence of movements of people and tightened security) points to its more complicated contradictions: 'In Beirut, a double imaginary of openness and closure is constantly reproduced and negotiated: Beirut is at once a city-refuge and a city-battleground.' These contradictions play out via low-profile daily activities, not only through spectacular and surprising events. In a similar manner, it could be argued that heritage destinations are also constituted through social and spatial practices that are inherently contradictory, and which are reinforced and challenged through ongoing activities that may not be particularly striking, but which nonetheless reinforce certain notions of place and identity.

Emotion is a key component that interweaves tourism, politics and these ongoing banal geographies. The sight of a valued landscape being bulldozed for the construction of an all-inclusive hotel or the reification of a divisive political figure through the erection of bronze monuments and plaques in a space that was part of a local playground can instigate a range of feelings and reactions. Protests – written, oral, imaginative and physical – which challenge (and reinforce) the dominant and contested use of tourist spaces illustrate the ways in which we *feel* landscapes (tactilely and intellectually). Heritage destinations therefore embody emotional geographies that tell complicated and visceral political narratives, and are also part of representations that may focus on specific sites while also being embroiled in broader discussions of power and space.

Explicitly engaging and analysing affective and emotional geographies is a critical component in moving towards more diverse understandings of heritage and place. This is also true for mediated geographies of heritage. In a thought-provoking paper exploring 'the emotional responses and affects of exhibitions', Crang and Tolia-Kelly (2010, pp. 2315–16) stress the need to address 'heritage sites as occasions for doing and feeling, of connecting different sensations, representations, and thoughts'. Although focusing their discussion more directly in the context of the British Museum and the English Lake District, Crang and Tolia-Kelly's attention to audience/visitor affective/emotional responses has relevance for understanding the varied mediated geographies, such as cinema spaces, film locations, television broadcasts, tourist destinations, art exhibitions, golf courses, living rooms, travel agents and brochures and local council meeting spaces. In my analysis of the two selected films and their wider reception, I will provide examples of the ways in which their geographies and affects encompass a range of emotional responses, activities and spaces.

The circularity and mobility of places, stories and identities are intricately interwoven with representations of heritage and emotions. This movement is not, however, effortlessly free-flowing, but is often slowed down, interrupted or diverted. As Crang and Tolia-Kelly (2010, p. 2316) continue:

> Heritage sites differentially enable and arrest the circulation of objects, people, emotions, and ideas ... Attending to the differential circulations and thickenings acts as a check on analytic elisions of ontological monism, seeing everything as the same substance, with an implied universalism affect.

For example, certain national, race, gender or place-based stereotypes may become more 'sticky' and less straightforward to shift when a limited range of images are reproduced and mapped onto specific locales and feelings for particular social groups. To challenge the variable viscosities of heritage representations, we need to be attentive to the diverse ways in which we experience place and how 'objects, people, emotions, and ideas' are encouraged to circulate, while opening up opportunities and spaces to depict and unearth stories, feelings and voices which are often 'stuck', marginalised or ignored. Paul Gilroy (1995) undertakes such a project of recovery through the image of the ship in the context of the Atlantic and the experiences of the Middle Passage, illustrating the significance of artefacts, such as people, objects and ideas, which are intertwined with this material and cultural symbol of circulation. Ships and coasts have often encapsulated movements and feelings of loss, escape and 'no return' – struggles which are just as relevant today. In relation to heritage tourism, coasts can also be viewed as being sites of additional artefacts: heritage, affect and emotion. And, as the case study films below illustrate, in more recent times, convergent media – for example, the varied formats of television, print media, radio, feature films, internet campaigns – have become increasingly intertextual in their circulations and can be utilised to reinforce/challenge particular identities or emotional (dis)connections with place. Media representations feed into each other while enabling the circulation of specific place images and identities, but not necessarily on an equal footing or with comparable economic or political support.

You've Been Trumped and *Jamaica for Sale* provide succinct examples of challenging universal responses to tourism developments by charting a path through conflicting emotions and practices that circulate around coastal developments, claims to local heritage and upbeat popular media depictions of tourism. The films themselves have not elicited uniform responses, being both praised and criticised to varying degrees by a range of politicians, developers, residents, journalists and academics (although their overall reception has been substantially positive, both from filmmakers and the wider public). The documentaries are incomplete and partial renderings of contested landscapes – as all stories are – but they are nonetheless important for complicating and contesting fixed notions of coastal heritage and the affective realms through which they are received. Just as museums are being pushed to re-think the ways in which we experience heritage materially *and* emotionally in varied contexts and forms, these films encourage other filmmakers, journalists, planners, residents, academics and entrepreneurs to reassess the ways in which our concepts of, and feelings towards, heritage, development and sovereignty are interwoven with mediated geographies and geographies of media (Pan and Ryan, 2013, and a range of examples highlighted in Mains *et al.*, 2015b).

Heritage tourism is, therefore, emotive and geopolitical, and requires an understanding of how localised routines and activities may reinforce and challenge dominant representations of space. A critical component of this fluid understanding of heritage destinations is the ways in which media representations are created and mobilised in relation to tourism landscapes. Given that visual images are such a central feature of the recognisability and success of heritage tourist destinations (Waterton and Watson, 2010), it is logical that media in multiple forms play a significant role in our understanding and negotiations of tourism. As I have noted elsewhere (Mains, 2015; 2008), tourism-related media coverage provides us with a certain type of mobility, but one that is often narrowly limited and with a circumscribed audience in mind. Media images of place, including coverage of heritage destinations, are frequently mobilised to appeal to viewers' emotive sense of patriotism, justice, nostalgia, loss or hope. While media depictions produced by tourism developers, and others seeking wider public support for new projects, may appear overly generic and simplistic at first view, they offer a few strands of a densely woven story. This media tapestry of tourism also includes the intersecting stories of those who counteract dominant narratives, and who may strategically deploy diverse media formats in order to reach a more varied audience and to harness emotional power.

While there has been a significant amount of scholarly research exploring the ways in which tourist destinations and place brands are developed, there has been less discussion focused on the ways in which critical media may provide alternative maps to these hotly contested landscapes. *Jamaica for Sale* and *You've Been Trumped* provide an opportunity to interrogate the theoretical ideas outlined above, while also suggesting a need to broaden the concept of heritage to include multiple 'visions' of landscape/land use through both everyday and exceptional negotiations of place.

A tale of two coasts: *Jamaica for Sale* and *You've Been Trumped*

Jamaica for Sale and *You've Been Trumped* are documentary films that complicate popular narratives of tourism as escape, opportunity and pleasure. Although made a few years apart, and on different continents, both films map out the challenges – limited economic resources, unequal political recognition, environmental degradation and entrenched ideas of 'development' (to name a few) – faced by residents who live in tourist destinations. The films resonate through a desire to connect with voices that are often missing from mainstream discussions of place promotion and landscape change, while paying attention to the importance of specific locales.

Jamaica for Sale builds on ground laid by an earlier film, *Life and Debt* (2001), a documentary by Stephanie Black that explores the impacts of Jamaica's spiralling international debt payments, neo-colonial trading relationships and externally oriented tourist landscapes. Like Black, Figueroa and McCauley strategically use music to capture a distinctive sense of place, and to emphasise the emotional impacts of overzealous construction, labour exploitation, poor governmental decision-making and increasingly endangered coastal environments. Throughout *Jamaica for Sale*, the directors give voice to residents who have been forced to relocate their homes as part of coastal 'clean up' processes, had their access to beachfronts blocked off, been confounded by government and large-scale business resistance to community organisations and witnessed ever-increasing rates of coastal pollution. Figueroa and McCauley provide support for local observations through the use of expert testimony in the form of interviews with academics working on related topics, legal professionals challenging the alleged (in)actions of the Jamaican government and non-governmental professionals who hope to provide more informed environmental and social awareness on policy processes. Images of idyllic unpopulated beaches are situated in stark contrast to polluted water, vast all-inclusive hotel structures and striking construction workers. The film illustrates the power of representation, with examples ranging from mid-twentieth century archival films exploring training for tourist-related employees and speeches by younger incarnations of political leaders praising the potential benefits of tourism, tourist promotions showing plans for a reconstructed and repackaged 'Historic Falmouth Cruise Port' harbour development on Jamaica's North Coast (Port of Falmouth Retail Centre, 2015), advertising and the (unfinished) construction of luxury condominium high-rise holiday homes owned by the Palmyra Resort and Spa – which was later placed in receivership (*Jamaica Observer*, 2011). *Jamaica for Sale* illustrates, however, that narratives counter to the assumed positive message of tourism development are also an important part of Caribbean landscapes.

Although at first glance they may appear to explore quite divergent landscapes – a blustery northeast coast Scotland with oil as the leading light in its economic landscape versus a widely popular Caribbean shore with well-established package tours and a tropical climate – Anthony Baxter's *You've Been Trumped* demonstrates the convergence of exclusionary narratives across these apparently disparate locales (as well as the convergence of media formats: feature and documentary film, tourism advertising, television news, painting, etc.). Set in the Menie Estate in Aberdeenshire, Baxter's film begins with an excerpt from another celluloid depiction of idealised rural coastal life, Bill Forsyth's iconic Scottish film *Local Hero* (1983). The latter film depicts the story of a small coastal village which becomes the recipient of the attention of a US developer keen to transform the area into a tourism hotspot in spite of its environmental credentials and alternative plans to create a protective marina. In Forsyth's film canny locals strategise to maximise their potential economic windfall. Ultimately, however, when it is discovered that an elderly resident who lives on the beach actually owns (and refuses to sell) it and the US tycoon at the head of the interested multinational finally comes to visit the village and succumbs to the setting's charms (pursuing an enduring fascination with observing the *aurora borealis*) the village's unique character is saved for another generation. While Forsyth's film challenges popular stereotypes of the naïve villager and depicts their surreptitious mapping of their coastal plans with irreverent humour, the villagers in *You've Been Trumped* face a far less sanguine and responsive Goliath (although humour is also utilised and unwittingly provided by one of the film's central characters, Donald Trump).

As the film commences, Trump International Golf Links has begun construction of a 'luxury' golf course, and in the process cleared and rebuilt existing sand dunes which were considered by many to be categorised as a Site of Special Scientific Interest (SSSI). Throughout the course of *You've Been Trumped*, Baxter juxtaposes the bombastic television appearances and newspaper statements made by Donald Trump and his colleagues (largely family members) in defence of his development with the increasingly concerned and thoughtful reflections of local residents who are dealing with the repercussions of these activities (light-blocking mounds of earth, eviction threats, broken pipes, public slander). In a similar style to Figueroa and McCauley, Baxter builds up a picture of the changing coastal and social landscapes over time, and he himself becomes a key figure as the police illegally arrest him for filming in the area. A mixture of interviews, reportage, animations and landscape shots is intermittently

From Menie to Montego Bay 187

interspersed with scenes from *Local Hero* – possibly as a reminder that documentary can be stranger than fiction (and can converge in even stranger ways).

Both documentaries have received critical acclaim and many screenings at international film festivals. *You've Been Trumped* has been shown to a range of European and US/Canada audiences (being based in the UK, Baxter has been able to tap into existing broadcasting networks and his film was screened on terrestrial British television on BBC Two and later BBC One). *Jamaica for Sale* has also screened on national terrestrial television in Jamaica and has been screened widely throughout the Caribbean and in other parts of the Americas, while being used as part of pedagogical exercises in a range of educational institutions (Bailey, 2010). Both films chart paths through highly contested and iconic landscapes, and in the following section I will look at these conflicts in closer detail.

Heritage, development and violence: affecting landscapes

Heritage, development and violence are three themes that emanate from *Jamaica for Sale* and *You've Been Trumped*. They are themes that may not appear so obviously interconnected in other settings, but which coalesce and collide sufficiently in these moving image formats to pose questions about the ways in which we communicate, feel and design our relationships with coastal environments. While discussions of heritage have tended to emerge from the fields of tourism studies, cultural geography and historical/critical landscape studies (see for example, Nash, 1996; Smith *et al.*, 2009; Tolia-Kelly, 2007, 2008), debates around development have historically focused more on the context of economic geography, postcolonial studies (for example, critiquing traditional notions of development as a linear – and positive – process), health and migration studies. Discussions addressing the geographies of violence have, quite reasonably, focused on topics in the area of political geography (for example, territorial and ethnic conflicts), feminist theory (gendered violence and human rights) and cultural geographies around race and representation. Pain (2009) also points out the importance of contextualising fear in a less hierarchical manner, while unearthing the opportunities for resistance and change. For the purposes of this section, this brief overview is necessarily an over-generalisation of the research that has been conducted in these areas – but it is helpful for thinking about what commonalties and tensions there may be in relation to how we think about emotion and place. In particular, in terms of emotions and the bodily resonances represented and experienced, tourist destinations exemplify the contradictory ways in which feelings about heritage can play out.

Jamaica for Sale and *You've Been Trumped* were selected as the focus for this study precisely because they provide provocative and timely examples of the ways in which broader concepts, such as development, can be brought into sharper relief and situated adjacent to other problematic landscapes that may function on different scales (for example, blocked village pipes, regional council meetings that result in residents possibly facing eviction, national legislative decision-making processes that challenge/reinforce individual/business legal rights, and local/national/international news media images that support/contradict particular environmental/business/resident causes). The selected films, and their mediated emotional geographies, provide a dynamic context through which to explore these almost unwieldy issues (heritage, development, violence) in a focused and in-depth manner while providing the opportunity to situate these specific representations within wider socio-spatial relations.

In what ways, therefore, do heritage, development and violence interconnect in *Jamaica for Sale* and *You've Been Trumped* through representations and productions of affective geographies? In the two films heritage is depicted as both central and contested in relation to tourist developments taking place in coastal areas. For entrepreneurial developers seeking to build large-scale all-inclusive hotels (*Jamaica for Sale*) or extensive golf resorts (*You've Been Trumped*), recreating past landscapes is seen as a way to include previous activities and features, but in a selective and managed form that enables the integration of new facilities while minimising (or erasing) features (informal settlements and non-compliant residents) that fail to fit within these 'idyllic' settings. Both films show examples of such nods to the past, including the re-creation of neighbourhoods, craft markets, sand dunes and a harbour. The films cast critical eyes on the means by which heritage is viewed as a bonus in the promotion of destinations, but problematic when it limits the ability to construct a new building, waste treatment plant or artificial sand dune, or to minimise a 'troublemaking' voice challenging such plans and their environmental impacts.

Although thousands of miles apart, the Highlands and coastal areas of Scotland and the North Coast of Jamaica (and Caribbean archipelago) have a long history of romanticised depictions in travel literature, including early tourism promotional photography and, later, feature films.[1] These destinations have been painted as quaint, scenic and/or somewhat 'wild' landscapes, situated in the past (relative to other more urban/ 'cosmopolitan' populations) via a neo-orientalist gaze that positions, and often continues to place, these potential tourist destinations in relation to external desires ready for visitors' consumption (Sheller, 2003; Thompson, 2007). If visitors/colonialists/entrepreneurs arrive at these distant shores only to find they do not quite match the geographical imaginary that has framed their journey and anticipation, a process of selective vision and hearing frequently ensues. The non-linear village streets, informal businesses or small family residences become incorporated into a *tabula rasa*, on which the cartography of modernity (in this context, tourism in the shape of all-inclusive hotels and golf resorts) charts its map. And it is not only physical structures that are utilised to stamp a new authority on the landscape – the struggle to define, lay claim and 'develop' coastal tourist destinations also takes place in the repertoires of representation and discourses of development (Escobar, 1995). Reflecting on Escobar's critique of development narratives, Reid-Henry (2012) notes, 'For Escobar, development amounted to little more than the west's convenient 'discovery' of poverty in the third world for the purposes of reasserting its moral and cultural superiority in supposedly post-colonial times'. In the context of Trump's language, lobbying and high-profile construction activities, along with those of many international finance organisations, foreign-owned hoteliers and extractive industries in the Caribbean, it could be said that the discovery of heritage and tourism has been strategically aligned with pre-existing development agendas (the Aberdeenshire links are hailed by Trump as a means of recapturing Scotland's golfing heritage, but in a way that fits within a drive to commercialise an environmentally unique area). As Marie Freckleton, an economist at the University of the West Indies-Mona, notes in an interview during *Jamaica for Sale*: 'The present development model seems to be based on expansion of tourism, and ... dependence on remittances. This is not a sustainable development strategy, in my view.' Creel (2003, p. 3) also notes:

> Coastal areas worldwide are major destinations for tourism, which represents the fastest growing sector of the global economy. Tourism dominates the economy of some regions and small island states; for example, tourism constitutes 95 percent of the economy of the Maldives and is the country's only source of hard currency.

Although written more than a decade ago, Creel's comments still resonate, as tourism continues to be viewed globally, and by Scottish and Jamaican governments, as a key income generator. While the economic contexts of the two countries show significant contrasts – Jamaica is a relatively recent postcolonial nation struggling with one of the world's highest debt to GDP ratios and relatively high levels of violent crime, while the Aberdeenshire coast is an area in which the North Sea oil industry has led to the area having among the highest average annual incomes in the UK (as well as rapidly increasing housing costs) – representations of coasts as sites for potential development, security and sustainability prevail in both.

The exclusionary development discourses interrogated in the films ignite impassioned responses within the mediums themselves, as well as among viewing audiences (to which I will turn to later). The documentaries illustrate that during their production the filmmakers faced a range of cavalier attitudes towards already existing material and social landscapes. These landscapes embody the *banal* and *personal* heritage of coastal areas, and the dismissive responses directed to these places and the film's directors and concerned residents galvanise and catalyse vociferous, angry and shocked responses. In tandem with the films, the process of filmmaking therefore becomes an active part of heritage storytelling practices. In both films, depictions of expensive golf and all-inclusive hotel resorts as expressions of heritage that are more desirable (and 'developed') than present-day landscapes, are called into question and met with incredulity, particularly given that the large majority of the local population will not be able to physically or economically access these sites. Anger, a sense of betrayal, and indignation are expressed by Menie Estate residents Molly Forbes, Michael Forbes and Susan Munro, as well as by residents in the local fishing industry and construction workers in Jamaica who rely on access to unpolluted beaches and shorelines in order to maintain their livelihoods. Indignation is not only the preserve of relatively low-profile and working-class residents: academics, non-governmental organisation leaders, artists and musicians coalesce to form a diverse, if under-funded resistance. During the course of the films (and beyond, via planning meeting protests, exhibitions, film screening question-and-answer sessions, etc.), a range of emotions – anger, frustration, empathy, joy, resistance and determination – emerge as part of these cultural landscapes and in other settings in which the issues raised by the films are discussed and acted upon (see, for example, the blogsite and activities created by the *Tripping Up Trump* (2015) activist group). Humour is also used to challenge dominant views in support of large-scale developments and limited critical coverage in local and national news. *You've Been Trumped* includes a segment from a student comedy night held in Aberdeen, during which one of the participants refers to Trump International's proposed activities, including the attempted removal of Menie Estate residents, as a 'twenty-first century Highland Clearance'. Although the statement is received with laughter, the joke is an attempt to undermine the positive promotional media being distributed by the golf developers and local press (whose limited and largely uncritical coverage is also highlighted). In public screenings and question-and-answer sessions following viewings of *You've Been Trumped*, awareness of this legacy of elite land ownership and control has also been raised as a source of anger and an enthusiasm for land reform, as well as the need for accountability. The Clearances reference also relates to the violence of forced removal of Highland populations in the past and to transatlantic slavery, the latter also explored in *Jamaica for Sale* when the construction of a new tourist-oriented harbour in Falmouth fails to engage with the pain and problematic history of processes of exploitation and racism (Figueroa, 2010).

Violence and displacement, on a smaller scale, resurface in the form of compulsory purchase orders in *You've Been Trumped* and in the destruction of small wooden-built houses that house hotel workers and local residents who are unable to afford title deeds or mortgages in *Jamaica for Sale*. As one woman, on the verge of tears, talks about the destruction of her house on Jamaica's North Coast, the hard work involved in establishing that home and the loss of her documents, it becomes increasingly clear that violence and affect come in various forms. Footage of a bulldozer knocking down homes where people had lived until earlier that day is also shown, along with fishermen recounting how the mangrove swamp was blasted with dynamite in order to construct a canal.

Violence in the form of bulldozers also reappears in *You've Been Trumped*, as the natural undulating landscape of the Aberdeenshire dunes is flattened or turned into more symmetrical versions, and embankments of sand and earth are built to block out the views of nearby householders. Walls and fences appear in *Jamaica for Sale* (Figure 9.1), blocking off beach access and creating what Mimi Sheller (a participant in the film) refers to as a 'spatial fix', which dramatically limits the possibilities for diverse forms of land use.

In both films, the heritage of subsistence fishing, diving, agriculture and family visits to the beach (shown through a range of older film footage, family photographs and stills), set alongside sweeping classical/electronic music, makes for an emotive landscape and one imbued with a sense of loss.

From Menie to Montego Bay 191

Figure 9.1 A still from *Jamaica for Sale* showing a wall around the construction of an all-inclusive hotel that has cut off beach access to local residents.

Photograph: Wendy Lee; in Figueroa and McCauley, 2008.

Violence appears in both films not only in a physical sense, but also through specific discursive practices that situate people as 'out of place' while attempting to exclude them from decision-making processes. In a discussion with journalists, Donald Trump responds to comments he had previously made (and went on to repeat several times), about Michael Forbes, a local resident who worked on the land and refused to sell it to the golf developers:

Journalist: Does calling Mr. Forbes and his property names … how does that help?
Donald Trump: … See, I happen to be a very truthful person. His property is terribly maintained. It's slum-like. It's disgusting. He's got stuff thrown all over the place. He lives like a pig. And I did say that…

In this dialogue, the landscape of a farm – where Michael has worked and lived for several generations – is no longer depicted as a valued part of long-standing coastal agricultural and family traditions, but is reframed for popular consumption (by someone who is relatively unfamiliar with the area and viewing it through developers' eyes) as a wasted space, with residents who are no longer described as people but as animals – or as 'a pig'. This dehumanisation of existing heritage landscapes and the people for which they have importance is a neo-colonialist process of erasure, an attempt to undermine dissenting opinions and justify highly contested developments on much-loved dune landscapes. It is also a depiction that has received the ire of many Menie residents and audiences beyond Aberdeenshire.

This corporate narrative is not, however, all-encompassing. Visible expressions of anger and disbelief are translated into several acts of resistance in *You've Been Trumped* and *Jamaica for Sale*: first, in the form of the documentaries themselves and the ensuing broadcasts, media coverage and community discussions; second, in the production of art, music and other media addressing the topics raised (for example, the art exhibition by David McCue held in Michael's shed and included in *You've Been Trumped* as a way of bringing together support for local residents); and third, in the creation of spaces through which the roles of critical residents and journalists are re-appraised and highlighted via the on-screen appearances of, and poignant interviews with, environmental activists, retired residents, academics, small-scale business operators, fishermen, shop workers and investigative reporters (for example, Anthony Baxter and John Maxwell). Anger, fear and frustration can, therefore, be transformed into a heritage of creative resistance and can be utilised to interrogate the narrative fictions of resort design plans, predicted job figures and improved housing security. Creative representations and community activism may inspire unforeseen connections, for example, with hope, bravery and joy – David McCue's series of paintings are an innovative example of this catalytic process, where 'local' residents become their own heroes and those of national and international audiences. Film, music, art and personal commentaries help to facilitate these empathetic and affecting connections with people, heritage and places that have often been ignored. Mainstream press and television coverage of current events and changing landscapes is contradictory and shifts over time, and this may partly come as a response to provocation from these resistant and creative challenges to dominating privileged voices that have previously received disproportionate (and uncritical) media attention. The attempted 'naming and shaming' of Michael Forbes by Donald Trump failed to erase him from the Menie cultural landscape and, combined with *You've Been Trumped* and increased public awareness of Menie Estate activities, Forbes' resistant presence was positively recognised through his widely publicised confirmation as 2012 'Top Scot' (part of the Glenfiddich Spirit of Scotland awards; Drysdale, 2012).

These filmic examples suggest that heritage embodies several competing forms: a long-held (perhaps nostalgic) personal connection to past places and activities; a particular story of a place connected to famous figures or events; and an opportunity to 'brand' destinations with historical connotations. There is also a fourth aspect to heritage's depiction in the films – one which spurs angry emotions and frustration for residents shown in the documentaries – and that is corruption. This latter aspect of heritage seems at odds with the idea of conservation and the celebration of culture, but several examples illustrate that questionable governance procedures and illicit deals are central: the documenting of town hall meetings (where residents face time constraints on comments or planning decisions have already been made prior to the meeting); the re-calling and overturning of decisions made by local councils; high-profile press events (where wealthy developers control who asks questions and offer 'attractive' participants employment – highlighted succinctly when Donald Trump is shown offering a job to Miss Scotland on their first meeting); and physical interactions and injuries incurred by those challenging proponents of tourist developments (the serious injury of workers at a hotel construction site in Jamaica and Anthony Baxter's arrest during filming). Illicit decision-making processes and the unequal sway of wealthy interests become increasingly visible. Although documentary (and feature) films are themselves depicting a specific version of particular events, they can simultaneously highlight and problematise the explicitness of partisan interests: in the two locales explored, partial stories of landscapes and heritage become part of the films' larger narrative development and wider public reception.

Feeling coastal heritage: beaches, communication and sovereignty

The opening scenes of *You've Been Trumped* include a now-infamous one from *Local Hero*, showing two men relaxing on a wall at the shore of a Scottish seaside village (filmed in Pennan, Aberdeenshire), a whitewashed guest house behind them: the location can be read as epitomising a quaint, peaceful, rural coastal landscape (and has since become a tourist destination for those interested in Bill Forsyth's/Scottish films). This image contains several elements that are returned to during this section: beaches, communication and sovereignty.

Following the view above, the camera looks out to the shoreline. On the left of the frame an older man stands in a red telephone booth, napping and/or waiting for a call. Gordon – the local hotelier, lawyer and go-to guy for all things needing fixing in the village – strides into view, as Victor and Mac, who are sitting and talking on the wall, turn and stand up to welcome their friend as he strides in front of the camera:

Victor: Breakfast ready Gordon?
Gordon: We've got a problem.
Mac: What?
Gordon: The Beach. Ben's beach.
Mac: What's the problem?
Gordon: The problem is it really is Ben's beach. [telephone in call box starts ringing] He owns the shoreline. Four miles of it: from the grass down to the low tide mark. I found it in the parish records when I was checking out some title deeds.
Victor: Can he prove it?
[The man in the call box becomes aware of the phone ringing and turns to answer it.]
Gordon: We can't steal the beach from him Victor, it's his.
Mac: We'll have to buy it from him.

(Forbes, 1983, cited in Bailey, 2010)

From this early scene in *You've Been Trumped*, the image then fades to the title: 'Aberdeenshire. 2010'. We then see a sweeping view of the Aberdeenshire coastline as Anthony Baxter's small red car is driven on the country road (the latter vehicle being a recurring image suggesting alternative mobilities and place-based narratives), and dramatic music is accompanied by the voiceover of a local news report stating: 'Donald Trump has arrived in Scotland to talk about his plans for what he claims will be the world's greatest golf course.'

The beach is a critical landscape and site of emotional enlightenment and tension in both *You've Been Trumped* and *Jamaica for Sale*. Although dynamic (as we can see with shifting sand banks and tidal cycles), these physical settings are emblematic of landscapes that have existed during an extended period of geomorphological and cultural history. Despite the rapid changes brought on by transforming land uses during the nineteenth and twentieth centuries, beaches are represented in the films (and popular media) as a relatively consistent, if contested, feature of coastal locations (Law et al., 2007). Indeed, it could be argued that it is both the dynamism and the presence of beaches that gives them such a unique sense of place, and a feeling of connection to a recognisable (if shifting) material setting.

As well as a significant proportion of Scotland and Jamaica's populations living within coastal areas, many residents of both countries are regular visitors to the seaside. These visitors, combined with international tourists and coastal residents, exist in tandem with the beach areas being represented in mainstream popular media as key recreational destinations. As Figueroa and McCauley (2008), Baxter (2011) and Creel (2003) point out, however, such an intensity and diversity of coastal activities can have both positive and negative effects:

> Tourism can offer some environmental benefits, such as greater appreciation of the value of natural resources. In the Caribbean, for instance, diving tourism has helped raise awareness about the need for reef conservation. But tourism can also have harmful effects ... It can lead to unsustainable coastal development as infrastructure is built on the shoreline to accommodate tourists. In the Caribbean, official estimates say that 70,000 tons of waste are generated annually from tourism activities.
>
> (Creel, 2003, p. 3)

Viewing the degradation of beach landscapes, such as the examples shown in *Jamaica for Sale* and *You've Been Trumped*, can be an emotive experience, evoking a sense of loss and nostalgia for past pleasures and the hopes for future generations. As White (2009, p. 1) reflects in relation to childhood experiences of beaches and holidays: 'It is a powerful image, not just a visual image, but something more fully and more sensually realised, something perhaps, that relates to the simpler sensual pleasures of childhood.' Beaches symbolise landscapes of social and environmental heritage: fishing, spending time with family, being able to physically enjoy being in the sea, and feeling the sand and sun on your skin and the temperature and buoyancy of the water. They are also inbetween spaces – they provide a tangible, aural and material connection between land and water – and are problematic spaces to bound (despite efforts to do so, as demonstrated by some coastal construction sites). Coastal areas become complicated regions in which to determine state-territorial and commercial boundaries and also point to nation-states' arbitrary claims to sovereignty.

Sovereignty is bound up in the notion that you have, and are recognised as having, the authority to govern and lay claim to a particular territory. Challenges to sovereignty can be seen in the beachscapes of Jamaica and Scotland. For example, Spanish developers of large-scale resorts are seen as challenging the authority of Jamaican officials through their ability to sidestep legal environmental frameworks, lack of accountability and attempts to limit accessibility in beach areas. Trump International has also been seen as challenging the rights of Scottish citizens through threats of eviction, construction in environmentally protected areas and cutting off residents' access to water. In an interesting development, Donald Trump found himself on the receiving end of debates about sovereignty delineations when his recent attempts to halt a proposed windfarm development several miles off the coast from his golf resort (because it would 'spoil the view' from the links) were dismissed by the Scottish judicial system amid criticisms that the developer was using his wealth to undermine domestic renewable energy policies (Carrell, 2014). In a response to earlier critics, a statement from Trump International was released to the press: 'We will not allow the Scottish Government or any other party to undermine what we have created at Trump International Golf Links and look forward to progressing the development further once this battle is behind us' (*The Herald*, 2013).

These debates about who should determine land use and access to beaches are controversial and emotive, and call into question the role of multinational corporations (also demonstrated in the earlier feature film *Local Hero*, mentioned above). Esther Figueroa and Diana McCauley play with this emotional and affective attachment to nationhood from the outset of their documentary. As *Jamaica for Sale* begins, a black screen appears and the Jamaican national anthem is played. In Jamaican cinemas the national anthem is played before each screening, and usually audiences will stand to recognise this practice, with the voiceover stating: 'Ladies and gentlemen, please stand for the national anthem of Jamaica.' While the anthem is played, a short film showing Jamaican sites or new construction activities may also be screened. Given this tradition, it appears as if a similar request is being made of the documentary's audience; however, as the music progresses and images appear on the screen, the images that are shown are not of new schools or hospitals being constructed or the faces of smiling children playing, but of zinc fences blocking off access to hotel construction sites, piles of rubble along coastal areas, workers digging, a congested dusty coastal highway overloaded with trucks hauling rubble, gated resorts, barbed wire fences around upscale condominium developments, a 'Scenic View' sign that points to a newly constructed wall, cruise ships docked safely behind fenced-off areas and, finally, a 'No Trespassing' sign with a view of the sea in the background. From the start of the film, viewers, particularly those who have stood up at the start of the anthem, are asked to call into question the assumed familiarity with the musical and visual imagery of nationhood. This may be accompanied by a feeling of discomfort and awkwardness, as well as a reflection on what is being brought into question: the trustworthy, bold and optimistic visions of the people and its government. It is a representational practice that involves an embodied mediated geography: the anthem and accompanying visual imagery may instigate physical (standing) and emotional (pride, nostalgia) responses. These audio-visual representations are familiar, but also a trick (and tricky). This film opening plays with habitual practices and assumptions, and in so doing displaces audience members by encouraging them to contemplate performances of nationhood and feelings of attachment and pride.

The cinema, therefore, comes into focus as a space where our experiences of the taken-for-granted can be challenged through unexpected emotional geographies. Reflexive and interactive engagements have been demonstrated following many screenings of *Jamaica for Sale* and *You've Been Trumped*. Having attended several public viewings of the films in Jamaica and Scotland, respectively – some of which have been followed by a discussion session with the director(s) and people featured in the films – there has been a consistently passionate, infuriated and concerned response by audiences. The films instigate an emotional residue that travels with people, as highlighted by the Tripping Up Trump (2015) and Jamaica Environment Trust (2015) activities instigated in the respective coastal locations as a response to the films and other related ongoing developments. As Law *et al.* (2007, p. 157) note (in relation to another coastal-based film, *The Beach*), this reflexive gaze 'brings into focus the viewer's relation to the environmental damage publicized by a range of journalists, fans and activists across various mediums'.

Cinemas are not the only communicative/mediated spaces that become reconfigured through public screenings. Within the films themselves, phone boxes, mobile telephones, televisions, radios and laptops connect people through time and space (where they may be seen as representative of feelings of excitement and anticipation, for example when waiting for a critical telephone call), or highlight (spatial and socio-economic) disconnection and distance (conveying, for example, refuge, sadness and loneliness – see Mann, 2015). This reformulation and representation of everyday spaces can also be seen through the reconfiguration of Michael Forbes' shed into an impromptu gallery and space for re-articulating new 'local heroes' in which people come together and where bodies, emotions and communication are integral. Like new forms of museums, such creative rethinkings and counter-representations of heritage require and inspire an openness to a diversity of experiences: '… intelligibility is reliant on the audience's openness towards an emotional and embodied encounter that orientates towards experience rather than making objects and text legible. The usual heritage encounter is counteracted in this event' (Crang and Tolia-Kelly, 2010, p. 2323).

Conclusion

> This is what will be etched into this land forever (Donald Trump in *You've Been Trumped*, 2011)

> How does one map a place that is not quite a place? How does one draw towards the heart? (Kei Miller, 2014)

In his poetry collection *The Cartographer Tries to Map a Way to Zion* (2014), Kei Miller explores a conversation between a colonialist cartographer and a 'rastaman' in Jamaica. Shifting between 'standard' English and Jamaican Patois, the discussion unfolds to chart out divergent worldviews that start to become slightly less distinct as the conversation progresses. The cartographer becomes significantly less expert and unshakeable in the certainty of his representations, and the rastaman is an eloquent spokesperson for sophisticated affective maps. In one telling exchange, the rastaman contemplates the 'official' mapmaker's reliance on the selective aspects of the material and visual:

> iv.
> The rastaman thinks, draw me a map of what you see
> then I will draw a map of what you never see
> and guess me whose map will be bigger than whose?
> Guess me whose map will tell the larger truth?
>
> (Miller, 2014, p. 19)

The final question above is particularly relevant for analysing the ways in which events unfold during *Jamaica for Sale* and *You've Been Trumped*. The films provide an opportunity to unearth the hidden cartographies of affect and emotional ties to place, while connecting and promoting a dialogue with those who control what heritage is promoted, valued and protected. The films illustrate, while also encouraging (as seen through later discussions, classroom exercises and exhibitions), an exploration into feelings, connections and familiarities that may not be observed at first glance, but which are keenly felt. Images of the absoluteness of large-scale tourist developments, depicted in stylised designers' plans and generalised press releases as politically neutral, come into sharp relief as part of wider dominant development discourses. Although the reconstruction of heritage and its associated material and social settings may be described as reasoned and emotionally detached, closer examination, such as those explored in this chapter, teases out the ways in which discourses about, and representations of, heritage and tourism are grounded in specific feelings about people and place. This grounding in emotion is also central to media, places and people who resist exclusionary tourist practices. These practices and representations are part of processes through which emotion can be both a constraining and enabling device; activities and images reflecting, reproducing and challenging affective geographies can impel action and more diverse representations (and, as Anderson (2006) suggests, may offer the possibility of hope).

Despite statements of developments, or heritage sites, being 'etched into the landscape forever', *You've Been Trumped* and *Jamaica for Sale* suggest that material and social landscapes can never be completely fixed, bounded or controlled: like coasts, heritage landscapes are relentlessly dynamic and prone to unexpected disturbances.

Note

1 The representational links between both locales can also be seen in the photography produced by the Dundee-based firm Valentine and Sons during the nineteenth century in an effort to document and promote travel to these 'exotic' locations (Salt and Mains, 2015).

Bibliography

Anderson, B. 2006. Becoming and being hopeful: towards a theory of affect. *Environment and Planning D: Society and Space*, 24: 733–52.

Bailey, M.J. 2010. Interview with Esther Figueroa, filmmaker of the documentary 'Jamaica for Sale'. http://jamaicans.com/documentaryjamaicaforsale/ (accessed 14 June 2014).

Baxter, A. (dir) 2011. *You've been Trumped*. UK: Montrose Pictures and Bell Rock.

Carrell, S. 2014. Donald Trump loses legal challenge to windfarm near his Scottish golf resort, *The Guardian*. www.theguardian.com/world/2014/feb/11/donald-trump-loses-windfarm-scottish-golf-resort (accessed 1 July 2015).

Crang, M. and Tolia-Kelly, D.P. 2010. Nation, race, and affect: sense and sensibilities at national heritage sites. *Environment and Planning A*, 42: 2315–31.

Creel, L. 2003. Ripple effects: population and coastal regions. PRB – Making the Link 1–8. www.prb.org/pdf/RippleEffects_Eng.pdf (accessed 14 June 2015).

Drysdale, N. 2012. Top Scot Michael Forbes demands an apology from Alex Salmond, *STV News*. http://news.stv.tv/north/203990-top-scot-michael-forbes-demands-an-apology-from-alex-salmond/ (accessed 7 July 2015).

Escobar, A. 1995. *Encountering development: the making and unmaking of the Third World*. Princeton, NJ: Princeton University Press.

Figueroa, E. 2010. *Historic Falmouth*. Jamaica: Vagabond Media. www.youtube.com/watch?v=bGYWI7Iy5Pc (accessed 7 July 2015).

Figueroa, E. and McCauley, D. (co-dirs) 2008. *Jamaica for sale*. Jamaica: Vagabond Media and Jamaica Environment Trust.

Forsyth, B. (dir) 1983. *Local hero*. UK: Goldcrest Films.

Fregonese, S. 2012. Between a refuge and a battleground: Beirut's discrepant cosmopolitanisms. *Geographical Review*, 102(3): 316–36.

Gilroy, P. 1995. *The Black Atlantic: modernity and double-consciousness*. Cambridge: Harvard University Press.

Harrison, D. and Hitchcock, S. (eds) 2005. *The politics of world heritage: negotiating tourism and conservation*. New York: Multilingual Matters.

Jamaica Environment Trust. 2015. Jamaica for sale. *Jamaica Environment Trust*. www.jamentrust.org/films/jamaica-for-sale.html (accessed 7 July 2015).

Jamaica Observer. 2011. Palmyra Resort confirms receivership. www.jamaicaobserver.com/business/Palmyra-Resort-confirms-receivership (accessed 7 July 2015).

Johnson, N. 1999. Framing the past: time, space and the politics of heritage tourism in Ireland. *Political Geography*, 18(2): 187–207.
King, T. and Flynn, M.K. 2014. Heritage as urban regeneration in post-apartheid Johannesburg: the case of Constitution Hill, in J. Kaminski, A.M. Benson and D. Arnold (eds) *Contemporary issues in cultural heritage tourism* (pp. 101–16). New York: Routledge.
Law, L., Bunnell, T. and Ong, C.-E. 2007. The beach, the gaze and film tourism. *Tourist Studies*, 7(2): 141–64.
Mains, S.P. 2008. Island village, tourist world: representing cultural geographies of Jamaica. *Caribbean Geography*, 15(1): 1–13.
Mains, S.P. 2014. Fieldwork, heritage and engaging landscape texts. *Journal of Geography in Higher Education*, 38(4): 525–45.
Mains, S.P. 2015. From bolt to brand: Olympic celebrations, tourist destinations and media landscapes, in S.P. Mains, J. Cupples, and C. Lukinbeal (eds) *Mediated geographies and geographies of media* (pp. 329–48). Dordrecht: Springer.
Mains, S.P., Cupples, J. and Lukinbeal, C. (eds) 2015a. *Mediated geographies and geographies of media*. Dordrecht: Springer.
Mains, S.P., Cupples, J. and Lukinbeal, C. 2015b. Introducing mediated geographies and geographies of media, in S.P. Mains, J. Cupples and C. Lukinbeal (eds) *Mediated geographies and geographies of media* (pp. 3–19). Dordrecht: Springer.
Mann, B. 2015. The film that makes me cry: Local Hero. *The Guardian*. www.theguardian.com/film/filmblog/2015/mar/12/the-film-that-makes-mecry-local-hero (accessed 1 July 2015).
Miller, K. 2014. *The cartographer tries to map the way to Zion*. Manchester: Carcanet Press.
Nash, C. 1996. Reclaiming vision: looking at landscape and the body. *Gender, Place and Culture: A Journal Feminist Geography*, 3(2): 149–70.
Pain, R. 2009. Globalized fear? Towards an emotional geopolitics. *Progress in Human Geography*, 33(4): 1–21.
Pan, S. and Ryan, C. 2013. Film-induced heritage site conservation: the case of Echoes of the Rainbow. *Journal of Hospitality and Tourism Research*, 37(1): 125–50.
Poria, Y., Butler, R. and Airey, D. 2003. The core of heritage tourism. *Annals of Tourism Research*, 30(1): 238–54.
Port of Falmouth Retail Centre. 2015. Welcome to historic Falmouth cruise port. *Historic Falmouth Jamaica*. www.hfcport.com/ (accessed 7 July 2015).
Reid-Henry, S. 2012. Arturo Escobar: a post-development thinker to be reckoned with. *The Guardian*. www.theguardian.com/global-development/2012/nov/05/arturo-escobar-post-development-thinker (accessed 30 June 2015).
Salt, K. and Mains, S.P. 2015. Transnationalism along the colour line: race, place and belonging in the Caribbean and Scotland. Presentation to the landscapes and lifescapes: material spaces and stories of connection between the Caribbean and the Scottish Highlands, 1700 to the present Symposium, Inverness.
Sheller, M. 2003. *Consuming the Caribbean: from Arawaks to Zombies*. New York and London: Routledge.
Smith, M., Davidson, J., Cameron, L. and Bondi, L. (eds) 2009. *Emotion, place and culture*. Farnham: Ashgate.

The Herald. 2013. Trump launches legal action against Scottish Government's approval of wind farm. www.heraldscotland.com/news/home-news/trump-launches-legal-action-against-scottish-governments-approval-of-wind-farm.1368719259 (accessed 1 July 2015).

Thompson, K.A. 2007. *An eye for the Tropics: tourism, photography, and framing the Caribbean picturesque*. Durham, NC: Duke University Press.

Timothy, D.J. and Boyd, S.W. 2006. Heritage tourism in the 21st Century: valued traditions and new perspectives. *Journal of Heritage Tourism*, 1(1): 1–16.

Tolia-Kelly, D.P. 2007. Fear in paradise: the affective registers of the English Lake District landscape re-visited. *Senses and Society*, 2(3): 329–51.

Tolia-Kelly, D.P. 2008. Motion/emotion: picturing translocal landscapes in the Nurturing Ecologies Research Project. *Mobilities*, 3(1): 117–40.

Tripping Up Trump. 2015. Tripping Up Trump: a popular movement against using compulsory purchase for private profit. www.trippinguptrump.co.uk/ (accessed 8 July 2015).

Trump, D.J. 2014. Greetings from Donald J. Trump. *Trump International Golf Links*. www.trumpgolfscotland.com/ (accessed 27 January 2015).

Waterton, E. and Watson, S. 2010. Introduction: a visual heritage, in E. Waterton and S. Watson (eds) *Culture, heritage and representation: perspectives on visuality and the past* (pp. 1–18). Farnham: Ashgate.

White, R. 2009. A short history of beach holidays, in S. Hosking, R. Hosking, R. Pannell and N. Bierbaum (eds) *Something rich and strange: sea changes, beaches, and the littoral in the Antipodes* (pp. 1–19). Kent Town: Wakefield Press.

10 Touching time

Photography, affect and the digital archive

László Munteán

Digital technology permeates every walk of our lives and plays a crucial role in the way we preserve, remember and forget the past. The concept of heritage as a fixed entity under institutional protection has given way to the reconceptualisation of heritage as a much broader terrain (Lowenthal, 1985; Smith, 2006). In her seminal work, *Uses of Heritage*, Laurajane Smith declares that heritage is to be perceived 'not so much as a "thing", but as a cultural and social process, which engages with acts of remembering that work to create ways to understand and engage with the present' (2006, p. 2). This new conceptual framework has paved the way to the recognition of the Internet, new media and digitisation as new areas of social, cultural and scientific activity where heritage is being produced. In 2003 UNESCO adopted the *Charter for the Preservation of Digital Heritage*,[1] which safeguards resources that are either 'born digital' or are digital surrogates of material heritage. The Charter refers to this latter category as 'digital documentary heritage', which entails the creation and dissemination of copies of documents that, in material form, were previously accessible to only a few. The proliferation of online archives and databases released by museums, libraries and cultural institutions attests to the democratic potential of this form of digitisation.

Digital archives of analogue photography collections constitute a substantial group of digital documentary heritage. Online photography archives, such as Getty Images and the National Archives' photo collection, have made the bulk of their analogue photographic collections available on the Internet. The democratic potential of digital documentary heritage, however, exceeds the function of disseminating analogue material in digital form. The ubiquity of the Internet in our lives has also given rise to archives and databases that operate at the grass-roots level, independently of institutions whose collections are automatically recognised as heritage. In this chapter I will use as a case study a Hungarian online archive of private photography, *Fortepan* (www.fortepan.hu),[2] to demonstrate how the digitisation of analogue images can lead to the production of new photographic heritage. More specifically, I am interested in how this archive operates as a platform for affective engagements with the new photographic heritage that it catalyses.

I will use affect in the sense of atmosphere, a quality which, in architectural terms, entails 'an exchange between the material or existent properties of the place and our immaterial realm of projection and imagination' (Pallasmaa, 2014, p. 20).

Atmosphere is what touches us viscerally, a quality that we perceive as outside of us, and yet we register it through a shared emotional disposition. If atmosphere 'emanates' from buildings, landscapes, objects and even groups of people, I will argue that the architecture of *Fortepan* is an atmospheric environment that, beyond nostalgia, caters to the fetishisation of the object world depicted in the images.

Atmosphere, heritage and meaning

Drawing on the foundational work of Deleuze and Guattari (1987, 1994) and, more recently, Brian Massumi (2002), the concept of affect has been adopted and reworked in non-representational theory (Anderson, 2006; Pile, 2010; Jones, 2011; Thrift, 2008), and has gained wide currency in the field of geography. In this discourse affect is distinguished from feeling and emotion as an intensity that is transpersonal, inexpressible and non-cognitive, as opposed to the pre-cognitive state ascribed to feeling and the cognitive state of emotion that ultimately lends itself to be expressed in language. Taken up recently by architectural theorists and introduced to the field of geography through the work of Ben Anderson (2009), atmosphere is a concept that activates these categories, while disavowing a rigid distinction between them.

In architectural discourse atmosphere is approximated as an affective entity that allows us to emotionally relate to space. In Juhani Pallasmaa's words, it is 'the overarching perceptual, sensory, and emotive impression of a space, setting, or social situation ... Paradoxically, we grasp the atmosphere of a place before we identify its details or understand it intellectually' (2014, pp. 20–1). Atmosphere, in this sense, resonates with the non-cognitive quality of affect, which, as Peter Zumthor contends, we apprehend through our 'emotional sensibility'; he defines this as 'a form of perception that works incredibly quickly, and which we humans evidently need to help us survive' (2006, p. 13). Paradoxically, we experience atmospheres as 'something quasi-objective', as Gernot Böhme puts it; 'Yet they cannot be defined independently from the persons emotionally affected by them; they are subjective facts' (2014, p. 43).

Following Böhme's and Mikel Dufrenne's earlier conceptualisations, Ben Anderson concludes that atmospheres:

> are impersonal in that they belong to collective situations and yet can be felt as intensely personal. On this account atmospheres are spatially discharged affective qualities that are autonomous from the bodies that they emerge from, enable and perish with. As such, to attend to affective atmospheres is to learn to be affected by the ambiguities of affect/emotion, by that which is determinate and indeterminate, present and absent, singular and vague.
>
> (2009, p. 80)

These ambiguities that are at work in atmospheres play crucial roles in the act of investing buildings, locations, landscapes and objects with meaning. Atmosphere, Anderson writes:

discloses the space-time of an 'expressed world'—it does not re-present objective space-time or lived space-time. It creates a space of intensity that overflows a represented world organized into subjects and objects or subjects and other subjects. Instead, it is through an atmosphere that a represented object will be apprehended and will take on a certain meaning.

(2009, p. 79)

Expressed worlds therefore cajole us into a realm that feels as though it is emanating from the environment around us and is at once registered through our emotional sensibility, as Zumthor observes. They open up an alternative spatio-temporal realm that cannot be articulated in semiotic terms, yet they are key to the meaning with which that particular environment or object will be invested by the beholder.

In their work on the role of affect in heritage sites, Kate Gregory and Andrea Witcomb evoke Walter Benjamin's notion of the dialectical image as an instance of shock laden with possibilities of insight as a quality akin to Anderson's use of the term 'expressed world' that unfolds within atmospheres. Looking through a Benjaminian lens, while exploring different manifestations of nostalgia in two historic buildings, Gregory and Witcomb view heritage sites as 'possible worlds which work through affective, corporeal and imaginative engagement to develop historical understanding' (2007, pp. 264–5). In what follows I will first contextualise *Fortepan* within new discourses of heritage and the archive and then explore its architecture as an atmospheric landscape of private photography that engenders an expressed world for affective engagements with the past.

Towards a new photographic heritage

The 2003 UNESCO Charter was followed by the 2005 convention on cultural heritage (Council of Europe), which grants agency to non-institutional actors, such as *Fortepan*, by acknowledging their increasingly active role in shaping what constitutes heritage. This sea change in the framing of heritage not only helps us theorise the cultural practices at work in people's relation to *Fortepan*; it also offers a platform to revisit digitisation and the role of new media technologies in heritage-making. Until Emma Waterton's (2010) commentary on the role of online communities in heritage-making and Elisa Giaccardi's (2012) edited volume *Heritage and Social Media: Understanding Heritage in a Participatory Culture*, this topic had received little academic attention. Building on Henry Jenkins *et al.*'s (2009, 2013) analysis of the structure and operation of social networks, which he calls 'cultures of participation', Giaccardi argues that 'Social media create infrastructures of communication and interaction that act as places of cultural production and lasting values at the service of what could be viewed as a new generation of "living heritage practices"' (2012, p. 5). These practices are multi-layered and polyvalent, entailing the negotiation of a multitude of voices and interests all of which weaken curatorial power (Silberman and Purser, 2012, pp. 13–14).

Instead of reinforcing the idea of an institution behind the online interface, *Fortepan* is an archive that owes its existence to the Internet and is rooted in cultures of participation where preservation is not antithetical but complementary to the idea of sharing. Launched in 2010 by Miklós Tamási and Ákos Szepessy, *Fortepan* features amateur photographs from the period 1900 through 1990. Initially a collection of digitised photos gleaned from house clearances and flea markets, the site went viral almost overnight and has since inflated to more than 40,000 images, thanks to private and institutional contributions. Family celebrations, reunions, holiday trips, home interiors, streetscapes, cityscapes and landscapes across the twentieth century make the exploration of the site an exhilarating, if not addictive, experience. Tamási and Szepessy regard *Fortepan* as a project 'aimed at the creation of a new archiving paradigm, based on openness, dialogue and participation, which stands in contrast with the traditional values associated with archives, such as discretion, permanence and immutability' (Fotofestiwal, 2013). Besides harmonising with Smith's formulation of heritage as a dynamic social and cultural process, the conviction to embrace instability and nonlinearity as consequences of openness reflects Arjun Appadurai's notion of archives as sites of 'the collective will to remember' (2003, p. 17).

Partaking in this collective will is not without sacrifices. By lending their albums to the curators donors officially waive their right of private ownership of their donation, thus expressing their consent to the licence of Creative Commons CC-BY-SA-3.0 under which *Fortepan* operates.[3] Though a number of public collections refrain from granting public access to their collections through *Fortepan* (Tamási, 2011), a growing number of Hungarian village, city and district archives, and even foreign institutions such as the Library of Congress and the National Archives, have contributed to the collection. There is, however, no hierarchy of donors on the website. The names of individuals and institutions that contribute to the site are listed in alphabetical order, in the same font size. This democratic arrangement illustrates what David Atkinson, in relation to re-conceptualisations of heritage, describes as an instance of 'democratizing memory' (2008, p. 382).

This is not to say, however, that the curators of *Fortepan* have no power over the content of the archive. Although curators in the new archiving paradigm operate increasingly as 'facilitators rather than authoritative scripters and arbiters of authenticity and significance' (Silberman and Purser, 2012, pp. 13–14), Tamási and Szepessy exercise full control over the selection of the material that gets digitised. For all the plurality of amateur photographs that the curators endorse, the analogue material is filtered through several criteria set by the curators themselves. The chronological scope of the archive, for instance, is one such filter. The sharpness of incoming analogue images, as Tamási mentions in an interview, is another. They particularly like photos of the first part of the twentieth century, which can be digitised in much better quality than snaps from the 1980s (introblog.hu, 2014). Therefore, it is within this curatorial framework that any democracy of photographs is possible (introblog.hu, 2014; Fotofestiwal, 2013). Consequently, once donors agree to the terms of Creative Commons they also accept that some of their photos may not be included in the archive. Also, donors agree that their pictures will appear without captions. If *Fortepan* attests to a collective will to remember, it does so as a bottom-up initiative that is nonetheless not without curatorial control. The site's reputation as an invaluable repository of the past grants *Fortepan* the legitimising force to turn private photography into new photographic heritage.

Touching photographs

The transformation of photographs from analogue/private into digital/collective raises the question as to what kind of affective engagements family photography elicits once taken out of the context of the family? If the exposure of private pasts serves a collective will to remember, what is it, then, that is worth remembering in these images? Ever since its inception, photography has been regarded as a gateway to the past. In his iconic work on photography entitled *Camera Lucida*, Roland Barthes writes passionately about the relationship between the photographic image and its referent. The photograph, he contends,

> is literally an emanation of the referent. From a real body, which was there, proceed radiations which ultimately touch me, who am here; the duration of the transmission is insignificant; the photograph of the missing being, as Sontag says, will touch me like the delayed rays of a star.
>
> (1993, p. 80)

Barthes' fascination with the photograph as an ontological trace of the past evokes what Charles Sanders Peirce in his theory of signs describes as an 'indexical' relationship between sign and its referent (1955, p. 106).

Barthes' celebration of the indexical power of photography pertains to the analogue images of his time. With the hindsight of more than three decades that have passed since the publication of *Camera Lucida*, Barthes' fascination with photographic indexicality may appear naïve if we consider that the photograph (both analogue and digital) is like any other means of representation – a construct, rather than an imprint, involving the photographer's choice of the frame and the physical and chemical processes that are at work in its creation. Still, for all our awareness of its mediated nature, we succumb to photography's lure as a trace of reality. As Elizabeth Edwards puts it,

> [t]he transparency of the photograph to its referent has been one of its most cherished features. Culturally, despite rational realizations that photography can 'lie', the photograph has been viewed, especially in its vernacular forms, as a window on the past.
>
> (2009, p. 332)

In other words, if there is anything like 'photographic truth', it lies in our emotional disposition towards the photographic image, rather than in the image itself (Kember, 1998, p. 31).

Understanding indexicality as an affective, rather than an ontological, property of the photograph allows us to re-read Barthes' use of the word 'touch' as a signifier of affective engagement with the photographic image. The ability of photographs to touch the viewer as traces of the past turns them into what Edwards calls 'surrogate memory' (2009, p. 332), denoting our inclination to impose our memories on them. Besides their potential to activate sight, Edwards emphasises their material texture, form and smell as integral qualities of analogue photographs that significantly influence our emotional responses to them as objects of memory and as relics to be preserved in albums or shoeboxes. These are the repositories from which the bulk of the images in *Fortepan* come.

Family photos constitute a particularly large group in the archive. The most salient feature of family photographs, as Gillian Rose (2010) argues, is that they play key roles in practices and rituals aimed at reinforcing togetherness, harmony and a sense of cohesion within the family. Sharing and displaying these photographs follows culturally embedded choreographies wherein they serve as cues for the telling of stories of the past and keeping the memory of distant or deceased family members alive. These practices of memory are informed by what Marianne Hirsch calls a 'familial gaze', a set of deeply engrained norms and codes of conduct that reinforces dominant ideologies of the family.

> Its content and even its mode of operating may be variable, but what doesn't change is that this ideal image exists and can be identified, and that it has determining influence. Within a given cultural context, the camera and the family album function as instruments of this familial gaze.
>
> (2012, p. 11)

Edwards' emphasis on the materiality of these images is helpful in seeing them as objects with their own biographies (Kopytoff, 1986) and exposing what happens to them when they are no longer held within the private realm of the family. Once their signifying function as reference points for sharing and reliving memories fades, they 'return to the ordinary, indeed disposable object, the detritus of material culture, as they cease to have meaning for the living beyond a generalized "pastness"' (Edwards, 2009, p. 334).

The house clearances and flea markets that Tamási and Szepessy continue to frequent signify the stage at which analogue images enter the realm of ordinary objects in Edwards's sense. Devoid of the rituals and narratives that anchored them as mnemonic objects onto the private realm of the family, they are now ciphers of a past claimed by a collective will to remember. The quality of indexicality that we ascribe to photographs does not diminish with their extraction from family relations. On the contrary, unmoored from family space–time, they become empty husks of the memories and stories that they once triggered. In this relation, what Edwards calls pastness is in fact an affective engagement with the past that is simultaneously intimate and impersonal. The will to remember, in Appadurai's sense, is exercised as a sensation of pastness without remembering. The memories and stories that they were meant to recall within the family linger on as a haunting absence, an auratic residue of unknown pasts that nonetheless determine what we sense as an atmosphere emanating from the pictures. The sensation of this residue is akin to what Anderson describes as an 'expressed world' that 'creates a space of intensity' (Anderson, 2009, p. 79). It is through this space of intensity, fuelled by our engagement with the photographs as imprints of inaccessible personal pasts, that pastness becomes an atmospheric quality where the object world of the images takes centre stage.

Bewitched by objects

In the previous section we have seen how the atmospheric quality of pastness is felt amid the audible silence of personal memories and stories when beholding images on *Fortepan*. What is this atmospheric quality's relation to the heritage value that the archive legitimises? Ultimately, what does new photographic heritage consist in? Barthes' terminology, which he employs to read images in *Camera Lucida*, serves as a point of departure in addressing these questions. As with the visitor's relation to the photographs on *Fortepan*, Barthes is an outsider to the personal histories of the images included in *Camera Lucida*. He studies their compositional features and he is attracted to their historical backgrounds with which he is familiar. He introduces the term *studium* to indicate this general aspect of engagement with the photograph. The studium, he writes, is 'an average affect', which entails an 'application to a thing, taste for someone, a kind of general, enthusiastic commitment … but without special acuity' (Barthes, 1993, p. 26). This general interest is ruptured by another aspect of experiencing photographs, which he calls *punctum*. The punctum is an 'accident which pricks me (but also bruises me, is poignant to me)' (Barthes, 1993, pp. 26–7). This accident is a kind of 'intensity' which can be triggered by a certain detail, but, more importantly, it occurs in the realisation of the irreversible passage of time arrested by the camera (Barthes, 1993 p. 96). The studium and the punctum can be conceived of as two affective modes; the former recalling and reinforcing pre-existing schemes, patterns and stereotypes, which the latter, in turn, undermines.

To demonstrate these two aspects of engaging with photographs, Barthes takes Alexander Gardner's 1865 photo-portrait of Lewis Payne, a young man sitting in shackles leaning against the prison wall, sentenced to death for attempting to assassinate the American Secretary of State, W.H. Seward:

> The photograph is handsome, as is the boy: that is the *studium*. But the *punctum* is: *he is going to die*. I read at the same time: *This will be* and *this has been*; I observe with horror an anterior future of which death is the stake. By giving me the absolute past of the pose (aorist), the photograph tells me death in the future. What *pricks* me is the discovery of this equivalence.
>
> (p. 96)

Here, the punctum arises from what Barthes registers as photography's indexical power. The impending death of Payne which he contemplates uncannily matches a photograph that he analyses in his book but does not show: the photograph of his recently deceased mother captured as a young child. The absence of the photo indicates, first, that for Barthes it also reveals an anterior future that bespeaks the irreversible death of his mother. Second, the average affect of the studium, which attracts him to Gardner's photo in the first place, is presumably missing in the case of the photo of his mother. There is no studium to absorb the punctum; the image

208 László Munteán

speaks to the past, not pastness. As an atmospheric mode, pastness, then, can only arise if the realisation of '*this has been*' is filtered through an average affect of interest, enthusiasm, recognition, liking or disliking. The absence from *Camera Lucida* of the photograph of Barthes' mother alludes to those images that people withhold from sharing on *Fortepan*, even though the site is devoid of stories that would contextualise the images. The absent presence of these stories and familial relations, as we have seen in the previous section, pulls viewers into an intimate world to which they remain outsiders. More than an average affect, this simultaneous sensation of interiority and exteriority is key to the experience of pastness as an atmospheric quality.

The affordances of *Fortepan* add new aspects to the pastness. The curators' choice of the year 1990 as an endpoint to the project indicates the temporal proximity of the political changes of 1989 as a historical landmark on the one hand and the dusk of the pre-digital era in photography on the other. More importantly, however, this time interval situates the project at a distance from the vantage point of the present. The ability to move back in time is enhanced by the timeline as one of the most prominent affordances on the site (see: www.fortepan.hu). Resembling an architectural ruler, it invites the visitor to enjoy time travel as one of the site's affordances. The distance that enables Barthes to construct readings of the images included in his book is thus reflected in the gap *Fortepan* establishes between the past and the present.

In order to examine the affective potential of the archive's affordances, allow me to choose a photograph from the archive at random (Figure 10.1).

I click on the year 1906 and I look at a photograph of a little girl standing at the gate of a building, which occupies the whole field of the image. The information below the photo indicates that it is number 88 Soroksári út, which appears on my cognitive map as one of Budapest's old industrial areas, now scattered with new office buildings and department stores. Apart from time, location and the name of the district archive that donated the photo, no other information is available on the site. In stark contrast to the girl, the gable stone features a bull's head, possibly denoting the building's relation to the nearby abattoir. As I study the façade exposed by the image, I take pleasure in reading the advertisement of a shoemaker. The gate envelops the girl, whose contours are blurred by slow shutter speed, which lends her a fairy-like appearance. Next to her I discern what seems to be a dog coming out of the building, but almost completely blurred. I am touched by the camera's inadvertent registration of movement and, as I look into the row of interconnected courtyards, I am carried away by the realisation that this *has been* and that everything I see in the photograph is gone by now. Instead of leaving me with pain or a sense of personal loss, this realisation fills me with the joy of discovery of this mundane moment of everyday life more than a hundred years ago. The temporal distance and absence of familial ties to the girl and the event allow a plethora of visual representations and stories of *fin-de-siècle* Budapest to pervade my imagination in the form of 'imagined memories' (Huyssen, 2003, p. 17).

Because of the indexical power that I confer upon the image so far removed from the present, I pay special attention to the advertisements, the details of plaster

Figure 10.1 Fortepan/Ferencvárosi Helytörténeti Gyűjtemény.

decorations and the courtyards in the background as though they were archaeological finds. The girl's posture and dress are part and parcel of my experience of material culture unfolding in the picture. Her story unbeknownst to me, I see her as a metonym of the place and time she inhabits. To me she belongs to the world of advertisements and plaster decorations that unfold as I delve into the photo.

This heightened sense of materiality is enhanced by an affordance of *Fortepan* which allows registered visitors to inscribe photographs with keywords, thus introducing new reference points to navigating the collection. If these keywords indicate what visitors have their 'eye out for', they indeed attest to the privileged role of objects as user-generated coordinates. Hats, coats, gates, ruins, swords, airplanes, uniforms, brands of cars and household appliances, to name only a few, have emerged as principal landmarks for navigation. In Luigina Ciolfi's terms, these tags are 'social traces' which entail 'ideas, opinions, physical trajectories and collaborative practices that embody the presence, activity and agency of multiple participants, and that can be represented in perceivable traces (e.g. visitor comments), in curatorial choices, and in the information on display' (2012, p. 73). While visitors who tag the images do not collaborate with each other, *Fortepan* has an official blog[4] where members share and discuss their research on the stories behind the pictures. Bloggers often engage in fierce debates as they try to identify the locations depicted in photographs where no such information is available.

These social traces are important indicators as to how visitors look at photographs in the archive. These traces illustrate their enchantment not so much with the materiality of the photographs that the archive renders in high-quality digital form, but with the material world which the photos indexically capture. Even though the materiality of that world is doubly removed (first analogically, then digitally) from the visitor to *Fortepan*, instead of generating detachment, the dematerialised image heightens the sense of place and materiality and instils a 'co-constitution of the visual and the material' which Gillian Rose and Divya P. Tolia-Kelly call 'ecologies of the visual' (2012, p. 4). The reason behind this sensation has to do with *Fortepan*'s method of displaying photographs. Not only do the photos appear in a uniform format regardless of their original size and texture, but their frames are also cropped, which invites viewers to look *through*, rather than *at*, them. To return to the photograph of the girl standing at the entrance of a house in Budapest, my inclination to look *into* the row of interconnected courtyards unfolding behind the gate attests to this spatio-material dimension. What *Fortepan* affords the viewer, then, is to experience the imagined materiality of the world beyond the photographs rather than merely the material texture of the photographs themselves. This experience is further enhanced by the fact that the images are positioned centrally and rendered large enough to dominate the computer screen, thus functioning more as windows than as surfaces.

Far removed from the present, the world of objects that unfolds on *Fortepan* is engaged at an affective register that both entails and exceeds nostalgia. In Dylan Trigg's words, *'nostalgia's grip on materiality becomes complicit with the imagined alteration of the past.* What this means is that memory, imagination, and place conspire together, serving as homogeneous platforms for the nostalgic body to impose and identify itself' (2012, pp. 199–200, emphasis in original). Although the joy I felt over perusing the details of the century-old photo of the gate may indeed be informed by my inscription of imagined memories of idyll onto a place that I see today as anything but idyllic, my fascination with objects, which, as the tags attest, is not simply my 'condition', also differs from nostalgia.

I am *bewitched*, as the Latin origin of the term *fascination* suggests, by the object world, an affective state devoid of the element of melancholy and longing central to nostalgia. In Barthesian terms, this fascination consists of the average affect of studium, which is not so much unsettled as it is enhanced by the punctum of time experienced as 'this as been'. Pastness, which we identified as an atmospheric scaffold governing our engagement with *Fortepan*, is thus anchored on an archaeological fascination with the material world that feels within reach on the computer screen.

Although I started this section by looking at a single photograph, the archive's online affordances invite the visitor to undertake a journey, rather than look at singular images. *Fortepan*'s interface is far from being as elaborate as the 3D walkthroughs of prominent museums (Cameron and Kenderdine, 2007; Kalay *et al.*, 2008), but central to the experience of navigating this minimalistic virtual landscape is that the visitor cannot map the extension of the archive, because only one image is made visible at a time. Although a recent upgrade shows other photographs of the same sequence in a small format under the timeline, the whole sequence is never revealed. But instead of generating a sense of loss, it is as though new streets of the city unfold one after the other, beckoning the walker to explore further and try out new paths (via keywords). Therefore, the keywords not only put the photographs into virtual drawers to facilitate searching but they also turn images into sequences that unfold one after the other as visual (and material) narratives. Walking, in this sense, constitutes an affective engagement not so much with the photographs *per se* as with the sequences in which they appear.

Despite the curators' initial aversion to displaying images of death (introblog. hu, 2014), photos of dead soldiers from the world wars, as well as evidence of lynching during the 1956 Revolution against Communist dictatorship, show a change. By the same token, photos that show the Krakow ghetto and Hungarian Jewish labour servicemen in Transylvania depict them in friendly conversation with their military guards. The tone of these photographs may even be uplifting, were it not for the viewer's knowledge of the Barthesian anterior future that brought horror to these servicemen soon afterwards. There are relatively few such images on the site, which is partly due to the fact that amateur photographers did not have the skill, the capacity and access to document what some of their professional peers could, either by skill or by accident. Amid the unfolding sequences such images turn fascination with objects into a confrontation with the human body reduced to the world of inanimate objects. Exposure to the violence of death turns pastness into past, fascination into punctum and objects without personal histories into a life brought to a violent end.

With this we have come full circle back to Appadurai's notion of the archive as a site for a collective will to remember and the question as to what is it about *Fortepan* that is embraced as heritage. According to Appadurai, the archive is no longer restricted as a site of collection and preservation, but is also used for the collective negotiation of what to remember and how to remember. In the case of *Fortepan*, the latter manifests itself in the archive not simply as a collection but as 'the product of the anticipation of collective memory' (Appadurai, 2003, p. 16).

The former constitutes the fluid body of amateur photographs filtered through curatorial choices. However, as the affective engagements afforded by the archive reveal, this new photographic heritage is valued less for the skill and artistic quality of the images and more for the richness of the material world that they depict.

In this section we have seen how the online affordances of *Fortepan* reinforce the indexical power of images and how this sensation of immediacy engenders the fetishisation of material objects imbued with the atmosphere of pastness. I will now turn to particular uses of *Fortepan* images once they are downloaded from the site.

Re-implacing photographs

The free use and circulation of images downloaded from *Fortepan* attest to Appadurai's notion of the archive as a site that is 'itself an aspiration rather than a recollection' (2003, p. 16). The number of blogs and online magazines that use *Fortepan* to illustrate articles is increasing. Over the twenty-four hours prior to this writing, the online news portal *index.hu* used an image from the archive to illustrate an article about the Treaty of Vienna in 1938 (Kolozsi, 2014). Four hours previously the online women's magazine *Nők Lapja Café* used *Fortepan* to demonstrate street fashion in the 1930s (15 magyar utcaidivat, 2014). In 2011, a blogspot dedicated to the discussion of recent books teamed up with *Fortepan* to organise a literary competition in which candidates wrote short stories, essays and poems inspired by *Fortepan* images (Fényképek bűvöletében 2014). Another site linked to *Fortepan* is dedicated to transforming stereophotos from the early 1900s into gifs so that their three-dimensional effect can be enjoyed without red and blue glasses.[5]

In order to further problematise the prominent role of materiality in engaging with *Fortepan* online, I will now focus on a project based on identifying the locations where images from the archive have been taken. Launched in 2011 by photographer Zoltán Kerényi, the project is called *Window to the Past*.[6] Initially a method of documentation employed in geographical surveys, rephotography has been used to demonstrate changes in a landscape by repeatedly photographing it from the same perspective over a period of time.[7] The popularised version of this practice features a large variety of techniques of superimposing images.[8] While there is no official name for this trend, my use of 'rephotography' as opposed to 'superimposed photography' to describe Kerényi's project acknowledges not simply the element of overlaying images but also the act of identifying the site of the original image which would in turn result in a new photograph.

The first phase of rephotography entails searching for the location shown in the old photograph and, once discovered, the effort of assuming the perspective that matches that of the image. Things that are recognisable in the photograph and surroundings play a crucial role as reference points – landscapes, streets, buildings and certain household fixtures and furnishings tend to remain stationary and may have outlasted human lives. After the image is mapped on its original setting, space is experienced temporally as though in a virtual archaeological excavation.

The materiality of space thus unfolds as the empty husk of the event that has transpired there. Consequently, the Barthesian reading of the photograph as proof of 'this has been', which, as we have seen, is an affective quality reinforced by *Fortepan*'s affordances, gains a new dimension in the first phase of Kerényi's project. Once he tracks down the material site of the old photograph, the Barthesian statement reads: this has been *here*. 'Hereness' is the experience of the indexical relation between material space and its photographic representation. Through the photograph's recycling into space, objects in the photo matching up with those in the physical environment are revealed as witnesses of the passage of time. The attention paid to objects here can be conceived as remediation of the fetishisation of materiality evidenced by the object-related keywords in *Fortepan*. For Kerényi, however, what is at stake is literally to re-implace[9] the old photograph. No matter how mundane these details may be, rephotography renders them time capsules where the past lingers on. To signify their potential to connect two temporal planes, I call these auratic objects time-bridges that ensure the sensation of the materiality of the past in the present.

While the first phase of rephotography involves spatial engagement with the old photograph, in the second phase hereness is transferred to the new photograph. Once captured by the new photograph, time-bridges transform the experience of 'here' into 'there'. Unlike in scientific surveys where the old and the new photographs are placed next to each other, in popular versions of the trend the old image blots out the terrain behind it and the time-bridges can only operate along the border of the old and the new picture. It is along this border that spatial continuities and discontinuities between the two temporalities unfold. This liminal zone between old and new, photographic and material – as I will demonstrate through one of Kerényi's collages – is crucial to the experience of the relationship established between photography and place.

We have seen how the digital removal of their frames deprives the photos in *Fortepan* of their materiality and at once enhances their potential to foreground the materiality of their referent. When working on the final collage and using digital technology to superimpose the old photo on the new one, Kerényi further trims the old image so that it matches seamlessly with the dimensions of the new one. In the final result, however, Kerényi simulates analogue frames by adding white lines to the old images, thus creating the effect of conventional photographs. Doubly removed from their analogue versions, the old photographs are finally granted material appearance through digital manipulation. As a result, they look as though they are literally superimposed on the new ones, evoking a glimpse of the past that took place there. As much as they separate the old from the new, the white frames allow our eyes to fill in the missing details and perceive continuities rather than discontinuities. Meticulous as he is, imperfections in matching the image with the background are precisely what Kerényi avoids. In fact, one of the remarkable characteristics of his montages is that they are for the most part seamless. As we have already seen, the absence of background information unmoors the characters from intimate family ties, memories and stories, which allows the viewer to perceive them as integral parts of the world of objects that unfold in the

214 *László Munteán*

photographs. Through rephotography they are allowed to re-enter the spaces they formerly inhabited and invest those places with a ghostly presence. Nonetheless, the more meticulous the superimpositions, the more we see figures rather than persons. The hierarchical division between the animate and inanimate dissolves into a democracy of objects and figures.

To see how Kerényi's project engages with and readjusts the atmospheric register of *Fortepan* images, I turn to one of his recent pictures (Figure 10.2).

Here, the old image Kerényi chose to rephotograph was taken in 1957 at a lookout point on Gellért hill, which offers a panoramic view of Budapest. Once this photo is superimposed on the new image of the same location, the iron railings, their stone footing, the pattern of the stone pavement and the horizon in the background serve as time-bridges connecting the temporal layers of the photographic palimpsest. The transition is so smooth that, were it not for the simulated white frame around the old photo, the black and white of the past would flawlessly morph into the colours of the present. Devoid of individual history, the figures come forth as historical characters: part and parcel of the world of objects and the panoramic scenery around them. As such, they serve aesthetic purposes insofar as they appear as a symmetrical counterpoint to the present-day tourist posing for a photograph. Moreover, the theme of the photographer being photographed, which underpins Kerényi's composition, metaphorically stands for his work as a rephotographer and adds a witty twist, which Kerényi applies in other works of his as well.

Figure 10.2 Zoltán Kerényi, Window to the Past, ablakamultra.hu.

Source: Zoltán Kerényi.

In the previous section we saw how pastness as an atmospheric quality caters to fascination with the world of objects as a subjective response whereby atmosphere is apprehended. The meticulous care with which Kerényi identifies and digitally manipulates time-bridges attests to this object-fetish. Instead of investing space with the sentimentality of social ties, he utilises the intimacy of private photographs as a choreographic device. The events that the original photos depict are relevant to him as long as their background is recognisable and traceable. Reinforcing the cult of objects facilitated by the affordances of *Fortepan*, instead of personing spaces, Kerényi spatialises persons by using them as props in order to assert the historicity of an urban setting. Apart from praising his skill of superimposition, the countless comments that Kerényi's project receives on the Internet bespeak a fascination with locations and objects that have weathered the stormy decades of the twentieth century. If *Fortepan*'s affordances invite viewers to view photographs as windows to a bygone world, Kerényi's *Window to the Past* materialises this potential. The tags on photographs in the archive serve as potential time-bridges in rephotography. As such, his project not only demonstrates a possible use of new photographic heritage; it also attests to the significance of details in the built environment embraced as heritage through affective engagements with the photographs.

Conclusion

The reconceptualisation of heritage as a cultural process increasingly defined by Internet-based participatory communities has been instrumental in giving due recognition to such bottom-up initiatives as *Fortepan*. By way of tracing the multi-layered transformations through which analogue photographs pass once admitted to the archive, this chapter has framed *Fortepan* as a catalyst of new photographic heritage that engenders affective engagements that, in turn, play a crucial role in defining what this new heritage consists of. In employing the notion of atmosphere as an affective quality apprehended emotionally, as though emanating from the environment, I observed the digital environment of *Fortepan* as an atmospheric space with performative potential. For what the archive offers is not simply digital surrogates of analogue originals but a fluid collection of private photographs that, presented without narrative context, exude the atmosphere of pastness. In the absence of narrative scaffolding, pastness envelops visitors in the form of fascination with the material referents of images, evidenced by the collective urge to identify objects and localities as traces of bygone times. To reiterate Anderson's formulation, it is through this fascination that atmosphere is apprehended and invested with meaning (2009, p. 79).

If the archive, in Appadurai's view, is 'an instrument for the refinement of desire' (2003, p. 24), meaning is more an affective than a semiotic quality. Approaching *Fortepan* as a locus of affective engagements has been helpful to reveal that the new photographic heritage that is being produced through the archive as a legitimising force goes beyond the appreciation of private photography *per se*. The public fascination with objects and places underlines the material referents of the images as potential uses of heritage, as exemplified by Kerényi's rephotography project. Besides re-implacing images in their environment, such a project reveals *Fortepan*'s potential as intermediary heritage, namely a repository of photographic reference to be applied to renovation and reconstruction projects as well as a platform for affective engagements with natural and urban landscapes as prospective heritage sites. The increasing number and variety of online communities and research projects that reference *Fortepan* as a source attest to this potential.

Notes

1 www.unesco.org/new/en/communication-and-information/access-to-knowledge/preservation-of-documentary-heritage/digital-heritage/concept-of-digital-heritage/
2 Fortepan got its name from the Hungarian photography equipment supplier called Forte, which produced the most widely used negative film in the years after the Second World War (www.fortepan.hu/?view=fortepan).
3 https://creativecommons.org/licenses/by-sa/3.0/
4 http://fortepan.blog.hu/
5 http://fortepan3d.tumblr.com/
6 www.ablakamultra.hu
7 Rephotography is mainly associated with the work of Mark Klett and Camilo José Vergara. *Thirdview* is a contemporary rephotography project dedicated to the mapping of the American West: www.thirdview.org/3v/home/index.html
8 Launched in 2011, Taylor Jones's site *My Dear Photograph* was among the first to popularize the technique: http://dearphotograph.com
9 I am borrowing this term from Edward Casey, who uses it to describe an aspect of re-experiencing places as part of the sentiment of nostalgia (2000, p. 201).

Bibliography

Anderson, B. 2006. Becoming and being hopeful: towards a theory of affect. *Environment and Planning D: Society and Space*, 24(5): 733–52.
Anderson, B. 2009. Affective atmospheres. *Emotion, Space and Society*, 2: 77–81.
Appadurai, A. 2003. Archive and aspiration, in W. Maas, A. Appadurai, J. Brouwer and S.C. Morris (eds) *Information is alive: art and theory on archiving and retrieving data* (pp. 14–25). Rotterdam: NAi Publishers.
Atkinson, D. 2008. The heritage of mundane places, in B. Graham and P Howard (eds) *The Ashgate research companion to heritage and identity* (pp. 381–95). Farnham: Ashgate.
Barthes, R. 1993. *Camera lucida*. London: Vintage.
Böhme, G. 2014. Urban atmospheres: charting new directions for architecture and urban planning, in Christian Borch (ed) *Architectural atmospheres: on the experience and politics of architecture* (pp. 42–59). Birkhäuser: Basel.
Cameron, F. and Kenderdine, S. (eds) 2007. *Theorizing digital cultural heritage: a critical discourse*. Cambridge, MA: MIT Press.
Casey, E. 2000. *Remembering: a phenomenological study*. Bloomington, IN: Indiana University Press.

Ciolfi, L. 2012. Social traces: participation and the creation of shared heritage, in E. Giaccardi (ed) *Heritage and social media: understanding heritage in a participatory culture* (pp. 69–85). London: Routledge.
Council of Europe. 2005. *The framework convention on the value of cultural heritage for society*. www.coe.int/t/dg4/cultureheritage/heritage/Identities/default_en.asp (accessed 18 December 2014).
Deleuze, G. and Guattari, F. 1987. *A thousand plateaus*. London: Continuum.
Deleuze, G. and Guattari, F. 1994. *What is philosophy?* London: Verso.
Edwards, E. 2009. Photographs as objects of memory, in F. Candin and R. Guins (eds) *The object reader* (pp. 331–42). London: Routledge.
Fényképek bűvöletében – az örökség. *Miamona Könyveldéje*. 2011. http://miamonakonyveldeje.blogspot.nl/2011/12/fenykepek-buvoleteben.html (accessed 19 December 2014).
Fotofestiwal. 2013. *Fortepan*. www.fotofestiwal.com/2013/en/events/Fortepan/ (accessed 18 December 2014).
Giaccardi, E. (ed) 2012. *Heritage and social media: understanding heritage in a participatory culture*. London: Routledge.
Gregory, K. and Witcomb, A. 2007. Beyond nostalgia: the role of affect in generating historical understanding at heritage sites, in S.J. Knell, S. MacLeod and S. Watson (eds) *Museum revolutions: how museums change and are changed* (pp. 263–75). London: Routledge.
Hirsch, M. 2012. *Family frames: photography, narrative, and postmemory*. Cambridge, MA: Harvard University Press.
Huyssen, A. 2003. *Present pasts: urban palimpsests and the politics of memory*. Stanford, CA: Stanford University Press.
introblog.hu. 2014. A *Fortepan* egy bőrönd, amiből mindenki kivehet (interview with Miklós Tamási). www.youtube.com/watch?v=41zZdqWlkH4 (accessed 18 December 2014).
Jenkins, H., Ford, S. and Green, J. 2013. *Spreadable media: creating value and meaning in a networked culture*. New York: NYU Press.
Jenkins, H., Purushotma, R., Weigel, M., Clinton, K. and Robison, A.J. 2009. *Confronting the challenges of participatory culture: media education for the 21st century*. Cambridge, MA: MIT Press.
Jones, O. 2011. Geography, memory and non-representational geographies. *Geography Compass*, 5(12): 1–11.
Kalay, Y.E., Kvan, T. and Affleck, J. (eds) 2008. *New heritage: new media and cultural heritage*. London: Routledge.
Kember, S. 1998. *Virtual anxiety: photography, new technologies and subjectivity*. Manchester: Manchester University Press.
Kolozsi, Á. 2014. Nyanyákkal foglaltuk vissza a Felvidéket. *Index.hu*. http://index.hu/tudomany/tortenelem/2014/12/18/nyanyakkal_foglaltuk_vissza_a_felvideket/ (accessed 19 December 2014).
Kopytoff, I. 1986. The cultural biography of things: commoditization as process, in A. Appadurai (ed) *The social life of things* (pp. 65–91). Cambridge: Cambridge University Press.
Lowenthal, D. 1985. *The past is a foreign country*. Cambridge: Cambridge University Press.
Massumi, B. 2002. *Parables for the virtual: movement, affect, sensation*. Durham, NC: Duke University Press.

Pallasmaa, J. 2014. Space, place and atmosphere: peripheral perception in existential experience, in C. Borch (ed) *Architectural atmospheres: on the experience and politics of architecture* (pp. 18–41). Birkhäuser: Basel.

Peirce, C.S. 1955. Logic as semiotics: the theory of signs, in Justus Buchler (ed) *Philosophical writings of Peirce* (pp. 98–119). New York: Dover.

Pile, S. 2010. Emotions and affect in recent human geography. *Transactions of the Institute of British Geographers*, 35: 5–20.

Rose, G. 2010. *Doing family photography: the domestic, the public and the politics of sentiment.* Farnham: Ashgate.

Rose, G. and Tolia-Kelly, D.P. (eds) 2012. *Visuality/materiality: images, objects and practices.* Farnham: Ashgate.

Silberman, N. and Purser, M. 2012. Collective memory as affirmation: people-centred cultural heritage in a digital age, in E. Giaccardi (ed) *Heritage and social media: understanding heritage in a participatory culture* (pp. 13–29). London: Routledge.

Smith, L. 2006. *Uses of heritage*. London: Routledge.

Tamási, M. 2011. Mediating archives: the Open Society Archives: photographic collection and the *Fortepan* initiative. Conference presentation at Archeologica fotografii. http://vimeo.com/25736024 (accessed 18 December 2014).

Thrift, N. 2008. *Non-representational theory: space, politics, affect.* London: Routledge.

Tizenöt magyar utcaidivat-fotó a harmincas évekből. *Nők Lapja Café*. www.nlcafe.hu/oltozkodjunk/20141219/magyar-utcai-divat-harmincas-evek/ (accessed 19 December 2014).

Trigg, D. 2012. *The memory of place: a phenomenology of the uncanny.* Athens: Ohio University Press.

UNESCO. 2003. *Charter on the preservation of digital heritage.* http://portal.unesco.org/en/ev.php-URL_ID=17721&URL_DO=DO_TOPIC&URL_SECTION=201.html (accessed 18 December 2014).

Waterton, E. 2010. The advent of digital technologies and the idea of community. *Museum Management and Curatorship*, 25(1): 5–11.

Zumthor, P. 2006. *Atmospheres: architectural environments surrounding objects.* Birkhäuser: Basel.

11 Commemoration, heritage, and affective ecology
The case of Utøya

Britta Timm Knudsen and Jan Ifversen

The event

At 3:25.22 pm on July 22, 2011, a bomb placed in a white Volkswagen Crafter parked near the office of the Norwegian Prime Minister Jens Stoltenberg and several other governmental buildings at Grubbegata, in the center of Oslo, explodes, killing seven people and shattering several buildings. An hour and a half later Anders Bering Breivik arrives at the ferry landing Utøyakaia in Tyrifjorden, 600 meters from Utøya, the island where the youth organization of the Norwegian Labour Party (AUF) is hosting its yearly summer camp. Disguised as a policeman, he lands on the island and immediately begins a shooting spree, searching for victims around the small island and hitting them at close range. Through his brutal action he transforms the peaceful summer meeting into a death camp in which the trapped are subjected to the terror regime of a perpetrator disguised as law enforcer. Only 72 minutes later does he give himself up, without resistance, to a terror unit of the Norwegian police. At this point he has killed 69 people and wounded another 66. People with boats from neighboring mainland Sørbråten rescue as many as 250 of the young participants fleeing from the island.

Terrorism is inextricably linked to its immediate broadcasting and spectacle in electronic and digital media, producing all kinds of emotions able to mobilize spectators around fear, anger, hate, sorrow, pity, etc. Due to their immediate broadcasting and dissemination via digital and mobile media—produced by victims, bystanders, occasionally positioned witnesses—the events become "*lieu de mémoire*", constructing a cultural memory for future generations (Erll, 2011; Frosh and Pinchevski, 2014). The affective and emotional dimensions of broadcast and digitally disseminated images of terrorism have to do both with their immediacy, positioning viewers as virtual bystanders to an event they can follow in real time (Bolter and Grusin, 1999; Grusin, 2010; Marriott, 2007), and with the aesthetic quality of the images, for example a particular softness and discretion in the images producing pity (Boltanski, 1999) or a rawness and deframed characteristics that enhance the feeling of authenticity (Knudsen and Stage, 2015a; McCosker, 2013).

Often terrorist attacks on a huge scale hit places of transit (train stations, airports, planes, as was the case in Madrid in 2004, London in 2005, Karachi in

2014) or global iconic sites (Olympic Games 1968, WTC and Pentagon 2001), or appear as acts of retribution, as in the attacks on the offices of satirical magazine *Charlie Hebdo* in Paris in 2015. The symbolic dimensions of certain sites are often targeted, but the ways in which they influence near-neighbors are often overlooked. The terror attack on Utøya differed from other incidents by being more extended in time and in the way that it took place in a natural environment that came to play an important role in the development of the action. The attack's being known as the Utøya attack (and not the Labour Party attack, for example) indicates that the island/nature itself is under attack and that the site itself plays a significant role. The terror attack on Utøya had more of the character of a serial mass murder, with a killer in face-to-face contact with his victims at various places on the island. These places were well-known parts of the summer camp cartography, with names such as the Love Path, The Café Building, and the Pump House. Nature came to play a role in the perpetrator's project, just as it will do in an outspoken way in the monument, *Memory Wound*, that has been chosen to commemorate the event.[1] The peaceful natural context of the island is not only visited but also actively used by the evil perpetrator, something which makes the nature *bad* and thereby alters the affective ecology of the place, an issue that we will investigate throughout this chapter. The concept of affective ecology that we use, taken from Davidson *et al.* (2011, p. 6), is defined as a relation between bodies, places and affects:

> This philosophical understanding of affect as a form of allure or attention, which provides the emotional "glue" that drive bodies to assemble into collectives and by which objects are understood to participate in microgeographies possessed of a specific situational ethos—or what we might call an ecology [...].

Geographers and heritage and tourism researchers have for quite some time now taught us that places are socially produced and performed, and that our experience of places depends on entanglements between nature/materiality, technology (media in its broadest sense, such as running shoes, skateboards and mobile devices), and bodies placed in social situations and carrying social frames with them. The concept of affective ecology adds two valuable insights to this knowledge: on the one hand it highlights the emotional dimension of a body–place relation, and on the other it frames "the sources of attachments to place as operating through a series of affective virtualities: nostalgia, desire, and hope" (Davidson *et al.*, 2011, p. 6). In this sense, an affective relation to place is something that is a "sleeping" potentiality, fundamentally real but not necessarily realized. The focus on the emotional glue between bodies and places is important when we deal with places that are affected by heritage in various ways: what happens to places that are appointed World Heritage Sites; what happens to places that carry a difficult heritage[2] (e.g. Nürnberg in Germany or Oswiecim in Poland) or are forced into a 'difficult' heritage by sudden events such as terrorist attacks? The ways in which these places are changed through their new heritage status depends on the ways in which future generations relate to the heritage in question. Davidson *et al.* are inspired—like Deleuze—by philosopher Baruch de Spinoza's (1997 [1678]) view on affects, or "passions," as he calls them, as pre-conscious proto-social momenta that either are experienced as "joy (laetitia) … or sadness (tristitia)" (Davidson *et al.*, 2011 p. 5). The crucial point here is that Spinoza moves the focus from the content or mood of the affect to what it does to the body involved: does it empower or disempower the body in question; is the capacity of the body increasing as the result of the affect (joy) or is it decreasing (sadness)?

> The mind, as far as it can, endeavours to conceive those things, which increase or help the power of activity in the body (Spinoza, 1997, Prop. XII).
>
> When the mind conceives things which diminish or hinder the body's power of activity, it endeavours, as far as possible, to remember things which exclude the existence of the first-named things (Spinoza, 1997, Prop. XIII).

In our perception of affective ecology we stress the two dimensions of affects. Affects are both virtual capacities of places ready to be actualized depending on the bodies experiencing the places, and actual capacities of bodies to act and react. The relation between these two dimensions of affect occurs in various combinations (bodies whose capacities to act are increased by actualizing affects of sorrow, for example).

Immediately after the events in Oslo and Utøya on July 22, 2011, commemorative activities began to take place. In one third of all municipalities in Norway more than one million people attended "rose parades" that had been initiated through Facebook. The participants carried red roses and, according to Kverndokk (2013, pp. 141–2), this overwhelming manifestation of support was the result of a "convergence culture … where old and new media collide" (p. 142). To a large degree, citizens responded and mobilized through social media, and this first phase of the commemorating process was characterized by a high level of consensus, which is typical for the circumstances. To the many spontaneous activities of commemoration were added reflections on the creation of more permanent memorials.

We claim that the social consensus experienced in the immediate aftermath of a violent event is often followed by a period of dissensus in which different opinions and interests are manifested. This return to political normality will, however, be changed as it is "haunted" by the voices of the victims (Blackman, 2015). The public debate on commemoration and the practices performed will revolve around the new symbolic positions of the victims and the witnesses. As we will show, the dissensus emerging from these practices and opinions circulates around the function of witnessing and intervenes in the affective ecology of the crime scene and its surroundings. We investigate commemorative processes as they unfold in the period after the immediate grassroots responses and before official memorials have been created. This means our interest lies with those processes and practices that fall outside the official remit. Our study of commemorative practices is based on ethnographical material collected during a field trip to Utøya island in 2014[3] and on online information about grassroots memorials. To outline the different opinions on how to commemorate the event, public debate regarding the official commemoration has been followed through a discourse analysis of comments in newspapers from September 1, 2013 to September 1, 2014.[4] We thus analyze these commemorative processes both from a discourse analytical perspective and from a more-than-representational perspective, looking in particular at changes between bodies–place–affects in relation to the Utøya attack.

Changing affective ecologies

We begin by looking at the role of Utøya in the commemoration processes. The island became the centerpiece in the narratives and practices that emerged. The traces of suffering left in nature and the bodies had to change older narratives and attachments to the island. Through a study of the grassroots commemoration performed at the island of Utøya itself, we will examine how practices of commemoration and witnessing played a role in changing the affective ecology of Utøya from a place of love, joy, and political enthusiasm to a haunted place. Drawing inspiration from more-than-representational theories (Anderson and Harrison, 2010; Thrift, 2008) and phenomenological event theory (Romano, 2009 [1998]), we investigate the entanglements of geography, affect, and commemoration which together form the witnessing and the practices linked to grassroots memorials.

The proximity between the killer and his victims at various identifiable death spots, as well as the exposure of the victims' bodies not only to the bullets of the perpetrator but also to the particular nature of the tiny island, are crucial to its new affective ecology. The unpredictable actions of the perpetrator—apparently he selected his victims by what he deemed to be their typically leftist appearance (NRK, 2012)—combined with the role of the place itself, its materiality (the rocks, the cold water in July 2011 because of weeks of heavy rain), its buildings and hilly landscape, highlight the violent mixing of human and non-human agency in the unfolding of the event. The island changes in at least three ways in the aftermath of the terror attack. First, through this violent event, the island "gets" a difficult heritage, that transforms the place for eternity. The haunted-ness of the island becomes necessarily a part of its future. And this future is one of the key issues in the debates around the island. Second, a very concrete change happens, through the attack, to the affective ecology of this particular place. Utøya, hosting youth summer camps for social democrats, has been a place of love and political enthusiasm, but must now include grief and sadness in its future life. Below, we look at how this additional affective layer at Utøya causes conflict and puts the whole future of the island into question. Third, this event brings to the surface and disputes a rootedness in nature that touches upon Norwegian culture in general. Nature itself is involved and "wounded" by the event, and is part of the conflict that ensues in the local community.

Raymond Johansen, the general secretary of the AUF, expressed the following very typical reaction in the aftermath of the attack:

I remember to have said to some friends that after 22 July, Norway and myself would be changed forever. They responded that everything would certainly go back to normal. And they were right and I was wrong. But I still believe now, three years later, that this must change Norway'.

(*Dagsavisen* 02.09.2014: 20–1)

Events seem to imply a hope for political change, and sometimes they themselves are the change they imply (the French revolution, May 1968, the election of Barack Obama, etc.); but sometimes, while igniting a desire for change, they paradoxically conserve or even re-strengthen existing cultural and political patterns and power-geometries (9/11, the Copenhagen shootings in 2015).

In the literature on events we can distinguish between those authors who focus on a micro-perceptual level (Anderson and Harrison, 2010; Knudsen and Christensen, 2015; Massumi, 2009; Simmel, 2005 [1917]), claiming that events and change happen all the time, and authors such as Badiou (1989, 2007), Derrida (2007), Jay (2006, 2011), and Romano (2009 [1998]) who look upon events more as ruptures of a certain socio-political scale. Events cannot be understood in light of their prior context, but only from the posteriority to which they give rise, as they are themselves the origin of meaning for any interpretation. The desire for significant changes to occur follows any violent event. As we shall see below, the tendency to place grassroots memorials at sites of violent events is likewise an expression of a desire for political change or an expression of protest. In the Utøya case the event connects to the future destiny not only of the crime scene itself but also of its near vicinity, as it is touched by the project of the national commemorative monument.

In the shadow of the tragedy, the discussions revolved around whether the island must turn into a memorial landscape. The fate of the island was caught between three options: Either the whole island had to become a memorial, and then just become the site of the 2011 attacks; or all material traces of the event must be erased (as was the plan of the AUF just after the event, restoring the island to its state before 2011); or summer camps should continue with certain places marked as memorial sites at the island. The third option found favor and summer camps have been resumed as we write in August 2015. But the destiny of the island aroused ferocious debates, as did the plan to install the *Memory Wound* monument. Although placed in metonymical proximity to Utøya at Sørbråten and metaphorically replicating its vegetation, soil, and environmental surroundings, the monument is a manmade intervention into nature. Because the location is non-natural it is discussed and contested, not least among the local population. As we shall show, neighbors of the monument (and of the event) dispute efforts to transform the whole affective ecology of "their" residential place into "everybody's" place of suffering and sorrow. They dispute both the idea that the passion of sadness should overrule their feelings of belonging to this place and the local place's transformation into a global cultural common through the intervention of the monument (Gibson-Graham et al., 2013). We are not claiming that predominant voices in the local debate did not acknowledge the event; they simply refused to let it influence their own relationship with their local territory.

Witnessing

How places are changed following interrupting events is influenced by the decisions of social actors involved in the commemoration practices. What they decide is, to a large degree, dependent on how they are positioned and position themselves *vis-à-vis* the event. One position, the witness, is of particular interest. Through testimonies, the witness acquires a privileged subject position in narratives of the event.

Based on Paul Ricœur's theory of the witness (Ricœur, 1992), and adding theoretical perspectives from new media scholarship on memory (Frosh and Pinchevski, 2014; Erll, 2011; Peters, 2009), witnessing can be said to possess three fundamental features. First, witnessing is tied to the *body* of a survivor, such as, according to Ricœur, a victim or bystander who survives a crime. For Ricœur a witness is always an eyewitness. As new media scholarship on memory suggests, however, it is no longer possible to distinguish between a person who is physically present and observing what happened (the eyewitness) and a person who is directly affected by a mediated event and commemorative practices around it (a secondary witness). Memory thus becomes "globital"; Utøya becomes a "*lieu de mémoire*" that is simultaneously globalized and digitized, and this creates new affective logics (Reading, 2011, p. 299). Bystanders and media audiences who followed the extensive media coverage also become witnesses. These media witnesses who take part in commemoration as a global common appear in our Utøya case as the disputed category of "tourists." As witnesses they have the right to be affected, but as tourists they are excluded as "strangers" without any right to the event whatsoever.

Second, the testimony from a witness is at one and the same time above and within a political realm. When it is above the political, it adheres to the juridical and the theological realm. When the witness claims to tell the truth it is only the law that can be judged. When her body is fully invested in the event as a hero or a survivor, she appeals to the right cause (the theological). Only when the testimony moves into the political field, with its struggles and debates, will it be contested. The different fields of witnessing are turned into positions that are used strategically to legitimate arguments in a political debate. In the discussion on how to commemorate Utøya, the witness position is divided between different categories of eyewitnesses. On one hand, there are the survivors; on the other, those spontaneous rescuers who saved the many young people fleeing from the island. The local rescuers also invoke their status as eyewitness when they participate in public discussions. The government and local authorities are also witnesses, as are the artist and the artistic institutions. These statuses can be questioned and disputed politically, which happens when the category of eyewitness is claimed by various social actors in order to give arguments more weight in the political; for example, when they oppose the survivors and their families in discussions on where to place the national monument or the grassroots memorials, as we shall see below.

The third basic tenet in witnessing is that it is characterized by an inherent paradox and a double obligation for the witness: she/he has to tell what happened, but the magnitude of the event exceeds ordinary channels of expression and is not possible to represent in ordinary language. Hence any witness communicates affectively. The witnessing is thus caught between being obligatory and impossible. A witness is an affected body communicating a violent experience that can only be partly rendered. When it comes to communicating the event, the suffering has to be lived, felt, and transmitted from the body of the survivor to the body of another.

Experiencing the affective ecology of Utøya

Britta visited the island on Saturday, September 27, 2014, on one of the first occasions that the area was open to the general public. It was a glorious late summer's day, with high temperatures and sun as she glimpsed her first view of Utøya from the bus. There it was, the characteristic silhouette of the island. Even though she had prepared for the moment, it did produce a true micro-perceptual shock (Massumi, 2009). Due to the extensive mediations, the simple effect of recognizing the island triggered a strong reaction. The sad mood was set immediately.

Britta arrived at the jetty to wait for the small ferry, where she joined a group of eight other people waiting for the boat. At the same time a couple in their mid-fifties who were from the mainland, Sørbråten, facing Utøya joined the group. The local couple had been in their cottage just opposite the island at the time of the shootings, but they had been indoors (the weather was horrible that July) and had not noticed what was going on close by. They were only made aware of the attack when they turned on the radio (the only media available in the house) after having received a text message from their daughter in Oslo. In that sense they only learnt about the event through the media. As they informed Britta, they felt remorse and guilt *vis-à-vis* the victims and their heroic neighbors. Since they had never visited the island before, they wanted to join the group of visitors to "see for themselves" and "to pay tribute" to the victims, survivors, and their families.

A young man in the larger group apparently knew the place well and took the lead. Since the larger group did not seem to mind, Britta and the couple silently appointed him as "their" guide and followed him around the island. It turned out that he was a survivor who was back on the island with his family for the first time since 2011. This meant that, in a sense, Britta was the only stranger making use of the public opening.

Britta's encounter with the island was therefore formed through the testimony of this survivor and his witnessing body, as well as through the grassroots memorials. The "guide" took the group on a tour of the island following in the footsteps of the assassin, pointing out significant sites where the murders took place. Touring the actual sites of the murders in this way could be considered as practicing kinaesthetic empathy (Thrift, 2008), where sympathy for those who happened to be present at the island on that fateful day is created through mental projections and bodily gestures in the *in-situ* landscape. The tour began at the main house, with the iconic sign of Utøya written in red, before heading north toward the kafébygget (the Café Building) where 13 people were killed. Nearby is the place that is supposed to host the official memorial at Utøya. Significantly, the memorial site is located at a "non-place," as the guide chose to represent it, not having a strong identity or even a name. On the pathway toward the designated memorial site, there is a plaque that carries a message honoring the volunteers from AUF that joined the fight against fascism in the Spanish civil war. Thus victims, survivors, and all future members of AUF become part of a grand historic narrative of heroic fight against fascism. This message was the only reference on the island of a counter-narrative to "political" deeds. But it certainly provides the victims at Utøya with a strong political agency and gives historical meaning to their deaths.

From the proposed site of the local memorial, the group continued along the Path of Love, leading from the northern to the southern point of the island and passing by the Pumping Station. Many fell victim here when they tried to hide behind a small wooden cottage on a slope descending toward the sea. At the time of the massacre, these outdoor places appeared safer than buildings, the guide told the group in a low and matter-of-fact tone. He immediately added that this was not the case. By meticulously pointing out the exact locations where some survived and some were killed, he gave testimony to the complete randomness of the event. Through his testimony the random nature of which places were protected and which were not became terribly clear. In the end, everybody was left with the impression that the "nature" of the island did not provide protection, rendering as capricious and untrustworthy what the young people had once considered beautiful and romantic scenery.

The tour ended at the school building, Skolestua, in front of which the guide stopped talking; he only resumed once the three "intruders," the local couple and Britta, had left the group. He had been rather sparse in information about his own experience and his own survival. It was only on the small vessel, Thorbjørn, that ships passengers to and from Utøya (in fact, the same vessel that brought the perpetrator to Utøya on that fateful day) that the young man revealed his story. He was one of 49 people who survived Breivik's bullets by hiding in the school building. When he found himself at the same place during his return visit to the island, he was apparently too close to the indexical sign of his own survival and his own entanglement with the event to include others in his testimony. He needed a private space and a family gaze. Only by gaining distance to the *in-situ* sites could he tell the story of his own survival. The tension between bodily presence and the untold story could indicate the "freshness" of the wound, but it may also be a sign of the difficult transition from private to public in communicating this event. Compared to the death site of the World Trade Center that Britta visited in 2004, there are noticeable differences. At Ground Zero guided tours were the order of the day and the guides were all survivors, relatives of victims, neighbors, or otherwise close to the event, witnessing it to the global public (Knudsen, 2011). At Utøya survivors were only hesitantly bringing visitors along. The tension between private and public, as well as the acknowledgement of the site as a global common, is at stake here. So is the question of distance and proximity in the witnessing process. The same tension can be observed in the practices of setting up grassroots memorials, as well as in the public discussions on the official commemoration.

Grassroots memorials

Grassroots memorials serve as momentary manifestations that appear in the aftermath of a violent event causing sudden death to citizens. They will be particularly noticeable immediately after the event, but can continue for many years.[5] The erection of an official monument often marks a form of emotional and practical closure of an event. Grassroots memorials can be viewed as individualized forms of participation in the commemoration processes. Often they are also made in critical response to weak official manifestations. They take the form of 'spontaneous' shrine-making and can be seen as expressions of mourning and grief, "nonformal and not regulated by any institution, and [with] no traditional societal rules or mourning costumes" (Margry and Sanchez-Carretero, 2011, pp. 11–12), even though they often follow common templates and have a fairly standardized content. Often grassroots memorials are maintained by relatives, friends, or people otherwise affected who continue to lay flowers or 'objects' at the crime scenes or sites representing the attacked people. In recent terror attacks – the Twin Towers in New York in 2001, the Madrid train bombings in 2004, the London bombings in 2005, the attack on *Charlie Hebdo* in Paris on 7 January 2015 – unofficial public manifestations of grief and mourning in the form of grassroots memorials played a huge role both at the concrete crime scenes and also as local manifestations of global solidarity all over the globe.

The events of July 22, 2011 prompted a large number of spontaneous grassroots expressions of grief and mourning all over Norway. The rose parades have been mentioned, but there were also more memorials at Utøya and its vicinity. On the island a "value" tree was set up at the campsite to gather all sorts of statements in the form of scraps of paper with messages such as "We will fight for you and you will always be part of us."

Similarly, the places on Utøya where people were killed, such as the Café Building (kaféhygget), the Campsite (Leirplassen), the Pumping Station (Pumpehuset), and the Path of Love (Kjærleiksstien), were made into sites of spontaneous mourning. In the Café Building, where 13 people perished, there are still traces of what happened: bullet holes in the walls and furniture, patches of white paint to cover bloodstains on the walls. Grassroots memorials in the form of roses, photos, poems, cuddly toys, and hearts made of felt, granite, or candy are added to the traces (see Figure 11.1). In the Café Building the memorials were remarkable and memorable because they brought the event to life through their function as *hypermedia*: Not only did they communicate the horrors directly, but they also evoked kinaesthetic empathy with victims through a claustrophobic feeling of being trapped as the grassroots memorials pointed to and enhanced the experience of the concrete circumstances of the violent deaths.

Figure 11.1 Heart of Candy in the Café Building.
Photograph: Britta Timm Knudsen.

At the Path of Love—another significant crime scene—the self-appointed guide noted a copper plate with a name inscribed on it nailed to a tree and then explained: "you cannot prevent people from leaving marks and commemorating, but this copper-contaminated tree will probably die, so maybe it would be better to come up with something that does not have such consequences for nature" (September 28, 2014). His utterance is notable because it points to the shift from civic expressions and emotional engagement forming a democratic monument that "we" build together to a more contested space that distance to the event—in time and space—establishes. His comment on the unsustainable character of this grassroots memorial indicates that this shift has taken place. The pervading sadness of the affective ecology of the place is, in fact, enhanced through the testimony of the body of the survivor, and through the commemorative vernacular practices visible on the island.

Politicization of commemoration

In her article "Forget Trauma? Responses to September 11," Edkins (2002) enumerates four ways of responding to trauma: to securitize, to criminalize, to aestheticize, and to politicize. In her view, only politicization acknowledges trauma as an event that opens to a new future. The immediate response to 9/11 by the US government was to securitize and focus on the terrorist enemies. In Norway, the response was of a more political nature, in the sense that there was an enthusiastic call to mobilize the population around democratic values (Jenssen and Bye, 2013, p. 224). Analyzing Prime Minister Jens Stoltenberg's speeches in the days immediately after the event, as well as the speech he gave at the commemoration the following year, Jenssen and Bye point out that he responded by calling for "more democracy, more openness and more humanity" (p. 223). He avoided focusing on the perpetrator's links with the Norwegian extreme right wing and how to prevent similar events in the future. His aim was to be politically affirmative.

Stoltenberg's response was met with enthusiastic support in the first consensual period immediately after the event, but soon after the responses became political, in the form of a diversity of opinions and interests, most markedly in relation to the location of the official national monument. The discussions are interesting because they revolve around the question of whether and how the affective ecology of the island could be changed as a consequence of the event.

We have conducted a discourse analysis of the debate on the creation of a national memorial as it unfolded in newspapers from September 1, 2013 to September 1, 2014, the focus being to identify the different positions that were formed in the debate. The intensity of the debate followed the rhythm of the governmental decisions concerning the national process, the possibility of an *in-situ* monument on Utøya, and its potential design. As far as the national monument is concerned, the jury set up by Public Art Norway (KORO), the government's professional body for art in public spaces, decided on two projects designed by Swedish artist Jonas Dahlberg as the official national memorial. Many reacted strongly to his idea for a memorial at Sørbråten. Due to local protests, the government decided in April to postpone the date for the inauguration of the Utøya memorial by one year, to July 22, 2016.

Two main features characterized the debate. First, it focused almost completely on how Utøya as a place should be included in the commemoration of the event. Second, it immediately involved participants that had been personally involved in the event and circulated around the importance of witnessing. The main arguments revolved around whether the location chosen by Dahlberg was the right one. Discussions on the design and aesthetic form of the memorial were secondary to the controversy around the location. Local newspapers—in particular *Ringerikes Blad*, which covers the region where Sørbråten is located—played a prominent role as channels for the debate.

For obvious reasons, no one contested the idea that commemoration should be conducted at a national level. The event had already been embedded in a strong national frame in the immediate reactions. The nation as a whole played a prominent symbolic part in the commemorative discourse. As one commentator noted when arguing in favor of a memorial close to Utøya: "For us this is the most important to settle. This is a place where those touched by the event can come, but it is also a place for the nation as a whole" (*Aftenposten*, 22.07.2014). It is, however, significant how few commentators chose to refer to the nation in arguing for their case. The debate mainly turned around a controversy between local and non-local issues.

The other important symbolic position in commemorative discourses is obviously taken by the victims. Whatever the opinions in the debate, the position of the victim as the main symbolic subject is untouchable. But the question of who can be regarded as victims and who are entitled to speak in their names remains open. Just as we have victims, we have a perpetrator. There is, however, hardly any mention of the perpetrator in the debate. At the time of the debate around the memorials, Breivik is not present as a person anymore, nor are the evil forces he incarnated mentioned. What is left is the pain and suffering, or the "wound," to use the metaphor that most commonly appears in the debate.

Next to the main symbolic positions we have the more concrete speaker positions represented in the debate. The people living in the place where Dahlberg's memorial will be erected—the neighbors of Utøya, as they are most often called—make up one position. The families of the victims speak from the position of being most closely related to the victims and to the event. They also speak through the national support group, The Supporting Group 22, which was established shortly after the event. The survivors who experienced and witnessed the horrors at close range make up a third position with direct connections to the event and to the place. Those involved in saving people fleeing from the island can also make claim to the status of eyewitnesses. Because many of the survivors as well as the victims were related to AUF, this organization takes a position in the debate far beyond that of a political party, due in addition to the fact that it is the formal owner of Utøya. Finally, we have all the rest—the bystanders who also witnessed and were affected by the event through the media.

Let us have a closer look at how these positions formed the debate. The neighbors were by far the most prominent. Their arguments against the location of the memorial went along two lines. First and foremost, they did not want it because it would impact on their daily life. They claimed that a memorial "outside their living room windows" would constantly remind them of the event. The term "re-traumatization" was used several times to describe how they as neighbors would go on suffering when living next to the memorial. In fact, they came close to arguing that a memorial in their neighborhood would not allow them to forget, which almost sounds like a contradiction in terms.

They were, however, well aware that a position of neighbors complaining about disturbances by visitors—more pejoratively described as "tourists turned toward terrorism"—was a difficult one to take. Therefore, they referred to themselves not as neighbors (which was a term mainly used by journalists commenting on the matter) but as witnesses: "We do this to safeguard our mental health and to maintain our right to live a worthy life despite the fact that we live where we live, saw what we saw and heard what we heard on the 22nd of July 2011" (*Ringerikes Blad*, 14.08.2014). To strengthen the argument, some of the commentators criticizing the location of the memorial even spoke of themselves as victims: "Ustranda Vel (the association established by the neighbors) has made enough sacrifices."[6] Getting close to the symbolic position of victim is an important way of bolstering the speaker position in order to take it out of the political realm and place it within the theological one. There is thus a situation of double closeness at play in the argument. The speaker who, in her or his own capacity, was involved is a victim of the event, and furthermore bound to it by the physical proximity to the island.

The case of Utøya 231

Efforts among the neighbors to establish a privileged speaker position went even further. Several comments related to their status as "saviours" or "boat heroes." A newspaper article carried the heading: "the neighbors and saving heroes are fighting to repress the 22nd of July" (*Bergens Tidende*, 14.06.2014). Although those speaking in the name of the neighbors only indirectly alluded to their status as heroes, there is no doubt that the double positioning of both victim and national hero, about to be betrayed by the national authorities, made their case stronger, particularly in their protest against the decision of the government. But despite these intentions, the comments also portrayed the neighbors as people with a right to privacy: "The people living along Utstranda [the part of the small town of Hole facing Utøya] are informing about big traffic problems, impudent tourists, pollution and a general misbehaviour that requires a lot of patience from the locals" (*Bergens Tidende* 14.06.2014). Here the problem is hardly about being haunted by bad memories. It is noticeable that outside visitors are portrayed as "tourists" with no legitimate right to be present. In this framing the tourist denotes a purely negative position, which together with the national authorities is seen as threatening local rights to the place. In the case of the national authorities, the dispute can be reframed as a case in which the government did not include the local population in their decision.[7]

Discussions relating to the neighbors' rights gained momentum when representatives of the local association secretly moved a grassroots memorial created immediately after the event to a location further away from their private homes (and next to the highway crossing the area) (see Figure 11.2). Although the neighbors

Figure 11.2 Grassroots memorial with view of Utøya.

Source: Ingvild-Anita Velde, Norwegian Broadcasting Cooperation, by kind permission.

claimed to have transferred it to another place (and used the same arguments as in the case of Dahlberg's memorial), their action raised protests. The spokesperson for the national support group was quoted as saying that it was "incredibly sad to witness the lack of respect for the intuitive memorial […] that mourners, families and others treated with respect. Notice what has become of it after the 'neighbors' moved it. They should be ashamed!" (*Dagsavisen*, 3.07.2014). From the perspective of this spokesperson, "neighbor" is certainly not a position that grants privileged rights to a memorial place, whether officially designated or spontaneously made.

The particular role of the island came up several times in the discussions on commemoration. It had already gained iconic status immediately after the event, and families and survivors visited it as part of the mourning process. There were discussions as to whether the island itself should be turned into a memorial. One commentator argued:

> tear down the buildings and leave Utøya in peace, so that those families in need of commemoration can go there. Utøya as a memorial is enough. AUF's ambition of building a "village" is absurd. Who would like to dance there?'
> (*Trønder-Avisa*, 14.07.2014)

The idea of transforming the entire island into a memorial—with no traces of previous activities—was based on the claim that it could never again be included in normal activities. In the words of a family member, "for us it is a cemetery. It is sacred ground" (*Oppland Arbeiderblad*, 19.07.2014).

The island itself as a non-human agent had attained a symbolic position. Not only should it be "left in peace"; it should also be given the status of an intimate memorial for the families. Some commentators directly criticized AUF for planning to continue the tradition of organizing summer camps. They claimed that the organization—despite its ownership—could no longer claim privileged access to Utøya, which was now a national concern or a global public good. Members of AUF did not refer directly to ownership, but still claimed that the island was part of their tradition. In a sense, they used arguments similar to those of the neighbors, in order to make Utøya theirs and to remove it from the discussion of memorials. But families also argued that Utøya should stay "a living island" where both traditional activities and commemoration could be practiced (*Dagen*, 24.07.2014). Neither those arguing for Utøya to become an untouched memorial nor those wanting to return it to former practices had any wish for an official memorial on the island. Although the discussion was fierce, the two opposing parties—AUF claiming property rights to the island and relatives of victims claiming the whole island as a memorial space—had to negotiate their positions.

In the end, the entire debate revolved around the right to speak in the name of the event and the place. Closeness to the victims (either as rescuer or a family member) and acquiring a status as primary witness made it possible to move arguments from the political to the theological realm. Survivors and families definitely had more legitimacy than neighbors, but the latter could also bolster their position by claiming their status as witnesses and their bodily investment. Only when they made recourse to their local rights did the political nature of their claim appear openly and become vulnerable. The single completely excluded position was that of the tourist, including all the secondary witnesses, who were thus left without any right to commemorate.

Conclusion

Immediate responses to tragic events are often characterized by strong feelings of consensus, as we witnessed in the aftermath of the terrorist attack on the French satirical newspaper *Charlie Hebdo* and the Copenhagen shootings in 2015. Millions of people forming silent crowds of mourning populate the streets of Paris, all over France, in solidarity marches in Europe, and in the rest of the Western world. Whereas silent witnessing prevails in the first commemoration phases after a violent event, we look at a period in a commemoration process where dissensus and witnessing stir up consensus and become political. In the case of Utøya we have thus emphasized the politicization of commemoration through growing disagreement as to how and *where* to commemorate the event. Local inhabitants or neighbors to Sørbråten dispute the specific location of the memorial because they fear that their place will be transformed into a *lieu de mémoire* and a global common that belongs to everyone, including those to whom they would prefer not to grant access. Even to the extent of removing a grassroots memorial, which is a rather extreme act, they strongly try to deny the inevitability of the event, the status of the place as a common, and the changes to the affective ecology of their locality.

We have likewise looked at the emotional dimensions of a difficult heritage site coming into being. In the case of Utøya we show how the question of commemoration radically alters the affective ecology of the island. A former place of joy, love, and enthusiasm is transformed into a heritage site as a consequence of the attack. The dispute as to what is to become of the island – returning it to normality, which means a rather nostalgic attitude toward the island, nearly denying the event; total preservation of the whole island as a memorial site (sadness); or co-habitation of former activities through the summer camps and the reality of the difficult heritage (sadness and hope) – runs through our material. It is important that heritage studies take into consideration these emotional realms that add to the identity politics we normally encounter in critical heritage studies, because they show us how landscapes are infused with emotions and how bodies in these landscapes react within and toward these emotions influencing the discussions around a future of a place. Such analysis will add other layers of signification to the discursive layers that need to be continued. In future heritage studies, it will be interesting to look at the relation between these two lines of interest (Knudsen and Stage, 2015b).

Danish survivor Kristian Kragh Lundø returned to Utøya on August 6 at the reopening of AUF's summer camps in 2015. The affective ambivalence of repulsion and attraction is clear in his reaction: "It is difficult. It is really getting to me.

I was frankly speaking not prepared for this. There is a battle inside me just now, I am enjoying it, I am looking forward to it, but on the other side it hurts" (6.8.2015 DR 1 21:30). After a day on the island, he stated: "I am even more convinced of my political work now. It is so right to be here" (7.8.2015, TV 2 News 20.45). We see in these two statements that the atrocity of the memories layered in the body of this survivor and his reunion with the haunted place produces, paradoxically, a feeling of empowerment and strength. Sad places do not necessarily make us sad: They can likewise increase engagement and will. The analysis of such emotional reactions toward heritage needs our full attention in the future of heritage studies.

Notes

1 www.theguardian.com/artanddesign/2014/mar/06/norway-massacre-memorial-jonas-dahlberg-anders-behring-breivik.,
2 "Difficult heritage" is Sharon Macdonald's term, taken from her book on Nazi relics in Nürnberg (2009); it follows the same line as Tunbridge and Gregory's "dissonant heritage" (1996), which also stresses conflictual views on heritage management, and it is followed by Joy Sather Wagstaff's 2011 book on 9/11, *Heritage that Hurts*.
3 Britta Timm Knudsen conducted a field study at Utøya on September 27, 2014. For several years, only families and members of AUF could gain access to Utøya. In autumn of 2014, the island was opened to visits from the public on some weekends.
4 We searched newspaper texts containing the following combination of search words: Minnesmerke AND Utøya, Dahlberg AND Utøya in the digital database, Retriever, www.retriever-info.com/da/?redirect=true, in the period mentioned. The search returned hits from around 200 newspaper texts, including many comments from readers.
5 A good example is the grassroots memorial created in the memory of Diana, Princess of Wales just above the Tunnel du Pont de l'Alma, her death spot in Paris, which always has fresh flowers, poems, photos etc. The particular thing about this grassroots memorial is that it hijacks an existent sculpture, the Flame of Liberty, a present the US offered France in 1989. The municipality has responded to this unofficial grassroots hijacking by putting up signs saying "This is not a memorial to Lady Diana." But the grassroots seem to win this battle.
6 In Norwegian to sacrifice, *ofre*, and *victim* (offer) are semantically closer than in English.
7 Survivors and families of the victims were represented in the jury that selected the winning project. This was not the case for the local community.

Bibliography

Anderson, B. and Harrison, P. (eds) 2010. *Taking place: non-representational theories and geography*. Farnham: Ashgate.
Badiou, A. 1989. *Manifeste pour la philosophie*. Paris: Seuil.
Badiou, A. 2007. The event in Deleuze. *Parrhesia*, 2: 37–44.
Blackman, L. 2015. Researching affect and embodied hauntologies: exploring an analytics of experimentation, in B.T. Knudsen and C. Stage (eds) *Affective methodologies* (pp. 25–44). Basingstoke: Palgrave Macmillan.

Boltanski, L. 1999. *Distant suffering: morality, media and politics.* Cambridge: Cambridge University Press.
Bolter, J.D. and Grusin, R. 1999. *Remediation: understanding new media.* Cambridge, MA: MIT Press.
Davidson, T.K., Park, O. and Shields, R. (eds) 2011. *Ecologies of affect. Placing nostalgia, desire and hope.* Ontario: Wilfrid Laurier University Press.
Derrida, J. 2007. A certain impossible possibility of saying the event. *Critical Inquiry,* 33(2): 441–61.
Edkins, J. 2002. Forget trauma? Responses to September 11. *International Relations,* 16(2): 243–56.
Erll, A. 2011. *Memory in culture.* New York: Palgrave Macmillan.
Frosh, P. and Pinchevski, A. 2014. Media witnessing and the ripeness of time. *Cultural Studies,* 28(4): 594–610.
Gibson-Graham, J.K., Cameron, J. and Healey, S. 2013. *Take back the economy: an ethical guide for transforming our communities.* Minneapolis: University of Minnesota Press.
Grusin, R. 2010. *Premediation: affect and mediality after 9/11.* New York: Palgrave Macmillan.
Jay, M. 2006. *Songs of experience: modern American and European variations on a universal theme.* Berkeley: University of California Press.
Jay, M. 2011. Historical explanation and the event: reflections on the limits of contextualization. *New Literary History,* 42(4): 557–71.
Jenssen, A.T. and Bye, K.H. 2013. When anger and sorrow was turned into openness and tolerance? Political oratory in the wake of the terror of 22nd of July, 2011. *Tidsskrift for Samfunnsforskning,* 54(2): 217–31.
Knudsen, B.T. 2011. Thanatourism: witnessing difficult pasts. *Tourist Studies,* 11(1): 55–72.
Knudsen, B.T. and Christensen, D.R. 2015. Eventful events: event-making strategies in contemporary culture, in B.T. Knudsen, D.R. Christensen and P. Blenker (eds) *Enterprising initiatives in the experience economy* (pp. 117–34). London: Routledge.
Knudsen, B.T. and Stage, C. 2015a. *Global media, biopolitics and affect. Politicizing bodily vulnerability.* London: Routledge.
Knudsen, B.T. and Stage, C. 2015b. *Affective methodologies.* New York: Palgrave MacMillan.
Kverndokk, K. 2013. Negotiating terror, negotiating love. Commemorative convergence in Norway after the terrorist attack on 22 July 2011, in C.A. Ingemark (ed) *Therapeutic uses of storytelling* (pp. 133–58). Sweden: Nordic Academic Press.
Margry, P.J. and Sánchez-Carretero, C. (eds) 2011. *Grassroots memorials. The politics of memorializing traumatic death.* Oxford and New York: Berghahn Books.
Marriott, S. 2007. *Live television. Time, space and the broadcast event.* London: Sage.
Massumi, B. 2009. Of microperception and micropolitics: an interview with Brian Massumi. *Inflexions: A Journal for Research Creation,* 3. www.inflexions.org (accessed 5 August 2015).
McCosker, A. 2013. *Intensive media. Aversive affect and visual culture.* New York: Palgrave Macmillan.
Peters, J.D. 2009. Witnessing, in P. Frosh and A. Pinchevski (eds) *Media witnessing: testimony in the age of mass communication* (pp. 23–41). Basingstoke: Palgrave Macmillan.
Reading, A. 2011. The London bombings: mobile witnessing, mortal bodies and globital time. *Memory Studies.* 4(3): 298–311.

Ricœur, P. 1992. *Lectures 3. Aux frontières de la philosophie*. Paris: Seuil.
Romano, C. 2009 [1998]. *Event and world*. New York: Fordham University Press.
Simmel, G. 2005 [1917]. *Fundamental problems of sociology*. New York and London: The Free Press and Collier-Macmillan.
Spinoza, B. de 1997 [1678]. *Ethics part III. On the origin and nature of emotions*. Project Gutenberg etext of the ethics. www.gutenberg.org/files/3800/3800-h/3800-h.htm (accessed 5 August 2015).
Thrift, N. 2008. *Non-representational theory: space/politics/affect*. London: Routledge.

Online articles:

23.4.2012, NRK, At jeg lever er like tilfeldig som at andre døde. Available online: www.nrk.no/227/artikler/breivik-husket-adrian-pracon-1.8092843. Accessed 5 August 2015.

News features on TV:

6.8.2015, DR 1, 21:30
7.8.2015 TV 2 News, 20:45

12 Social housing as built heritage

The presence and absence of affective heritage

Sophie Yarker

This chapter uses the lens of affective heritage to consider some of the implications of designating housing, in particular social housing, as cultural heritage for residents' and former residents' sense of belonging to place. In doing so it adds to debates within heritage studies that question traditional processes of heritage management and their tendency to obscure and undermine the cultural meanings of landscape and the built environment for communities (Waterton, 2005). It does this by drawing on empirical material generated during walking tours with residents and visitors of the unique case of the Byker estate, Newcastle upon Tyne, England. Through a lens of affective heritage, the memories, ordinary affects and affective atmospheres emerging during the walking tour open up gaps between what is valued as heritage by residents, and what is designated heritage by the authorised heritage sector.

In 2007 the Ralph Erskine-designed Byker housing estate became one of only two post-Second World War social housing estates in Britain to be given Grade II* listed building status. For its advocates, such as English Heritage, this recognised the estate's 'ground-breaking design (which has been) influential across Europe' and celebrates it as a 'pioneering model for public participation' (Pyrah, 2015). For many others, including a number of residents, however, the listing 'was neither here nor there' (Pendlebury *et al.*, 2009). Built as part of a national programme of 'slum clearance' housing redevelopments in the 1960s and 70s, the colourful and unconventional design of the estate has attracted architectural interest and divided local public opinion, leaving its architectural credentials often sitting in uncomfortable tension with its multiple indicators of social and economic exclusion. Byker therefore provides a unique and rich case study from which to explore the affective heritage and consequences of listed building status for local residents.

The theoretical framework of this chapter engages with a Thriftian approach to affect concerned with describing 'practices, mundane everyday practices, that shape the conduct of human beings towards others and themselves in particular sites' (Thrift, 1997, p. 127). Three important points require drawing from this statement. The first is a turn towards an emphasis on practice which allows us to think about affect as something that goes beyond precognition and to consider how affect becomes known through its movement between bodies. The second is an understanding of affect 'working' on our bodies through contagion. This is the ability of affect to circulate among people, and moves towards an understanding of affect that is interested in its potential capacities as 'broad tendencies and lines of force' (Thrift, 2004, p. 60). This is of particular interest to an understanding of how affective heritage moves between bodies during practices such as heritage walking tours. The third is the 'mundane everyday practices' of heritage in social housing, which diverge from much of the literature surrounding emotions and heritage, which often focuses on the spectacular and the overtly traumatic. Affect, then, can be understood as sensation linked to a person's environment, and serves as 'a sense of push in the world' influencing our behaviour (Dittmer, 2010, pp. 91–2). Here, a social and relational understanding of affect is useful for our purposes of understanding the relationship with built heritage and individuals. In this sense, affective heritage can be understood as the affective atmospheres (Anderson, 2009), ordinary affects (Stewart, 2007) and memory that prompt subconscious attractions and aversions (Dittmer, 2010) to the heritage environment around us. A brief discussion of these three main dimensions of affective heritage pertinent to this chapter will be given before a context of the debates surrounding the listing of post-war social housing is provided.

Affective heritage

As places where people live their everyday lives, the ordinary affects (Stewart, 2007), memory and affective atmospheres (Anderson, 2009) that make up affective heritage in this analysis offer an illuminating way of understanding the subtle nuances of peoples' complex relationships with place and the impact that heritage designation can have on our sense of belonging and local attachments. For Stewart (2007), ordinary affects are the sensations indicating that something is happening which should be attended to. This *something*, described by Stewart as akin to Williams's (1977) structures of feeling, provides palpable pressures within the ordinary, everyday experience of place. They are 'public feelings that begin and end in broad circulation, but they are also the stuff that seemingly intimate lives are made of' (Stewart, 2007, p. 2). Elaborating further, Stewart continues that they can be

> ... experienced as a pleasure and a shock, as an empty pause or a dragging undertow, as a sensibility that snaps into place or a profound disorientation ... they can be funny, perturbing or traumatic. Rooted not in fixed conditions of possibility but in the actual lines of potential that a something coming together calls to mind and sets in motion.
>
> (2007, p. 4)

Ordinary affects, then, are present states of potentiality and resonance which can provide potential ways of knowing. They become a window or a way into understanding how people experience place and how they negotiate and express that relationship.

Social housing as built heritage 239

For Duff (2010), it is this interplay between the experience of affect, or ordinary affects, and the capacities and propensities experienced in place that give form to what Anderson (2009) describes as affective atmospheres. Such atmospheres can be understood as a set of collective affects used interchangeably with mood, feeling, ambience and, more broadly, a sense of place: 'they envelope and thus press on a society from all sides with a certain force' (Anderson, 2009, p. 78). Atmospheres become an important way of understanding how affect is experienced between bodies and in place. Atmospheres allow affects to attach themselves in some way to the materiality of place – in this instance, listed social housing. Key for Anderson is the ambiguity between presence and absence held within affective atmospheres, and it is this which forms a central focus for this chapter in illustrating the use of affective heritage to understand the potential dissonance between authorised heritage discourse and the lived experience of built heritage.

This brings the discussion to the third element of affective heritage: memory. The importance of space, place and location for shaping memory has long been recognised by geographers (Blunt, 2003), and recalling memories of the way places used to be is a central part of how they are experienced in the present (Degan and Rose, 2012). In this way memory acts as a key register 'offering a sort of historical sedimentation' (Waterton, 2014, p. 76), producing an affecting response to heritage for the individual and also contributing towards affective atmospheres for the collective. The importance of memory becomes particularly heightened in a changing urban landscape where both individual and collective memories will be attached to the materiality of the built environment, such as to housing. Jones and Evans (2012) write of urban redevelopment that 'destroying material traces of sites with deep place associations resets the clock on the embodied relationship between the individual and that environment' (p. 2326). The empirical evidence discussed in this chapter refutes this statement. The material was absent; the homes, streets and back-lanes that formed part of the material backdrop of participants' biographies had gone; however, this absence served as a presence for the sharing of memories producing the affective heritage of Byker's redevelopment. This is, in many ways, unique to the heritage of social housing in an instance where people continue to live out their everyday lives in such heritage sites. Memory work is seldom about providing an accurate and authentic record of the past; it is in fact more often about seeking to establish a connection between place and personal biography. Therefore, nostalgic memories and an affective atmosphere of nostalgia, looking to the past and critiquing the present, become a key part of affective heritage, important for understanding the meanings and values associated with a myriad of heritage sites, not least of all those which are still residential dwellings.

Armed with the above conceptual understanding of affective heritage, this chapter demonstrates the important tensions of presence and absence (Anderson, 2009) in affective heritage and how this can reconfigure, and disrupt, people's relationship with place. The listing of buildings brings to our attention the authorised heritage discourse (Smith, 2006) of a place; what is valued, preserved and worth protecting. In valorising what is present, however, authorised heritage

240 *Sophie Yarker*

discourses can by extension hint at what is absent. It is through an analysis of affective heritage that these absences and the complex relationship with what is present are brought to the surface through memory, ordinary affects (Stewart, 2007) and affective atmosphere (Anderson, 2009).

Listing of social housing in Byker

The Byker estate is a place of complexities. On the one hand, as a predominately social housing estate, the community has become a byword for social deprivation and crime, and the Erskine redevelopment still plays a vivid role in the political imagination of many residents and former residents (Yarker, 2014). On the other hand, and because of the redevelopment in the 1970s, Byker is celebrated for its architectural credentials and hailed as one of the most influential pieces of architecture of its time. How these conflicting elements are made sense of by residents in the lived experience of the estate becomes the central curiosity of this chapter. This makes it a rich social milieu from which to explore questions of how affective heritage can be used to critique authorised heritage practices such as the granting of listed status to buildings.

The listing of a building is one of the most obvious, comprehensive and yet contentious ways in which the built environment can be subsumed under attentions of an authorised heritage discourse. As a professional discourse that validates and defines what is or is not heritage, the authorised heritage discourse, as defined by Smith (2006), posits heritage as innately material, aesthetically pleasing and an inherent 'good' for the community concerned (Waterton and Smith, 2010). In addition, there is often an assumption of the 'good' of preservation for future generations and an uncritical connection to community identity. These assumptions and values of authorised heritage have been roundly critiqued (Avrami *et al.*, 2000; Pendlebury *et al.*, 2009; Smith and Akagawa, 2009) and are further problematised by an analysis of affective heritage.

As a result of such critiques, those within the heritage industry have, in recent years, had to consider more fully the plurality of meanings of heritage and the values ascribed to the built heritage environment. This has led to a widening out in the attentions of the heritage sector to include more recently built heritage, bringing the listing of buildings into the post-1945 period. As a result of this enfranchisement, the welfare state and associated buildings built for the working class are now, more than ever before, being recognised as cultural heritage. For some, this marks a move within the heritage sector to begin to cast off its somewhat elitist persona and preservationist stance which, despite a succession of reforms, continue to shroud such work. However, as Pendlebury *et al.* (2004) argue, this has not been wholly accompanied by a fundamental broadening out of the heritage sector; as such, the relatively narrow set of assumptions around what is valued as heritage is constructed continues to persist.

The use of heritage in attaining the objectives of social inclusion has been a different purpose of the broadening of the attention of the heritage sector (*The Heritage Dividend*, 1999; *A Force for our Future*, 2001; *People and Places*, 2002);

however, these shifts in the focus of heritage often fail to adequately respond to its egalitarian ideals (Pendlebury *et al.*, 2004), with much of the heritage sector's work continuing to replicate the values and assumptions of those within the industry, albeit with a nod to those groups it was often deemed to exclude – social housing tenants being one such group (Smith, 2006). In short, the heritage sector appears to remain reluctant to relinquish its powers of definition and recognition when it comes to the designation of heritage value (Waterton and Smith, 2010).

Alongside such critiques there also remains the question of whether these perceived excluded groups wish to be engaged in the arena of formalised heritage in the first place (Pendlebury *et al.*, 2004). The decision to list the Byker estate was shrouded in an aura of community consultation; ultimately, however, the decision to list was instigated elsewhere, by the Department of Culture, Media and Sport. The exact details of how this particular post-war social housing estate came to be listed are difficult to pin down, with various parties and stakeholder groups holding different views on the matter. However, what does appear to be of consensus is that the impetus came from the need to preserve rather than any wider social motives towards inclusion or renewal. Research conducted with stakeholders of the estate leading up to the listing suggest local residents felt at best ambivalence and at worst suspicion and mistrust towards the decision to list (Pendlebury *et al.*, 2009). The feeling among many residents was that the decision to list came 'from outside', with particular references to middle-class tastes and values, and is important to bear in mind for subsequent discussions in this chapter. The feeling of interventions being 'done to' rather than being 'done for' the community was a recurring local structure of feeling in Byker (Yarker, 2014, drawing on Williams, 1977) and is inflected through many of the dimensions of affective heritage to be discussed later in the chapter.

Capturing affect in Byker

Capturing affect is, of course, a difficult question for researchers. In attending to the non-representational, Thrift (2000) has previously critiqued cultural geography's methodological timidity, leading Latham to call for 'new ways of engaging with how individuals and groups inhabit their social world through practice and action' (2003, p. 1993). Thrift advocates drawing on performing art in our methods in emphasising the role of practice and performance. To this end, Duff (2010) draws on de Certeau's understanding of 'pedestrian speech acts' (1984, p. 97) in walking to understand how people are affected by the ways in which they create 'novel pathways and alternative uses for the city's designated places' (Duff, 2010, p. 884). The value of walking interviews in researching the affective dimension of place has also been demonstrated by Jones and Evans' (2012) development of the practice of rescue geographies, used to capture the embodied relationship between communities and urban space. Interviewing while walking, they argue, allows for a 'placing' of the personal and demonstrates the active connection between the body, landscape and memory. The practice of walking itself allows oral history-type recollections to be connected to very specific geographical sites and allows an exploration of the practised and affective dimensions of belonging to place to be observed and explored. In trying to capture affect it, of course, becomes 'represented' in some form in our field-notes, transcripts or audio files. However, by moving through space and attending to the memories, ordinary affects and affective atmospheres encountered by participants, this research makes an attempt at discovering what may be more helpfully understood as 'more-than-representational' understanding of heritage (Waterton, 2014, drawing on Lorimer, 2005).

Walking interviews were therefore selected as the best methodology to capture the nuanced and embodied relationship between people and place. As part of the celebration of Byker's architectural heritage, resident-led walking tours have run on an informal basis since 2008, loosely associated with a programme of national 'Open Heritage' events which sees key sites of historical and social interest opened up to the public for tours and exhibitions.

The specific tours attended for this research ran with the aim of developing a self-directed walking tour 'designed for those who wish to explore the Byker redevelopment' (Byker Discovery Walk, 2012). The research informing this chapter accompanied a total of four walking tours undertaken in June–September 2012. The number of tour participants varied from five to twelve and was mainly made up of current and former residents, as well as a few local heritage enthusiasts usually drawn by the opportunity to explore an iconic local landmark. The tours were pre-planned and I joined as an interested party, outlining at the outset that I was interested in how residents responded to the listing of the estate. As such, the approach taken towards collecting data was fluid. My usual approach was to engage in informal conversation with one or two participants as we walked and at some stage ask if they were comfortable for the discussion to be audio-recorded on a Dictaphone. In all cases the response was positive, giving me an audio transcript of the conversations not just between myself and the participants, but also between each other and with the tour-leader. Of course, as multiple conversations often happened at the same time, the audio recorder could not pick up everything, so field-notes written up in as much detail as possible directly after the event became an additional source of 'capturing' the affect of the tour. It was from these notes that I was able to reflect on the emotive responses of the participants, as well as my own affective response. The sections that follow draw heavily on the empirical material produced by the walking tour research.

'Seeing and feeling through the gaps'

The physical act of walking through the streets of the estate, tracing the pattern of the redevelopment, served to highlight the very material changes in Byker's layout. During the redevelopment the urban layout changed dramatically. The architect and his team took advantage of the topography of the hillside to make the most of the enviable views both up and down the River Tyne. Linear rows of terrace flats that had previously run down the steep hillside to the river were replaced with houses facing into one another. Streets became squares; blocks faced into communal gardens and, for those higher up the hill and flats within the perimeter wall of the estate, looked out onto riverside views. Whether understood as reconceptualising the community's relationship with a deindustrialised river or as a privatisation of the views, the change in the community's familiar layout is still a matter of debate (Yarker, 2014). In a nod to continuity, however, many of the original street names were retained, although often bearing little relation to the pathways they had once marked. It was these linkages to the past which often served as prompts for those on the walking tour to be able to situate themselves within the now less familiar landscape.

During one walking tour I met Linda, a divorced woman in her late 50s who had moved back to the community to look after her ill mother. Linda was in the process of tracing her family tree and joined us on the walking tour as she was curious to learn more about the streets on which she – and previous generations of her family – had grown up. For Linda it was Kendal Street, the street where she was born and raised and where her mother and aunts continued to live after she moved away, which held the most memories of her time in Byker. Kendal Street, as she remembered it, was no longer there; yet the patterns of everyday life and emotional memories of her family having been there connected her to place and remained vivid in her memory:

> My mother and my aunties, they all lived on the same street, Kendal Street, that they were born on and they were in and out of each other's houses, their mother's house and what happened, they all got scattered all over, and honestly they broke their hearts. They all got moved to Walker etc. and they were now a bus ride, or two bus rides, away and it makes such a difference and they always said we don't care about hot water and we just want what we had. Because there was a whole debate about should we keep the houses and just refurbish? A friend of mine was a university student at the time and involved in the demolition and said they had great big beams, you know they were solid houses. They could have just modernised and given us inside toilets – it could have all have been done and you would have kept the community. And you get social problems, tear a community apart.

There was a physical absence in that Kendal Street, as Linda remembers it, was no longer there. This absence, in tension with the presence of the listed buildings of the redevelopment, produced for Linda a palpable pressure (Williams, 1977) of loss prompted by the memory of her aunts' 'heartbreak' and of a community torn apart. This affective heritage of loss and the sense of perceived injustice that accompanied it ignited for Linda a well-worn debate on the politics of the time; a time that saw Byker and working-class communities like it designated as slums and redeveloped within an agenda of modernisation and 'progress'. Linda's questioning of the legitimacy of the redevelopment – why did they not just refurbish instead of demolition – directly challenges the architecture that listing recognises as well as the authorised heritage discourse that listing values and ultimately protects. Within the listing, there is an absence of recognition of the 'heartbreak' experienced by families such as Linda's or of the value of community and desire for the continuity of close social bonds. Further into the walking tour, Linda discussed her relationship with Byker today:

> Now I am back and I love it, and I feel committed to it, this community and to providing the connections between people. I think it is almost like, this stage in my life I'm walking on the bones of my ancestors or something.

Here, Linda describes the ordinary affect of connection to her ancestors when walking the estate. The presence of her ancestors' spirits during her movement through the estate acts as 'fragments of past moments glimpsed unsteadily in the light of the present', to borrow from Stewart (2007, p. 59), which are 'snapped into place' by being physically back in the place where she and her ancestors had grown up. This affective heritage references the tangible elements of the built environment (the ground beneath her feet) along with the intangible ordinary affects and memories produced by being back in Byker. There is a haunted quality to Linda's experience of place: her ancestors and family memories of Byker are both present and absent in the affective heritage that the listing prompts, and the affective atmosphere of loss for Linda is charged with both personal reflection on the 'heartbreak' of her aunties and a wider-reaching sense of injustice on the part of the community. These are emotions and experiences not reflected in the authorised heritage discourse of the listing, and therefore the affective heritage of the built environment amplifies even further. Coming back to the present, Linda reflected on her reasons for remaining in Byker:

> I think it was reaching a certain stage in life where the kids are grown and my job ended in Manchester City Council, do I want to stay in Manchester the rest of my life? I love being near the sea here and I suppose the rest ... (pauses) everything is familiar, comforting memories from your childhood.

The identification of a familiar and comforting atmosphere would, at first, appear to be in tension with Linda's earlier emotionally changed critique of the architectural changes. When challenged on this point, Linda returned to her memories of what the area had previously looked like and drew on personal memory in recreating the familiar and comforting atmosphere she spoke of:

> It does (look different) I know, but I still ... just across this bridge here to the left is Albion Row and my aunties had a pie shop, so I always remember the pie shop, just full of smells, being in the kitchen. Even though it isn't here it is still in my head, I fill in the blanks.

Here, Linda demonstrates again the elements of both presence and absence in the affective heritage of Byker. The current architecture held little meaning for her; instead, she saw past it and through the gaps, filling in the blanks in her own mind of what used to be there and how this connected her memories of place with her sense of attachment to Byker. The presence of the listed 1970s architecture reminded her of what is absent: Kendal Street and her aunts' pie shop on Albion Row, with its kitchen full of smells. However, what is physically absent is not without affect. Linda drew on these affective atmospheres of comfort and familiarity and sensory memories of smell and sight to create a presence of Albion Road, as she can relate to it in her own mind. She was thereby able to re-connect herself to a landscape that, for the most part, was unrecognisable to her in its current physical form.

Claiming space

The performance of the walking tour itself proved a notable way in which those participating could lay claim to space, locating themselves within the landscape either as a source of authority and knowledge or by the remembering and marking of personal life events. For one older gentleman, Norman, the site of a mini-roundabout – on one of the few main roads cutting through the estate – became transformed for both himself and the rest of the group, as the site of his former home and place of birth. In laying claim to this space, he paused while one of the tour organisers took a photograph of him standing on the spot where the house once stood. In this instance it was very clearly the absent that was being valued as heritage by this gentleman, and any reference to the authorised heritage which had replaced the house of his birth was made in opposition to what it had physically removed.

Throughout the walking tour Norman spoke of there being 'nothing for him' in Byker anymore. A widower, with grown-up children and grandchildren scattered across the country, his expressions of loneliness were made all the more poignant with the taking of this solitary photo. However, for him, this was the moment in the walking tour at which he was able to assert himself back into the landscape and re-kindle a sense of attachment and belonging. This had little to do with the architectural heritage the walking tour was supposed to be attending to and, much like Linda's sense of 'walking on the bones of her ancestors', demonstrates the affective atmosphere of presence-in-absence that individual memories create and attach to.

Linda's and Norman's experiences of the walking tour speak to very personal feelings of being both simultaneously placed and displaced by the affective heritage of Byker, and the affective atmospheres that it produced for them. Both felt a pulling and pushing of 'affective force' (Duff, 2010) produced in certain spaces of personal memory within the estate. However, there were also examples of a collective remembering in place that sought to locate not only individuals in Byker but also a collective sense of identity and community which many felt had been lost at the expense of the redevelopment. This reminds us of the contagion of affect in its ability to circulate among people (Dittmer, 2010) and draws attention to the movement of affect between bodies. With specific reference to participants of heritage and architectural tours, Craggs *et al.* (2013) refer to such tours as being an 'interpretive community' (drawing on Fish, 1980). Extending this further, Reed (2008) suggests that members of a single interpretive community are distinguished by what they agree to 'see'. This agreement hints at the relational dimension to affective heritage but also its grounding in place. In the case of Byker, there appears to have been a tendency to agree to 'see', and therefore value, what was not actually physically there. This must be understood in the context of the politics of Byker as a former 'slum-clearance' redevelopment and site of opposition (architecturally at least) to the modernisation discussed earlier. It also hints, in agreement with Geoghegan (2013), at the sociability of such affect and emotional responses to heritage, and as a result the tour produces a sense of belonging and connection both to place and to those with whom they are partaking in the experience.

One such example of this was a discussion of the condition of the terrace housing which had previously dominated the area. Those participants of the walking tour who could remember the old housing in which they lived as children gleefully swapped experiences of weekly tin baths in front of the fire, ice on the inside of the windows in the winter and outside toilets. These objectively dour stories were told with great humour and affection, but mostly laden with a sense of irony and finished with the statement: 'but those were the good old days!' Of course, many of these storytellers were fully aware of the 'audience' to which this narrative was being performed, as many others on the tour would have had little or no personal experience of the conditions they were speaking of. Nonetheless, there was a sense of a collective identity being asserted based around a shared experience of living in 'old' Byker. Told in the context of urban renewal – the demolition and replacement of what was judged at the time to be sub-standard living accommodation (and sometimes 'sub-standard communities') – these stories take on a political meaning of opposition to the modernisation of an area which for many meant the death of a sense of shared community. To trace the societal changes in Byker back to the architecture is, of course, impossible; what is worthy of note, however, is that in this instance the new architecture of Byker comes to stand for much of what was lost in terms of community and collective identity. The redevelopment occurred at a time of significant social, economic and political change in the UK. However, many of the perceived negative consequences of this – an increasingly individualised society, inequality and decline of 'community values'–nevertheless become transposed onto the architecture, making the decision to preserve by listing even more contentious. A sense of loss circulated among the participants of the tour through the memories and affective heritage of the listed built environment. It was an affect which produced a sense of identity for many participants that united them in opposition to the redevelopment of the 1970s and, in part, the listing of 2007.

Vying for place

Agreement was not always the result of shared narratives during the tour. As well as shared memories of old Byker, participants on the walking tour at times disagreed with regard to how things had been, often resorting to mobilising 'authentic' memory in an effort to claim space. The listing includes all the Erskine

Social housing as built heritage 247

redevelopment buildings, as well as all of the original community buildings which were retained by the architect. The former funeral parlour, in which the Erskine development team set up an office, was one such building preserved presumably for its architectural importance, and served as a reminder of the participatory credentials and spirit of 'demystifying the architect' of the redevelopment. Now the home of the housing association rent office, a blue plaque on the side of the building remembers the 'humane and inspirational architect'. The tour stopped at this site and the tour guide started to explain its historical significance, narrating:

> This is the remnants of a corner of a whole street of houses, Brinkburn Street, and this was retained because it became the architect's office, so they were based in Byker during the redevelopment which is quite unusual.

This piece of architectural information was directly related to the participatory credentials celebrated by the authorised heritage discourse of the listing. It was, however, roundly dismissed by some of those on the tour, as talk instantly turned from what was present to what was absent and the memories this prompted. In this instance, the conflicting memories expressed by members of the tour produced an effect of vying for place among the participants, as each sought to assert their own recollection of the past in an attempt to lay claim to place and their attachments to it. The following 'friendly disagreement' over the local butchers demonstrates this:

> Janet: I was on the corner of Brinkburn and Kendal Street, my Mam and I lived above Howard the butchers.
> Raymond: Oh – I know which one you mean, it wasn't called Howards though. Janet: It was, it was Howards.
> Raymond: Oh right, there was another name though, because my Granny lived half way down the bottom end of Kendal Street.
> Janet: I remember, because when I got married the butcher lads sent me a telegram saying 'take advice from one who knows, tie your nighty to your toes'! and that got read out at the wedding! I was 18 years old, I was dead embarrassed!
> Raymond: I was talking to this old guy the other day from Killingworth, and he was talking about the butchers as well, but there was another name for it, I have never known it as Howards, I knew it as something else and I can't think of what it was.
> Janet: It must have been Howards, he was there for ages.
> Raymond: I know but I didn't know it as Howards…
> Janet: Maybe that was just his first name, maybe I knew him by first name.
> Raymond: Matches! That's it! Matches! I just had to dig a bit deeper. And I tell you why when you come down the street at certain times of the day, Matches, you smelled it before you say it, it used to stink as a butchers at times.

Janet: Well believe me, my bedroom was just upstairs and it used to vibrate because the fridge was underneath, but the worst day was Mondays because they used to boil the sheep's heads right outside the window downstairs and none of my friends wanted to visit me.

Raymond: You could come down Brinkburn Street, just half way down and you could smell the butchers.

This exchange between Raymond and Janet was conducted in good spirits, but for both of them there was a clear need to be 'right' and a need to be able to connect their own personal memories about Matches the Butcher to place. To do so, both drew on sensory memory: Janet described feeling embarrassed by the butcher boy's telegram and Raymond commented on the smells. These sensory memories produce affective forces, demonstrating Degan and Rose's (2012) suggestion that sensory engagements with place are mediated by memories of what is absent and that 'buildings, streets and squares may be seen, heard and smelt through memories of what was once there' (p. 3280). Despite the butcher's shop no longer existing, there was enough in the memories and affects circulating between the participants and the built environment for both Raymond and Janet to begin 'filling in the blanks', as Linda had done in relation to the former site of her aunts' pie shop. The exchange between Raymond and Janet demonstrates the need to be able to use affective atmospheres, created by memory, to locate oneself in place and to stake a claim in belonging there. Here, the concept of embodied remembering (Waterton and Watson, 2014) is helpful in explaining how the 'remembering' performed by Janet and Raymond in these spaces in Byker allows us to access and encounter the more-than-representational experience of affective heritage. Memory and heritage become linked with affective heritage through such sensory and embodied experiences of place.

As mentioned earlier, many of the stories and memories told by members of the walking tour of 'the good old days' and of Byker before the redevelopment could easily be dismissed by some as simple nostalgia, yearning for the past and retreating from the present. This would be to underestimate the value and meaning of such sentiments. As an affective register in the context of listed social housing, where there is a personal and a collective sense of loss and longing for home and community, it becomes pertinent to make brief reference to the question of nostalgia and to provide a discussion of how nostalgia, as an affective atmosphere and register, may be better interpreted.

Reflective nostalgia

An attempt has been made in recent years to rescue nostalgia from sinking into reactionary and conservative politics (Bonnett and Alexander, 2013) and, on this basis, there is an argument that nostalgia needs to be reinterpreted to take into account its potentially progressive and reflective nature. This is a critique that originates from the ground-breaking work of Fred Davis (1979) and his understanding of reflexive nostalgia as a more self-aware, critical concept. It is unsurprising that heritage tours of communities such as Byker, which have witnessed stark physical as well as cultural and social change during a particular period, should prompt such expressions of nostalgia. House museums have been explored within the heritage literature in a similar way, in that they 'collapse the present with the past' for the visitor and create an atmosphere of presence and absence (Gregory and Witcomb, 2007, p. 265). This was very much the experience for residents and former residents of the Byker walking tour – the physical presence of the listed redevelopment brought into sharp focus that which was absent. As nostalgia is often prompted by change, this causes it to be so often negatively conflated with a sense of wanting to 'hold on' to some idealised version of the past at the expense of looking to the future. As Hodge (2011) argues, heritage practices and scholars should engage more positively with the work of critical nostalgia and confront its social functions.

This would demand an understanding of the potential for irony within nostalgia – an understanding which would enable a more complex interpretation of statements such as 'but those were the good old days'. Such examples of remembering in place would amount to a critical-awareness ability to reflexively locate oneself alongside place, instead of being wholly within it. Therefore nostalgia of this sort should not be dismissed as reactionary and inward-looking, but seen as potentially outward-looking and progressive.

Working with this re-interpretation of nostalgia, it can be seen – in this instance – as playing a role in the claiming of space by an individual or group. Access to a collective form of nostalgic memory, for some of the walking tour participants, gave a sense of access to a privileged 'insider' position (Rowles, 1983) premised on knowledge and connection to 'old' Byker and 'the good old days'. Nostalgia, for the participants on the walking tour, was about far more than wallowing in 'sentimental denial' (Hodge, 2011, p. 120); rather, by participating in the telling and sharing of nostalgic memory, they were able to reassert their attachment to Byker and belonging in place *through* the use of nostalgia, despite the presence of an authorised heritage discourse of the redevelopment which stood in contrast to this. A reflective and critical use of nostalgia therefore becomes an important part of the affective atmosphere of heritage.

Conclusion: presence and absence in affective heritage

This chapter has demonstrated the importance of affective heritage in illuminating the gaps between the authorised heritage discourse of listed building status and the value ascribed by individuals and communities. In teasing out the different elements of affective heritage – using the elements of memory, ordinary affects and affective atmosphere – it has become clear how individuals can see and feel through the gaps of authorised heritage and create new spaces in which to reconnect themselves with the landscape and claim a sense of local belonging and attachment. Key to this understanding of how these tensions between authorised heritage discourses and community discourses are worked out is an understanding of both presence and absence in affective heritage.

The empirical findings presented in this chapter have sought to draw attention to the need to develop a more sophisticated understanding of the impact of listing social housing on the people who live within it, who may not share the same values of heritage that their homes have become enshrined within. In short, the listing of the built environment has complex and often unexpected impacts on residents' relationships with place, and the lens of affective heritage has been shown to highlight some of these tensions and provide a window into these relationships with place. This chapter showed how individuals were able to connect themselves back into a landscape which, in many ways, was quite alien to them, through affective atmosphere and memory. In this way authorised heritage discourses around the process of listing did not always have the effect of dislocating an individual from place, but did cause a re-working of belonging that drew on what was both absent and present in the affective atmosphere produced by heritage discourse and the built heritage itself. In particular, this chapter highlighted the need to think about how we understand nostalgia in people's relationship with the built environment, not as a straightforward plea for conservatism and a return to what was there before, but as a plea for a place that can support the image of self and memories bound up in a person's biography. Understanding the probing and working out of the tensions between absence and presence within affective heritage is key to this.

The exploration of presence and absence in affective heritage presented in this chapter is one which interprets the relationship as co-constituted and co-existing, rather than the two existing as opposing binaries (Jones *et al.*, 2012). It was through the absence of tangible buildings and streets, as well as the absence of intangible value attached to them though the listing, that members of the walking tour were able to be 'present' themselves in place and to locate themselves within the built heritage that surrounded them. It was also the very imposing presence of the listed architecture of the redevelopment, and its jarring with the heritage values of residents, that brought what was absent to the surface.

This co-constituted relationship between presence and absence is captured well by the idea of places being 'ghostly' (Maddern and Adey, 2008; Wylie, 2007) or haunted (Edensor, 2008; Mah, 2010) and substantiated by some of the narratives explored above. The participants in the walking tour made regular mention of people who were no longer in Byker, as well as to the absence of certain parts of the built environment. For Linda this was her mother and aunts and their pie shop, and for Norman his grown-up children and former home. Yet their presence was felt in a haunting affect for both participants, stirred up by the practice of the walking tour itself and the memories it brought back.

The haunting affect produced by the absence of certain features of the built environment has been explored by Mah in the haunting memories of an industrial past produced by industrial ruins of former shipbuilding communities. The presence of the 'ruination' of buildings connected to disused shipyards serves as a locus of 'living memories' for participants in Mah's research. Memories linger in such sites and the absence of conditions of their former use is palpable.

A fascination with the simultaneous presence and absence in such 'ruins' has attracted considerable academic and popular attention in recent years. The de-urbanised formerly great cities such as Detroit have become the poster children for the provocatively named trend in 'ruin-porn' and the transcending of sites for artistic and political purposes by urban explorers has attracted considerable attention and critique. These examples, however, demonstrate that discussions around presence and absence tend to privilege the visual at the expense of other sensory experiences (Jones *et al.*, 2012). The material presented in this chapter serves to develop these debates by contributing empirical findings of other senses and the more-than-representational being drawn upon in the experience of presence and absence. The absence of smells, for example, features in both Linda and Raymond's accounts of place, as does the absence of a 'friendly atmosphere' or a sense of belonging for Norman. These more intangible expressions of absence are evident in the affective heritage of place in Byker and form an important part of the dissonance between the lived heritage on the one hand and the authorised heritage of listed buildings on the other.

These tensions created in the participants of the walking tour illustrate what Stewart refers to as 'pressure points and forms of attention and attachment' (2007, p. 5) – ordinary affects which provide potential modes of knowing through 'fragments of past moments glimpsed unsteadily in the light of the present' (p. 59). The affective heritage experienced was fleeting, a glimpse into the past, yet made unstable by being viewed through a sense of today. The authorised heritage discourse of the listing acted as a powerful illuminator of anything (memories, affects and conflicting ideas of heritage) that did not align with the values enshrined in the listing. It was this which brought the tensions between what was present and what was absent to the surface, and it was this which needed to be reconciled within the participants as they sought to locate themselves within the authorised heritage landscape of the listed estate.

Listing buildings in which people still live and go about their everyday lives, and to which they have often strong and emotional attachments, will, without doubt, produce strong reactions and the potential for disagreement. Listing a building in a community with a history of locally disputed top-down intervention intensifies such a response. The dissonance in Byker between the value of heritage enshrined by the Grade II* listed status and the local structure of feeling around the redevelopment held locally has been highlighted and explored in this chapter through the conceptual tool of affective heritage. Listing a building which people call home serves to add another layer of interpretation onto a place already laden with conflicting meanings and values. In the case of Byker, this served to produce further complexities in the relationships of residents and former residents of the community. Affective heritage, attending to the affective atmospheres, ordinary affects and memory brought to life during a heritage walking tour, provides one route into understanding these complex relationships.

Bibliography

Anderson, B. 2009. Affective atmospheres. *Emotion, Space and Society*, 2(2): 77–81.

Avrami, E., Mason, R. and de la Torre, M. 2000. *Values and heritage conservation*. Los Angeles: Getty Conservation Institute.

Blunt, A. 2003. Collective memory and productive nostalgia: Anglo-Indian homemaking at McCluskieganj. *Environment and Planning D: Society and Space*, 21: 717–38.

Bonnett, A. and Alexander, C. 2013. Mobile nostalgias: connecting visions of the urban past, present and future amongst ex-residents. *Transactions of the Institute of British Geographers*, 38(3): 391–402.

Byker Lives. 2012. *The Byker Discovery Walk*. Developed by Colin Dilks, produced by Byker Lives.

Craggs, R., Geoghegan, H. and Neate, H. 2013. Architectural enthusiasm: visiting buildings with the twentieth century society. *Environment and Planning D: Society and Space*, 31: 879–96.

Davis, F. 1979. *Yearning for yesterday: a sociology of nostalgia*. New York: The Free Press.

de Certeau, M. 1984. *The practices of everyday life* (S. Rendell, trans.). Berkeley: University of California Press.

Degan, M.M. and Rose, G. 2012. The sensory experiencing of urban design: the role of walking and perceptual memory. *Urban Studies*, 49(15): 3271–87.

Dittmer, J. 2010. *Popular culture, geopolitics and identity*. London: Rowman and Littlefield.

Duff, C. 2010. On the role of affect and practice in the production of place. *Environment and Planning D*, 28: 881–95.

Edensor, T. 2008. Mundane hauntings: commuting through the phantasmagoric working-class spaces of Manchester, England. *Cultural Geographies*, 15(5): 313–33.

English Heritage. 1999. *The heritage dividend: measuring the results of English Heritage regeneration 1994–1999*. London: English Heritage.

Fish, S. 1980. *Is there a text in this class? The authority of interpretive communities*. London: Harvard University Press.

Geoghegan, H. 2013. Emotional geographies of enthusiasm: belonging to the Telecommunications Heritage Group. *Area*, 45: 40–6.

Gregory, K. and Witcomb, A. 2007. Beyond nostalgia: the role of affect in generating historical understandings at heritage sites, in S.J Knell, S. MacLeod and S.E.R. Watson (eds) *Museum revolutions: how museums change and are changed* (pp. 263–75). New York: Routledge.

Hodge, C. 2011. A new model for memory work: nostalgic discourse at a historic home. *International Journal of Heritage Studies*, 17(2): 116–35.

Jones, P. and Evans, J. 2012. Rescue geography: place making, affect and regeneration. *Urban Studies*, 49(11): 2315–30.

Jones, R.D., Robinson, J. and Turner, J. 2012. Introduction. Between absence and presence: geographies of hiding, invisibility and silence. *Space and Polity*, 16(3): 257–63.

Latham, A. 2003. Research, performance and doing human geography: some reflections on the diary-photography, diary-interview method. *Environment and Planning A*, 35: 1993–2017.

Lorimer, H. 2005. Cultural geography: the busyness of being 'more than representational'. *Progress in Human Geography*, 29: 83–94.

Maddern, J.F. and Adey, P. 2008. Editorial: spectro-geographies. *Cultural Geographies*, 15(3): 291–5.

Mah, A. 2010. Memory, uncertainty and industrial rumination: Walker Riverside, Newcastle upon Tyne. *International Journal of Urban and Regional Research*, 34(2): 398–413.

Pendlebury, J. Townsend, T. and Gilroy, R. 2004. The conservation of English cultural built heritage: a force for social inclusion. *International Journal of Heritage Studies*, 10(1): 11–31.

Pendlebury, J., Townsend, T. and Gilroy, R. 2009. Social housing as heritage: the case of Byker, Newcastle upon Tyne, in Gibson, L. and Pendlebury, J. (eds) *Valuing historic environments* (pp. 179–200). Farnham: Ashgate.

Pyrah, C. 2015. Planning and Development Regional Director for English Heritage, quoted at www.davidlammy.co.uk, cited in Glynn, S. *Byker Newcastle upon Tyne in standard colour*. Herne Bay: Categorical Books.

Reed, A. 2008. City of details: interpreting the personality of London. *Journal of the Royal Anthropological Institute*, 8: 127–41.

Rowles, G.D. 1983. Place and personal identity in old age: observations from Appalachia. *Journal of Environmental Psychology*, 3: 299–313.

Smith, L. 2006. *Uses of heritage*. London: Routledge.

Smith, L. and Akagawa, N. 2009. *Intangible heritage*. London: Routledge.

Stewart, K. 2007. *Ordinary affects*. Durham, NC and London: Duke University Press.

Thrift, N. 1997. The still point, in S. Pile and M. Keith (eds) *Geographies of resistance* (pp. 124–51). London: Routledge.

Thrift, N. 2000. Dead or alive? in I. Cook, D. Crouch, S. Naylor and J. Ryan (eds) *Cultural turns/geographical turns: perspectives on cultural geography* (pp. 1–6). Harlow, UK: Prentice-Hall.

Thrift, N. 2004. Intensities of feeling: towards a spatial politics of affect. *Geografiska Annaler, 86 Series B: Human Geography*, 1: 57–78.

Waterton, E. 2005. Whose sense of place? Reconciling archaeological perspectives with community values: cultural landscapes in England. *International Journal of Heritage Studies*, 11(4): 309–25.

Waterton, E. 2014. A more-than-representational understanding of heritage? The 'past' and the politics of affect. *Geography Compass*, 8(11): 823–33.

Waterton, E. and Smith, L. 2010. The recognition and misrecognition of community heritage. *International Journal of Heritage Studies*, 16(1–2): 4–15.

Waterton, E. and Watson, S. 2014. *The semiotics of heritage tourism*. Bristol: Channel View Publications.

Williams, R. 1977. *Marxism and literature*. Oxford: Oxford University Press.

Wylie, J. 2007. The spectral geographies of W.G Sebald. *Cultural Geographies*, 14(2): 171–88.

Yarker, S. 2014. *Belonging in Byker: the nature of local belonging and attachment in contemporary cities*. Unpublished Thesis, Newcastle University.

Part III
Practices

13 'Please Mr President, we know you are busy, but can you get our bridge sorted?'

Keith Emerick

In this chapter I will consider the difficulties that can arise between 'the public' and heritage practitioners over the ways in which heritage is perceived, used and managed. These conflicts and difficulties usually revolve around two themes: (1) what is heritage? and (2) how can you use it in the present? Two case studies will be presented that illustrate affective and emotional dimensions in heritage and the degree to which people are passionate about heritage matters. The two case studies operate at different scales, but are linked by the intensity and tenacity of the participants; they also reflect an intensity that is in contrast to the measured responses of the heritage professionals who are also 'players' in the two cases. I will conclude with a proposal for a different, more inclusive approach to heritage management that reflects the nature of such engagements.

The relationship between the public and heritage practitioners can be difficult. Quite often the two sides are presented as exactly that: two unambiguously distinct and opposing sides or forces in conflict. The reality, of course, is different; 'the public' is made up of numerous people who feel, believe and act in different ways and no two heritage professionals feel, believe or act in the same way. However, current conservation and heritage practice is shaped by a professionally (although not necessarily formally) agreed practice based on conservation charters, principles, conventions and guidelines inherited and developed from past practice. Prevailing national or international heritage legislation adds a further layer to practice and the manner in which expertise is exercised. Even if a heritage practitioner might believe one thing, she or he is constrained by what an institution, or the heritage law of that country, allows or instructs that person to do (Jones, 2009). The implication of this mass of self-referencing documentation, as observed by Laurajane Smith (2006), is that there is an Authorized Heritage Discourse (or AHD), and the effect of this discourse is to value, privilege and sustain a narrow expression of heritage based on expertise, 'the old', authenticity and fabric. Heritage – usually property – is defined as something inherited from the past to be handed on, unchanged, to the future. In practice there are two extremes of fabric – the iconic and the everyday. Practitioners are largely drawn towards, and are passionate about, the iconic, and position such places in national and international narratives in ways that are meant to engage a wider public in the same way. The everyday becomes second-class and diminished as 'local', but has its own passionate supporters who recognise and value such places as part of their lives and stories. However, the iconic also has 'local' meanings; these, though, are usually considered less 'legitimate' than any national narrative. Similarly, the local can be emblematic of bigger national and international questions. In each case, people and communities stand in some relation to heritage in a way that reflects their engagement with it.

The practical implication of this discourse for the heritage manager is that they have inherited a particular perspective about what heritage is and how it should be assessed and valued – and this is supported by a very specific language used in the descriptions of places, in heritage legislation and policy, in communications and by one practitioner to another (see the list description of Creets Bridge, Kirkby Malzeard – included in this chapter – and note the brevity and restraint of the language). Emotion plays little part in the process. In general terms, fabric is preferred above the more intangible meanings and 'associations', but the very idea that fabric can be used to tell a national story is itself an 'association'. However, the language of conservation and heritage management changes over time. As the heritage sector develops and reflects on its practice, the language can become more nuanced. An example of this is the growing awareness among practitioners that conservation and heritage have to be more inclusive, and, in reflecting multi-faceted and multi-cultural societies, have to address a broader spectrum of engagement. The practical outcome of this is that a wider range of places are identified or designated as 'heritage' places. But if the language can change, is the same true of practice? Does this really change? And is the language really changing, or is it subtly shifting to ensure that expertise retains its dominant position? It has been argued that heritage is akin to a mirror and everyone should be able to see themselves in that mirror (Hall, 2005), but do those 'new' communities of interest and prospective participants (let alone the 'old' communities) have the opportunity to manage their own cultural heritage, or are the choices and decisions still held by the same people?

The prevailing approach to heritage management is referred to as a 'values'-based approach, whereby the assumption is that conservation and change are not mutually exclusive. To decide how a place can be changed or elements retained, a place is assessed by means of the values attached to it by people. *The Australia ICOMOS Burra Charter* (Australia ICOMOS, 2013) identifies aesthetic, social, scientific, spiritual and historical as the principal value groups, but there are other values, and other organisations identify similar value groups but name them differently. The sum of the values is called 'the significance' or 'cultural significance', and ideally a values-based approach should lead to inclusive and participatory approaches and solutions in the management of heritage places. The values-based approach should also allow 'space' for emotional responses to a place. Aesthetic value can include the feelings of awe or terror generated by, for example, a Cold War nuclear warfare structure. However, inclusivity relies on practitioners being open to becoming one of a variety of stakeholders rather than the deciding voice and accepting a variety of views, feelings and perspectives as legitimate. A similar but still emerging idea is that of 'people centred conservation' (Vermeulen,

2006), which is linked to the idea of 'living heritage' (Court *et al.*, 2014). 'Living heritage' recognises that conservation and heritage management take place in living communities and stresses the need for partnership, participation and capacity-building in the search for local solutions. It is less clear what this means in practice and how such approaches deal with conflict and dissonance, although the idea of 'living heritage' seems to be coalescing around the intangible: craft skills, dance, performance and so on. The reality may be that 'living heritage/people-centred conservation' merely states the obvious about the manner in which people and community have been disenfranchised by the location of power in expertise; things that conservationists and cultural heritage managers have been slow to recognise, or have been happy to ignore.

How can these changes in language and principles be seen in actual cases, and how do they reveal the two main themes of this chapter: how can we use heritage places in the present, and what is heritage? The first case study looks at a dispute over a bridge in North Yorkshire, England, where there was also a clear distinction between the two main parties about how a heritage place could be used in the present. The evolving disagreement took an unexpected turn, but one that clearly demonstrated how passionate some people were about 'their' heritage. At the same time, it illustrated that the actions of heritage managers can be defined by short and long-term political considerations, by the careful assembly of facts and data and ultimately by pragmatism. The second case study considers an area of heritage frequently ignored by heritage practitioners: the young and their cultural heritage. Of particular note here is the anger felt at being excluded from cultural dialogue, and the way that this, in this instance, generated an effective and ultimately successful campaigning group.

Creets Bridge, Kirkby Malzeard, North Yorkshire: background

Creets Bridge (see Figure 13.1) is a small road bridge over a minor stream, to the east of the village of Kirkby Malzeard, on a minor road leading from the city of Ripon.

It sits in the Harrogate district of North Yorkshire, UK. Kirkby Malzeard is six miles to the north-west of the city of Ripon and five miles north-north-west of the Fountains Abbey and Studley Royal World Heritage site. It is also within the Nidderdale Area of Outstanding Natural Beauty (AONB), a designation placing the area at a slightly lesser grade of importance than a National Park, but a grade that is a material consideration in planning matters. The bridge is listed Grade II, a designation that marks it out as nationally important for its architectural and historical 'special interest' and identifies it for special consideration in the planning process. The bridge is narrow and relatively short in length. Approaching from the city of Ripon, vehicles descend a wooded hill, turn a sharp right-hand bend and are on the bridge before they see it. The parapet of the bridge is relatively low, looking like two low walls. The only immediately distinguishing feature is the two piers of stone with stone balls at their top at the village end of the bridge.

Figure 13.1 Creets Bridge following its reconstruction, seen from the village.

The official list description (its entry on the National Heritage List for England, NHLE) describes it as follows:

> Bridge. Late 18C Coursed ashlar stone. Single segmented arch with voussoirs and keystone. Walling breaks forward either side of the arch to form pilasters. Band, coped parapet.

The list description also includes the NHLE number, its National Grid reference, its local authority location (by County, District and Parish), the date when first listed (25 June 1987) and bibliographical references.[1] There are 374,081 listed buildings in England;[2] of those 2.5 per cent are Grade I ('of exceptional interest'); 5.5 per cent are Grade II* ('of more than special interest') and 92 per cent are Grade II ('nationally important and of special interest').

The village does not have a Conservation Area (a local authority designation identifying the part of the village that has special interest for its character and appearance, and which also merits special attention in the planning process), but it does have a Village Design Statement (VDS). The concept of the VDS (now replaced by Neighbourhood Plans) was conceived by the Countryside Commission for England and Wales (a statutory body, which, following several mergers, is now part of Natural England, a national government agency providing advice on natural heritage) as a tool to help reverse what they considered to be the principal threats to the English and Welsh countryside, specifically the standardisation of new buildings and poor design. The aim of a VDS is to be representative, to give a community a recognised voice in the planning process and enable a village community to manage change, whether the change is cumulative or large or small-scale additions and alterations through the identification of the qualities and features that make a village distinctive. Once completed, VDSs can be 'adopted' by the local authority as part of the Supplementary Planning Guidance – documents that inform larger-scale planning policy.

AONBs are a curious 'halfway house' between National Parks and undesignated landscape. Distinctive landscapes were defined as AONBs because people, largely the inhabitants of those landscapes, felt that the landscapes were identifiable and they cared about the retention of the particular qualities and characteristics that made those landscapes significant, but they can also be understood as operating as 'buffer zones' to National Parks. The concept of the VDS was also part of a larger Countryside Commission objective to give rural communities the opportunity to promote and sustain what was important to them, and this was achieved through a highly successful grant scheme called the 'Local Heritage Initiative'. Grants were made available for local projects and the conservation of those features that people considered particularly distinctive, emblematic or just loved at the local level, but were not recognised by national designation criteria or by national grant awards.

The Kirkby Malzeard VDS followed the accepted format and process: a collaboration between parish council (acting for the village community) and Harrogate Borough Council (as the planning authority). The content of the different elements of the VDS was established, with each section concluding with a set of recommendations: village context; its economic and future prospects; the character of the landscape setting; the relationship between village and surrounding landscape; settlement pattern and pattern of the village; character of streets and open spaces; the character of buildings; sizes, styles and types; height, scale and density; materials; architectural features; highways and traffic; footpaths; cycleways; street furniture and so on, ultimately presenting a 'design guide' for new additions to the village.[3] Creets Bridge is not mentioned in the main body of the text as a particular distinctive building or feature, because it is 'outside' the village and not visible from it. However, it is mentioned in the highways section of the VDS in the following terms: 'Creets Bridge on Ripon Road is particularly narrow with limited visibility.'[4] The draft VDS was approved for consultation on 13 September 2001 and ratified on 12 June 2002 by the Head of Planning Services, Harrogate Borough Council, with the Ward members.

On the night of 2 November 2000, Creets Bridge suffered some flood damage, resulting in scouring of the foundations to the eastern parapet; this caused distortion of the arch and a partial failure of the parapet. A temporary bridge was put in place to take the traffic off the bridge. Discussion began almost immediately about suitable solutions, including repair of the original, introduction of a lights system, converting the temporary bridge to a permanent one and using the opportunity provided by the partial collapse to dismantle the damaged section, rebuild and then widen the bridge, with the latter being the preferred position of the Kirkby Malzeard Parish Council, which argued that most of the village residents wanted a wider bridge.

Repairs were not forthcoming and by 2002 the condition of the bridge had deteriorated. In February 2002, a listed building consent application was submitted by the Highways section of North Yorkshire County Council for the dismantling, rebuilding and widening of the bridge. The proposal recommended the removal and reconstruction of the eastern half of the bridge using original stone as far as possible, and widening it from 4.9 metres to 6.5 metres. On 11 June 2002 (within days of the ratification of the VDS), the application was considered by the North Yorkshire County Council planning committee.

In the intervening period between submission of the application and its presentation at the planning committee meeting, various positions had been adopted. It should be remembered that this application was being considered before the introduction of a 'values-based' approach to planning matters, which came about in England in 2010 with the issue of *Planning Policy Statement 5: Planning for the Historic Environment* (DCLG, 2010), now replaced by the *National Planning Policy Framework* (DCLG, 2012). The chief difference between the *Planning Policy Statement*, the *National Planning Policy Framework* and what came before is that the two former documents adopt the current international conservation language of value, significance and harm, and thereby attempt to balance competing heritage values in order to allow the significance of heritage places to be defined and sustained. Applications for works to Grade II listed buildings were, and are, usually determined solely by the local authority. However, where it is proposed that a listed building should be demolished, it is the practice that English Heritage is consulted. English Heritage, the national body that manages the designation process, advises government on heritage matters and provides heritage advice to local authorities, but since 1 April 2015 has been divided into two organisations: a charity, the English Heritage Trust, which manages the national collection of historic places; and Historic England, the organisation providing statutory advice to government and local authorities on heritage matters.

By June 2002, there was a 116-signature petition opposing widening, six letters in support of the proposal and twenty-two letters opposing the widening. The supporters of widening argued that the opportunity should be taken to widen the bridge and thereby take account of the requirements of modern traffic, which would reduce the risk of accident to vehicles and the bridge, allow increases in traffic volumes and allow Kirkby Malzeard to be accessed by larger commercial vehicles. Objectors argued that demolition and widening would detract from the character and appearance of the Nidderdale AONB, increase traffic speeds and reduce road safety for drivers and pedestrians, adversely affect the character and historical value of the bridge and be time-consuming and expensive.

The various agencies and authorities made their positions clear: two parish councils (one of which was Kirkby Malzeard) supported the proposal, while English Heritage, the County Council's Heritage Unit, the Society for the Protection of Ancient Buildings (SPAB), the Ancient Monument's Society, the Council for British Archaeology and the Cyclists' Touring Club all objected. Their objections were that demolition, reconstruction and widening would compromise the architectural integrity and symmetry of the original design and would harm the intrinsic character and merit of the bridge, and, fundamentally, that such measures were not 'justified', failing the existing test of what was 'desirable' and 'necessary' in existing legislation (*Planning Policy Guidance Note 15: Planning and the Historic Environment*, DoE 1990).

English Heritage and others made the point that the bridge could be repaired and returned to its former form, but that the applicant had failed to justify why the bridge should be demolished and widened, and had not provided any evidence (such as analysis of traffic patterns) to support their arguments about the benefits of bridge-widening. The discussions continued in this vein for several months. At the June 2002 planning meeting, recommendations from Council officers (the staff of local authority specialists) were tabled, outlining that the listed building consent should be refused because the proposal was contrary to County Council policies, *PPG15* and the County Council's adopted Heritage Strategy. It was considered that the widening of Creets Bridge would adversely affect the special character and visual appearance of the listed building and its wider setting, adversely affect the character and ambience of the Nidderdale AONB and unacceptably affect the amenity of non-motorised road users. However, the Council members (the elected representatives) decided to ignore the recommendations and agreed the application to widen the bridge. The objectors' response to widening took some time to develop, but ultimately they challenged the decision and had it referred to the Government Office – the usual practice for an application that is submitted and determined by the same authority. Before considering the next phase of the unfolding argument it would be useful to review the respective positions and their implications.

Creets Bridge: the issues

In many respects, the events surrounding Creets Bridge were not unusual for the practising heritage manager; people agree or disagree with 'official' views and decisions and either escalate those disagreements (they might write to their local Member of Parliament) or accept the decision. The inequalities of representation are a frequent issue in such debates. The numbers of objectors and supporters stated above indicate that the objectors to bridge-widening were in the majority, but the elected Parish Council were clearly in support of widening. But is local debate always between or within the same group of people who are comfortable about writing letters of objection or support? Students of critical heritage studies are familiar with the idea that heritage is contested, but 'contested' is not a word that the cultural heritage manager would immediately recognise and understand. For the heritage manager, the issue is one of 'understanding'; once the importance of a place is understood by the public, then the reasons that underlie an 'expert' decision become clear.

However, this is far from straightforward. If we return to the official designation description of the bridge we can see that the language is 'clipped' and entirely that of the expert, the old, the authenticity of fabric and the aesthetic. The only element missing from this list is an idea of attribution – who designed and built it and who for – but any architectural historian reading this description would know immediately what the bridge looked like and where it would fit into a typology or architectural narrative. But there is no contextual information. What the description does not say is that the bridge is part of a wider eighteenth-century planned landscape associated with the Aislabie family (who created the Fountains Abbey and Studley Royal pleasure gardens) and also part of the designed approach to the eighteenth-century Mowbray House (immediately adjacent to the bridge), where the design of pilasters and ball finials on the bridge is also repeated in the design of the gateway entrances to the house. In addition, small, neo-classical bridges are a significant characteristic of many of the villages in the Nidderdale AONB. Similarly, there is no 'human' context, and no narrative. How big is the bridge? Is it longer or shorter than a standard single-decker bus (the bridge is about the same length as a bus)?

If we then look at the language used in the official objections, the stress is on architectural integrity, the reduction of symmetry, loss of intrinsic character and merit, while the proposed widening was not 'necessary' and was 'unjustified'. But who decides what is 'necessary'? The supporters of the widening scheme phrased their arguments in terms of common sense, local utility and the potential to make life different and provide a future for the village. They recognised that the bridge was designated and 'important' but could not see how, given the circumstance and the inevitable dismantling that would have to take place, the designation should stand in the way of an opportunity to carry out a 'sensible' modification of the structure.

If we look at the VDS, the introduction makes it clear that a VDS analyses the characteristic style of a village 'describing those qualities valued by its residents. Local knowledge, views and ideas have been gathered to make a statement about what will most benefit the development and prosperity of a village' (p. 1). It was also noted that the aim of a VDS was to give a community a voice in the planning process – so, interestingly, the 'official' documentation is pulling in two directions. The designation description, national agencies and amenity groups emphasise the position of Creets Bridge in a national narrative, but in a way that is only accessible to experts, while another set of documentation, responding to the need for inclusion and greater participation in the planning process, emphasises how local character can be part of development and change. The implication of this is that while there is a *wish* to think holistically and inclusively about cultural heritage, there is a lack of *will* to address the implications of what this means for heritage expertise and how heritage expertise meshes with other areas of expertise or interest outside of their own. However, language remains fundamental to any debate, and the language of the expert (whether on paper or in discussion) remains measured and calm and is part and parcel of the construction of rational argument. And yet heritage experts are passionate about their own fields of interest, but have learned to moderate that enthusiasm when dealing with the provision of advice because that advice is part of planning law. It may be the case that the reliance on the precise and controlled use of language, and the creation of specific definitions of professional terminology, has led to the creation of blind spots whereby the 'meaning' of a case or project can be lost: that heritage professionals fail to see the wood for the trees. To the supporters of the bridge-widening the issue was simple and clear, and the inability of others to see this made them more passionate about their case.

Both local supporters of and objectors to the widening scheme used the term 'our' when referring to the bridge, indicating that there was a sense of ownership, but the agencies and amenity groups continued to place the bridge in a national narrative, for which they were the custodians – so their idea of 'our' (if they had used the word) was a 'higher' level of ownership, and an ownership that assumed to be more even-handed, rationalist and elevated than the passionate disagreements that were happening at the local level. They claimed to speak for something that had no voice of its own; inevitably, the voice it was given 'voiced' the precautionary principle, rationality and national narrative, and inevitably, this 'voice' eclipsed the local sensibilities. But this was exactly the role and perspective that is sought for experts (usually by experts) and in this sense their position, particularly that of English Heritage as advisors to the local authority, was absolutely correct because it met the obligations of the existing legislation and their national remit. It could be argued that the supporters of the widening proposal had little or no empathy for the architectural merit and national importance of the bridge, but how had this been related to them in the past? And did national importance mean that the bridge had to remain as it was whatever the consequences and forever? Interestingly, the word 'authenticity' was rarely used (although 'integrity' was mentioned by the Council for British Archaeology in terms of 'architectural integrity'). This may be because the reality of the necessary conservation repair, whether that repair would be to dismantle the damage and then reconstruct it, or to dismantle the bridge and then rebuild and widen it, would inevitably reveal that 'authenticity' had already been lost.

Would a 'values-based' or 'living heritage' approach have produced a different result at this stage? If we use the value groups in the *Burra Charter* (Australia ICOMOS, 2013), we can see that the bridge has some historical and aesthetic value but limited social and scientific value; however, this does not necessarily produce a 'better' answer and does not help us address the conflict between the various parties. However, if a decision was taken to dismantle, rebuild and widen the bridge as part of a heritage craft skills initiative for young apprentices (for example), it could be argued that the social value would be enhanced, thereby creating a new history or new layer of story for the bridge in addition to providing valuable experience, although its historical and aesthetic value could well be diminished. Another way of looking at the balance of values is to accept that the historical value could be enhanced through the necessary archaeological recording related to the construction of the new bridge, which could reveal more of the story of the bridge during the dismantling and the insertion of new foundations in the river bed. The crux of decision-making then becomes a balancing exercise to see if it is possible to establish where the public benefit or social value of each possible solution might lie. Would a 'living heritage' solution, if there is such a thing, be different? If 'living heritage' includes craft skills, then it is possible that a craft skills rebuild would meet that agenda. But at the moment it is difficult to see where the balance of power lies in this approach or how it might address the conflict that existed between the groups and within a community.

Arguably 'living heritage' is about the tension that exists between the iconic (the things that experts recognise), the cultural meanings, experiences and stories behind the everyday (the places, activities and things experts do not see) and the everyday aspects of the iconic, so there should be a relationship with the Creets Bridge case. However, national heritage legislation did not and does not enable such an approach, even though it could be considered as having some similarity to the VDS because it aims to support and enable people to make decisions about and be involved in the management of their own cultural heritage. But tensions still exist, and this brings us back to the next phase of activity at Creets Bridge.

Creets Bridge: the next phase

Frustrations grew about the lack of progress in finding a solution for the bridge's repair. The local community felt that their lives were being made more complicated and could not understand why the referral system to the Government Office had to take place and take as long as it did, and why it might ultimately result in a public inquiry (a potentially long, expensive and complicated legal proceeding). Referral to the Government Office meant that the planning decision would be made by the then planning minister, Tony McNulty. As a consequence, the actions escalated over 2002, with letters being sent to local MPs and the then Prime Minister, Tony Blair, asking them to resolve the situation and end the delay.

In January 2003, it was revealed that one Kirkby Malzeard parish councillor had taken a very unusual and dramatic step. Several of the local newspapers (*Yorkshire Evening Post*; *Northern Echo*) broke the story that a parish councillor had written to the US President, George W. Bush, to ask him to bring pressure to bear on Tony Blair and the British government in order to resolve the lack of progress and push for a widened bridge: 'It seems to me he is the only one who can influence our Government at the moment', the councillor told the *Northern Echo* on 24 January 2003. It is perhaps worth refreshing our memory about the events of late 2001 and early 2003 to appreciate the political context of this letter.

American domestic and foreign policy was shaped by the events of 9/11 when terrorists hijacked four commercial aircraft and flew them into the World Trade Center in New York, the Pentagon in Washington DC and the Pennsylvania countryside. The fatalities numbered approximately 3,000 and it was the worst attack on American soil since the Japanese attack on Pearl Harbor in December 1941. The events of 9/11 led to the creation of the Office of Homeland Security in September and the definition of plans to defeat world terrorism, with the primary targets being Osama bin Laden and the al-Qaeda organisation. In October 2001, allied military operations (codenamed 'Enduring Freedom') began in Afghanistan. In January 2002, President Bush cited North Korea, Iran and Iraq as an 'axis of evil', intent on threatening world peace. In Moscow in May, President Bush and

Russian President Vladimir Putin signed a nuclear arms treaty, vowing to reduce the number of their nuclear weapons by two-thirds.

By July 2002, the simmering problems generated by the collapse of Enron and WorldCom resulted in new laws on corporate abuse. The Dow Jones Index dropped below 9,000, its largest one-day loss since 2001. In September 2002, President Bush and Prime Minister Tony Blair began to argue for military action against Iraq leader Saddam Hussein. In December, United Nations weapons inspectors conducted inspections searching for chemical and biological 'weapons of mass destruction', but did not find any. In March 2003, George Bush issued an ultimatum for military action giving Saddam Hussein forty-eight hours to leave Iraq. The deadline passed on 19 March 2003, and President Bush informed the world that America was at war with Iraq. In April, President Bush and Prime Minister Blair issued a joint address reassuring the Iraqi people that they would be able to live in peace, security and freedom in a post-Saddam era. This was followed on 1 May 2003 by a 'Mission Accomplished' declaration by President Bush, although Saddam's whereabouts remained unknown.

There are at least two ways of thinking about the letter to President Bush. One is to assume (certainly if you were one of the heritage managers involved in the case) that, at this dramatic international political juncture, such a letter was an act of pure madness and hysteria and exhibited a failure to understand how the planning system worked; the other is to consider that it illustrated exactly how passionate (and frustrated) people were about their heritage, and a simple, damaged, uncomplicated old bridge. Curiously, this rather simple act of letter-writing (although motivated by an intense feeling) illustrated that the expression of passion was something that perplexes – and even embarrasses – the heritage community, because their training and responses to an issue focus on the measured, the gradual and the controlled assembly of facts. It is not clear whether the letter to George W. Bush had any impact, but in March 2003 the planning minister confirmed that he would support the application to dismantle and widen the bridge. Archaeological, restoration and reconstruction work began later that year. The objectors were faced with an interesting problem in that they could have objected to the decision and asked for a public inquiry (PI), and the application would then have been determined by a government planning inspector. However, PIs are expensive and time-consuming, lasting several months where necessary. The outcome, of course, cannot be known, so the objectors had to be pragmatic and weigh up whether this was the case they wanted to 'die in a ditch for' (a frequently used term when considering PIs). If PIs are expensive, long and resource-hungry, is this the case you want to commit to? And while you are focused on this PI, the rest of your work is piling up, and nobody else is going to sort it out. If you lose, are you prepared to accept the damage it could do to the reputation of an organisation? What would a loss say about your powers of analysis and assessment and your ability to pick the right, winnable, case? Is a damaged Grade II bridge *really* worth all that disruption to an organisation, all that energy? Conservation never exists in a vacuum; political considerations are always there and the pragmatism of walking away from a case is more common than might be considered.

268　*Keith Emerick*

If we accept that heritage is about story and place and 'doing' rather than fabric, then the events of 2000 to 2003 at Creets Bridge have added to its meaning and story. If we look again at the value groups we can see that the archaeological work associated with the foundation and widening work added to its historical value, because it added more information. The original bridge was recorded. Perhaps the aesthetic value of the eighteenth-century creation was diminished, but it still sits in an attractive rural location above a tumbling, lively beck, so it still has an aesthetic and gives a sense of being 'old'. However, the social value of the bridge has been increased because of the events surrounding its future. If heritage is understood as something that is created in the present, then the debates over Creets Bridge illustrate that heritage continually expands, rather than diminishes. And, should people feel daring, they could push the link further and rename the bridge as the Creets Bush Bridge or the Creets Enduring Freedom Bridge. To others of course, it is A Bridge Too Wide.

Although all the participants at Creets Bridge understood that the historic fabric was a 'heritage asset', simply disagreeing about how to use it, the second case study considers an example of heritage as an activity. In this example, although one set of participants strenuously argued that the activity was cultural heritage, the heritage sector failed to see the relevance of the debate and failed to see the case as something that could inform ideas of heritage or heritage management practice. Of the two cases this is perhaps the closest to the idea of 'living heritage', but would practitioners recognise it as such?

Long Live South Bank (LLSB), London

If you were to ask which planning application in the United Kingdom generated the greatest number of objections, you might be tempted to think that it was a project such as a wind farm, a bypass or another piece of infrastructure in a green, rural and middle-class part of the country. The actual answer is quite surprising. The Southbank Centre in London is a large complex of buildings of arts and cultural activity on the south bank of the River Thames in the borough of Lambeth, between the Hungerford and Waterloo bridges. The complex is made up of three main buildings: the Royal Festival Hall (RFH), the Haywards Art Gallery (HAG) and Queen Elizabeth Hall (QEH). The complex had its origins in the 1950s with the Festival of Britain, a festival designed to show the world that Britain had recovered from the Second World War. A number of buildings were erected for the event, but only the RFH remains from that period; however, it was always intended to be part of a larger complex of arts and performance spaces and buildings. In the 1960s, the HAG and QEH were added. The National Theatre building is further east on the south bank.

Designed in reinforced concrete with some Portland stone and Derbyshire marble cladding, the buildings are in a brutalist style that has references to the work of Corbusier and modernist Scandinavian influence. The RFH is listed Grade I while the HAG and QEH, being part of the overall design concept, have 'Certificates of Immunity', which means that, although they are of national importance, the Secretary of State at the Department for Culture, Media and Sport will not list them.

The complex has a long, popular, wide riverside walk between it and the River Thames, stretching along most of the south bank. Of particular relevance in this example is that the complex also has an extensive undercroft, which has been used for the past forty years by a number of youth groups but specifically skateboarders, who consider the undercroft to be the oldest surviving skateboarding site in the world and the home and birthplace of UK skateboarding.[5] Originally defined as public land and managed under the now dismantled Greater London Council (GLC), the complex is a publicly funded private estate run by Southbank Centre Limited. As the estate took over, it began to modify the undercroft areas with the insertion of retail units, gradually pushing the skateboarders, BMXers and street artists into a smaller area of undercroft. There was no discussion or consultation with the skateboarders, but there were suggestions from the Southbank Centre that the spaces were being made unavailable and inaccessible for repair and maintenance reasons, suggesting that access would be allowed at some point in the future. Inevitably, the spaces remained empty for long periods of time before retail units were moved in, but the effect on the community of users was to slowly erode the home and culture of that community.

Events then escalated in March 2013 with a proposal from the Southbank Centre setting out its intentions to create a Festival Wing in the remaining portion of the undercroft beneath the QEH that would be both an arts and retail complex. Again, there was no discussion with the skateboarding and street art community and it was stated that a new, purpose-built skateboarding park would be constructed beneath Hungerford Bridge for their purposes. The public outcry against these proposals began immediately.

A group, 'Long Live Southbank' (LLSB), was formed to represent the community of young users – principally the skateboarders, BMXers and street artists – although it was always acknowledged by LLSB that the undercroft as a meeting place and the activities that went on there were to be understood as cultural heritage, as a cultural 'space' and as cultural heritage belonging to all. Petitions against the various iterations of the development proposal numbered 15,000 and 31,000 (the latter being the record for people presenting objections to a planning application in the United Kingdom), while the membership of LLSB quickly rose to 160,000+. LLSB and the LLSB website[6] made the most of the power of social media, with the latter full of stories, memories, news, links, images, posts, publications and comment – chiefly by the young users of the site, although big-name supporters (such as London Mayor Boris Johnson, TV presenter Jeremy Paxman, boxer Mike Tyson and artist Damien Hirst) were keen to be associated with the protest. LLSB were aware that they were up against a well-resourced, well-connected, experienced and powerful opposition (the CEO of Southbank Centre Limited was a former employee of the advertising firm Saatchi and Saatchi) who would play the 'culture' card – to their mind, skateboarding was not a valid use or activity, whereas classical music and gallery space would be beneficial to all, would attract tourists and a substantial revenue stream and was somehow 'more' democratic than the current use.

Despite this, LLSB were determined to fight for the continued use of the undercroft and, with the support of their legal advisor (planning lawyer Simon Ricketts of the firm SJ Berwin), made imaginative use of the existing legislation to oppose the development: having the undercroft recognised as an 'asset of community value' by Lambeth Council, and using the 2006 Commons Act to register the undercroft as a 'Village Green'. This latter initiative was rejected by the Council, but LLSB then responded with an application for a judicial review of that decision to reject the application. A protracted argument developed, with both sides presenting their case and arguments through a variety of media and spokespeople, but the core of the argument remained the same: the proposed retail and gallery units were of clear public benefit while skateboarding was ephemeral, set against the idea that the youth activities and the space were cultural heritage, inclusive and enabling.

When it looked as though the disagreements would continue, LLSB and Southbank Centre announced on 18 September 2014 that they had reached an agreement, securing the QEH undercroft

> as the long-term home of British skateboarding and the other urban activities for which it is famous. The agreement has been formalised in a binding planning agreement with Lambeth Council. In the agreement Southbank Centre agreed to keep the undercroft open for use without charge for skateboarding, BMX riding, street writing and other urban activities.[7]

Councillor Lib Peck, Leader of Lambeth Council, said:

> I'm pleased that Lambeth Council was able to work with both sides and find an imaginative solution to resolve this. Shared public space in London is precious and Southbank Centre is a great asset to the country's cultural life. This agreement is a sensible way of protecting both and we can all now look forward.[8]

Long Live Southbank: the issues

Although the Southbank/Undercroft case was not a case in which heritage managers had to become involved through the UK planning system, it raised a number of issues pertinent to heritage management and critical heritage studies. Of particular interest was the reference to what was effectively a hierarchy of culture, and this was intimately linked to the questions of class and age relating to both the asset and the people involved. Heritage managers and conservationists frequently make the case that heritage places are to be handed onto the future and that this is an altruistic responsibility towards succeeding generations. For this to happen it is essential, therefore, that heritage managers are able to explain to the young why old things matter. But why is the relationship only one-way? Why cannot older heritage managers understand that the young have heritage and heritage places that they feel strongly about, as well as their own perspectives on 'established' heritage places? There are perhaps several competing strands here: the tension between those who understand heritage as an active process in the present, those who think that the past has a role to play in the present (as an agent of growth) and those who see the role in the present as static and passive (the heritage place as tourism site). The latter two perspectives tend to understand heritage as old places, but the conservation of those old places is fundamentally about old values somehow shaping the future (Spennemann, 2011). In this respect the engagement with the young is about making the young understand why old things matter, rather than understanding (or even asking) what they value and why. If heritage is an active construction or process in the present, then the present is as valuable as the future and it is the juxtaposition of activities that generates engagement among people. And, indeed, 'age' remains a cornerstone of the AHD and how we think about heritage: a frequent question, asked wherever there is a system for designating heritage places as nationally or internationally important, is 'how old does something have to be before it can be designated'? But age also extends to who takes part in heritage; as previously stated, the heritage sector goes out of its way to convince the young to be interested in old places, but professionals identify the middle-aged and retired as the people most likely to be 'heritage' volunteers or their allies, largely because they are more likely to already appreciate the 'old' and the past.

For the LLSB community, the undercroft and the activities that took place there were their cultural heritage, and the heritage of the nation. They made a clear connection between what they were doing in the present, how they wanted the place to be used in the future and the way the place was used forty years previously as the home, birthplace and epicentre of skateboarding in the UK. It was important to them; they felt strongly about it. But outside of that community, few people in the cultural or heritage sectors understood the legitimacy of that perspective. Skateboarding and BMXing were presented by detractors as temporary and ephemeral; street art was vandalism and a forty-year history was largely irrelevant.

As various offers and counter-offers were put forward, LLSB were criticised by the developers for not accepting the seemingly generous offer of a purpose-built, up to £1m skateboarding park underneath Hungerford Bridge, but that site – as LLSB frequently made clear – was physically removed from Southbank and had no 'value' to them. Additionally, LLSB made it clear that the 'offer' was merely an expression of money equalling might and right, while the development proposal further eroded the concept of public and community space. And because LLSB were arguing their case from a cultural heritage perspective, they were forced into using the heritage and conservation language they thought was correct. Thus 'preservation' figures frequently in the web site: 'preservation not relocation', 'dedicated to preserving the Southbank Undercroft and its diverse creative community' and 'you can't move history'. At the core of what they were arguing was the certainty that the undercroft was dynamic and was about youth,

youth values and change and growth – positions contrary to the idea of 'preservation'. Fortunately for LLSB, the Southbank Centre group did not seize upon this mismatch between terminology and ideas and use it as means of undermining their argument by suggesting that LLSB were about 'freezing' the past, or were inconsistent in their arguments and direction. However, this use of language illustrates how pervasive a particular idea of heritage has become and the manner in which expertise has shaped heritage in the mind of the general public and forced them to adopt a particular language. But it is a matter of some concern that heritage managers did not see the importance and implications of the LLSB/Southbank debate and failed to offer their 'readings' of the relevance it had to, and how it might affect, practice. Conversely, it could be argued that the failure of the heritage sector to engage was probably advantageous to LLSB because heritage practitioners would have had to adopt the position (using the existing criteria) that skateboarding and the undercroft were not 'nationally important' and were not, therefore, cultural heritage. Many heritage practitioners and heritage agencies in England will roll their eyes at the idea of intangible heritage or state categorically that it does not exist in England (Smith and Waterton, 2009), but LLSB clearly proves that this is not the case and that the intangible is a fundamental part of place.

The Southbank Centre group made their arguments about the value of culture and the way in which LLSB was somehow exclusive but also denying access to an improving, classless and enduring idea of culture – although members of LLSB were quick to point out that, on a particular day, the cheapest item on offer was a 45-minute orchestral recital costing £9.50 per ticket, while much of the proposed gallery space was undeveloped or, where it was developed, was largely unvisited. But what the Southbank Centre group failed to see was the way in which the activity at the undercroft had been 'improving' for many of the users, and had politicised a community of the young. It was also the case that the cafes, restaurants, galleries and recitals just did not appeal to a large proportion of the population of the south bank and its environs. As the LLSB website put it:

> This is not about destroying the expansion of access to culture, this is about the questioning of the integrity of archaic hands that seek to stifle and silence culture. We are the voice of young people who stand for the future of artistic and cultural expression in an ever-changing world, for everyone. Not on paper, but in practice.[9]

Conclusion

Both the cases discussed here tell us something important about the way that people feel about heritage and how strong those feelings can be. They also illustrate something about how an institutionalised heritage system responds (or does not respond) to these feelings. Would a values-based approach to heritage have made a difference or offered a new perspective on events at South Bank? The skateboarders attached historical, aesthetic and social values to the space and the activities but the heritage sector did not see the debate as a cultural heritage matter, or recognise that there was such a thing as heritage of the young. Skateboarding was not cultural heritage, and even though a fifty-year-old building might be considered as heritage, the type of activity going on in the undercroft was not something heritage managers understood as 'legitimate'. It is likely that the undercroft to the QEH would have been perceived as something that detracted from the aesthetic value of the building, so its conversion to retail units and gallery space would have been seen as enhancing the values of the building. Would a living heritage approach have offered anything? Possibly. Skateboarding, BMXing and street art can be understood as 'intangible', and the way in which the skateboarding community managed their own protest and 'heritage' could be seen to align with the aims of the approach. But why did heritage managers avoid the debate?

If a values-based approach is founded on community participation and an understanding of what places mean to people, the outcomes should be that heritage managers write and speak clearly and distinctly, with the minimum of jargon, and are able to express why a place is important to us/the community now. But it should also mean that heritage managers become proficient in listening and become less embarrassed by, or at least less dismissive of, expressions of passionate support for particular cases, stories or issues. If the heritage sector wants people to be involved and committed to heritage issues, then passion and emotion have to be accepted as parts of that dialogue. Once you learn to listen and accept a variety of positions as valid, you can then draw out the meanings and associations attached to places. In England designation descriptions are being re-written in more accessible language, correcting the brevity and assumed objectivity of the technical language seen in the Creets Bridge description. Once official documentation was free of emotion, and specialists had to control their language, but now (certainly in England) documents are meant to communicate enthusiasm for heritage places and specialists are expected to be vocal advocates for their subject. But has the prevailing execution of a values-based approach changed practice and is the sector any closer to addressing the questions of what heritage is and how is it used in the present? Is heritage any less about planning law? In October 2014, English Heritage took the bold move to designate the 1978 purpose-built skateboarding park at Rom, East London as a Grade II listed building (see English Heritage, 2014). The site was designated for its historical, architectural and cultural values and is the first such site to be designated in Europe, and the second in the world. While this is an innovative approach to recognising youth culture, the test will come when management choices are required for the site: will it be allowed to change and who will lead on the management – the users of the site or the heritage professionals?

It seems that heritage management and conservation practice are still at an impasse where expertise is still about passing old fabric to the future, even though the need for pragmatism drives many solutions – as was seen at Creets Bridge. The international demand for growth means that heritage practitioners are coming under greater scrutiny to demonstrate that heritage is not a brake on growth, but is an attribute that can generate revenue – principally through tourism. The implication of this is that heritage managers have less time to do (or learn) the community aspects of their work, and the outcome is a focus on the iconic and the authority of expertise. For the public, trying to influence experts on matters affecting heritage places, the use of measured, calm, rational language referring to national policy and guidance remains the essential requirement. Emotion and passion in letters is misplaced. The pragmatism lies in the necessity to be able to say that this part of a place is significant and should be retained, but this part of the same place is less significant and can be changed. But does this mean that practice has to remain the same?

An approach that privileged 'story' over fabric and accepted that intervention was valuable might break the logjam of practice, but it would require a major shift in emphasis for practitioners to accept and understand heritage as a process rather than a product. A review of Creets Bridge could reveal that the pragmatic conclusion was avoidable if the case had been seen as an opportunity to create a new story for what was essentially a damaged structure. However, the blind spot in heritage practice relating to the young and young heritage revealed by the LLSB case is a cause for concern and needs further research. But it also offers great potential – to discover a heritage practice that is not only vibrant and meaningful, but also strongly felt, constantly renewing and thriving on juxtaposition and difference in the present.

Acknowledgements

Thanks to Jason Cains of LLSB for kindly providing comment on the draft text.

Notes

1 See www.english-heritage.org.uk/professional/protection/process/national-heritage-list-for-england
2 See www.english-heritage.org.uk/caring/listing/listed-buildings
3 See www.nidderdaleaonb.org.uk/Documents/VillageDesignStatement_KirkbyMalzeard.pdf
4 See previous link.
5 See www.llsb.com.
6 See previous link.
7 See www.llsb.com.
8 See previous link.
9 See www.llsb.com.

Bibliography

Australia ICOMOS. 2013. *The Burra Charter*. Sydney: Australia ICOMOS.
Court, S., Thompson, T. and Wijesuriya, G. 2014. Keynote address: a people centred approach: improving engagement and conservation and management capacities for cultural heritage, *Engaging Conservation*. https://engagingconservationyork.files.wordpress.com/2014/08/iccrom_2014_07_11.pdf (accessed 25 January 2015).
DCLG. 2010. *Planning policy statement 5: planning for the historic environment*. London: DCLG.

DCLG. 2012. *National planning policy framework*. London: DCLG.

DoE. 1990. *Planning policy guidance note 15: planning and the historic environment*. London: Department of the Environment.

English Heritage. 2014. *National heritage list for England*. http://list.historicengland.org.uk/mapsearch.aspx (accessed 25 January 2015).

Hall, S. 2005. Whose heritage? Un-settling 'the heritage', re-imagining the post nation. In J. Littler and R. Naidoo (eds) *The politics of heritage: the legacies of 'race'*. London: Routledge, pp. 23–35.

Jones, S. 2009. Experiencing authenticity at heritage sites: some implications for heritage management and conservation. *Conservation and Management of Archaeological Sites*, 11(2): 133–47.

Smith, L. 2006. *Uses of heritage*. London: Routledge.

Smith, L. and Waterton, E. 2009. *Heritage, communities and archaeology*. London: Bloomsbury Academic.

Spennemann, D.H.R. 2011. Beyond 'preserving the past for the future': contemporary relevance and historic preservation. *CRM: The Journal of Heritage Stewardship* 8(1 and 2): 7–22.

Vermeulen, S. 2006. Action towards effective people-centred conservation: six ways forward. *Policy Matters* 14, March 2006: 64–72.

14 Dark seas and glass walls – feeling injustice at the museum

Practitioner perspectives: Rosanna Raymond

An interview with Rosanna Raymond by Divya P. Tolia-Kelly

The artist Rosanna Raymond is a curator, performer, artist, and lecturer and was a member of the UK-based Polynesian community until she returned to Auckland in 2014. Rosanna is a New Zealand-born Pacific Islander of Samoan descent. Rosanna is also a published poet, writer and founding member of the *SaVAge K'lub*, with art works held in museum and private collections around the world. Rosanna has forged a role for herself over the past 20 years as a producer and commentator on contemporary Pacific Island culture in Aotearoa New Zealand, the United Kingdom and the USA, fusing traditional practices with modern innovations and techniques and specialising in work within museums and higher education institutions as an artist, performer, curator, guest speaker, poet and workshop leader. She has had solo exhibitions and made contributions to international collections over this time.[1]

Rosanna has collaborated with Divya P. Tolia-Kelly since 2005, in the form of interviews, conversations, joint performances and co-writing, and here talks through the role of representing Polynesian and Māori culture at heritage spaces such as the museum. Their conversation is about affect and heritage, beyond a theory-centred or ethnocentric account. The focus is on the role of affect within the heritage space and the role of emotion in shaping, producing, representing and articulating cultural heritage and values.[2]

Divya: Thanks Rosanna for agreeing to do this interview for our edited book on *Affect, Heritage and Emotion*. Much of my thinking about affect at the museum has been inspired by your work and as such we wanted you to have an opportunity to publish and share your experiences of curating, participating and engaging with heritage practices and places in light of emotions.

Rosanna: Recently, I've been working with about 16 different *savages*.[3] Yeah. It's an Asia Pacific Triennial. I'm building a savage clubroom. I've been working with the Queensland Museum. So, we're getting museum collections and putting them with our collections. Some of the negotiations there have been interesting – producing new pieces. It is like a huge exhibition inside a huge exhibition. Last night, I had to try and finalize all the drawings

	and ... and then even things like the customs to get into Australia is nuts. So, half of our stuff can't even be included because it's seeds and shells and feathers. They are being stopped at customs, who ask us detailed questions about the materials within the artworks. Then we're like – those are shark's teeth, and then we get these letters back, what sort of shark's teeth. It's just like…
Divya:	How do you deal with the customs regulations in this circumstance?
Rosanna:	Oh, it's really complicated. It's given us all a headache, all of us.
Divya:	So, are you having to make new objects … new 'fake' things to look like the originals? Or are you going to get a special dispensation?
Rosanna:	Well, we're trying to get special dispensation. Because we can't prove provenances in the way that they would accept, due to the nature of our heritage, some of it has been passed through families. We can't prove that Aunty gave me that, and so it's that classic case of yeah, some of these things are actually endangered, it's so frustrating. So to express our culture our expressive culture, certain things dictate the inclusion and aesthetics. Again, it's Indigenous people that lose out. Two of the girls in the savage club are from the Solomon Islands. Some of their regalia is all this beautiful shell, and it's got little dolphin teeth. It's really special to them and really … and we can't take them over to Australia because they can't prove when they came into the family, so now we're having to look at alternatives and also we cannot care for them. In this case, the museum is actually trying to help us but we're getting knocked back by a different sort of institutional madness [laughs]. Those two girls have lost some of their most important regalia. Without the evidence they cannot access it. It's interesting, the difference on working with a museum and an art gallery. Like, I've got this beautiful Kahu Kiwi, a kiwi cloak. I said I want to put it on a brace so that it's floating. In the art gallery, their conservators sit there and go, 'oh no, we have to confirm that'. It's like, the museum has just told you that you … that we can do that. So three months later, I get an answer. I think a lot of that is just very rigorous protocols within the institution…
Divya:	Yeah, but you're an artist, surely your professional view and integrity has to count for something?
Rosanna:	Thank you. Yeah. Luckily the curator I'm working with is fantastic.
Divya:	Obviously I've connected with you over a period of time because affect and emotion and the ways in which they play out in museum exhibitions are quite political and interesting. But I'd like you to say how you describe your work that you do and where you're at right now.

Rosanna Yeah, well, it's definitely been a journey through working through that actual museum space. I don't think if I'd connected with the museum space I probably wouldn't have thought so much about a lot of the issues that I explore now within that space. Yeah, so where I first, I think, sort of came into contact with the museum it was from the outside as a visitor looking through the glass. I was really trying to connect with my community whilst living in London. Then I sort of managed to get a little foot in the door and I started to actually be able to get behind, into the actual collections. That really opened up a whole new world for me. Through that I started to actually read a lot of what they were writing. I really started to take notice of it because I found so much of it problematic. From there I started to attend a lot of the conferences, where I realized I could actually challenge a lot of those notions. I still felt that gap between being an artist and the voice of the academy, and whose voice was noticed and how it was noticed. I think working with Ngāti Rānana[4] and the British Museum, London, sort of having a sanctioned presence, and actually forcing the museums to acknowledge that living dynamic, and the ceremony and the ritual was a really vital part of the actual collections. Not just bringing us on as the entertainment, you know, we like to sing a few songs and blow a few conch shells. I really felt like when the community, you know, when we're there physically rendering the ceremonies, it's such a highly charged space. A lot of the curators feel something; they feel it. Yet when they talk about it they go back to a very object-based training; you lose that emotion because of that preference of the western intellectual realm, and how information is dispersed as factual information about these communities. I've really started to want to challenge that space. That's when I really started to look at how the body – how the living body – living dynamic I call it – interfaces and collects as time and space so that it becomes a shared space with the ancestral past, and allowing it into the now. So that's where I'm at now, actually really making space, and through the body activating this spark of life that then we … that then we take out of the space. So that took me over probably five years of working inside the museums and struggling with a lot of issues to really hone what am I doing as an artist-activist.

Divya: How does that feel politically? Being an artist within the museum?

Rosanna: A lot of the feedback from my peers at home [in New Zealand], they were like, what are you doing there? Why are you making yourself into an artefact? Why are you even engaging with these people? They're such a waste of time [laughs]. This reflects how

Practitioner perspectives: Rosanna Raymond 279

there's so much animosity towards the museum space. It is seen as such an artefact of the colonial past. Actually, I'm really enjoying being able to challenge and change things. But talk about two steps forward and a couple back, in terms of negotiating space! How I get into the space, who invites me, what time I'm allowed. There are always so many constraints. It is such a highly negotiated space. It's still incredibly challenging. Some museums are great, and then you realize that they're not all like that and there's still so much work to be done inside the museum.

Divya: Can you just answer two things: when did you started working with the museum space and where? You said initially there was a gap. Can you just articulate what the gap is?

Rosanna: Yeah, well it's interesting because I was looking for my community when I was in the UK. It's something I didn't need to do in New Zealand because it's completely surrounded by it. There was very little need for me to go to a museum and stare at something behind a glass case, because a lot of the 'artefacts' we've all got in our homes. Over there I've got a hula skirt hanging up, I've got bark cloth on my wall, I've got greenstone. So I didn't need to actually rely on the museum. When I went away from my base culture the museum all of a sudden became a place where I knew that I could find parts of our history. I was quite shocked actually at how little we're represented. That's when I really did get a shock. Knowing how many of our cultural treasures are in these places. Let's take the British Museum for example: I knew that they had one of the biggest collections of Pacific Island cultural treasures in the world so I was really excited to go to this huge museum. You can imagine how I was quite overwhelmed with how little we are represented. There are a few little corners, and that was actually quite shocking to me. That's when I had to really work hard to try and find out, well, where are they? Then I realized that they were all in the storerooms. From there I had to build up a relationship. Luckily, I had an introduction through Professor Nicholas Thomas, who had written about the Pacific Sisters and the work that we'd done in New Zealand. So that was my in: an introduction from one of their kind. But it was very different when it was just Rosanna Raymond trying to ring them up and get in.

Divya: He becomes a sanctioning authority or gatekeeper?

Rosanna: It was like, who are you? What are you doing? It was only when I learned a new language like the word 'research', '*I want to research the collections*', rather than '*I would like to look at your collections*'. I mean, there was a sense of frustration. I saw my role as an artist, too. It was to actually ensure that a lot of my fellow peers actually knew that they had the right, that we all had

	the right, to actually ring up these museums and ask to see the collections that we are actually descended from. It's that invisible old sign, not welcome. So it was about understanding the set of protocols, because we've all got our protocols. We have cultural protocols that are in place including how we interact with the actual treasures once we're there. It still hurts my ears when I hear the word *artefact*.
Divya:	Artefact is a deadening term. A *violent* term.
Rosanna:	Yes, it hurts my ears. I still, I get the pain. It still hurts my ears when I see and hear the grass skirt. You just know that this grass skirt is actually made of usually hibiscus fibre, and it has a particular way that it is procured, that plus how many hours go into it. That it would have danced and swished around and made great noise and so all these little things just kept triggering more and more inside me as an artist to understand how I could bridge the gap in this dormant state that I saw them in and how we conveyed them as a living dynamic. Because I suppose Indigenous is, non-intellectual, it's very hard for people to associate indigeneity with contemporary. I think that is where the museum hasn't been particularly helpful, because they're constantly framing it/us within the past because that is their definition of their collections. They're from the past. So I feel that some of the responsibility of the modern museum is to bridge that gap and actually to look at the living community and how it is today. Not necessarily portray it because it's too – it's a very hard thing. But to at least have an engaged relationship with those communities for which they hold cultural treasures. It's like entering a relationship with somebody. You can't just have their bracelet and not deal with *them*.
Divya:	Just to rephrase, you're saying that the dynamic, the living, is what you want to ensure people see? But also that part of your work is to create these relationships between museums, the community, and give life to that relationship?
Rosanna:	Yeah, I think give it a living dynamic. I mean it's hard when you've got tens of thousands of cultural treasures from the Pacific, and you're based in London. But there are communities there. I think a lot of museums have to try a lot harder to connect with these communities, because our collections are everywhere. It's incredible. Little tiny museums up in Durham or somewhere all have a treasured fish hook or have a piece of greenstone. I think the relationship with the museum and the base country or people should be established. There should be a lot more inter-community, even inter-institutional, linking up so that you've got access to information. I worked with the Royal Albert Museum, Exeter,[5] and I came in and I researched the collections through my artist's eyes and sort of just – because of my connections

	with the homelands – I was able to give them websites, living artists, how to get in touch with different community groups. Even which books that they should have to read up on it. So, just simple things like that. My other thing is trying to get that particular culture to understand the role of guardianship, *kaitiakitanga* (guardianship) as opposed to ownership. Because we don't even own these concepts let alone these things. A museum will sit there and tell you that they own it, even though you made it, or your family gifted it. And therefore you can't even use a photo of it, or – so you get into all this crazy copyright ownership sort of issues, which is really contradictory, isn't it?
Divya:	How would you say that Polynesian or Oceanic culture gets represented within the sorts of spaces that you've been a scholar, artist or curator at? You said a bit about the deadening with the grass skirt, for example. You're seeing it differently in New Zealand/Aotearoa, being based in Auckland at the moment. Can you give us a feel of what's happening in terms of representation?
Rosanna:	Yeah, I think the representation is still very static. I find that you fix that by making sure that you have the living dynamic involved. Whether it's visible or whether it's even behind the scenes. The museum in Leiden[6] has an incredible amount of good practice. They wanted to purchase a Māori canoe so they did their research, they came to New Zealand, and they met up to purchase this canoe (or *waka*). The community's like, 'well, no you can't actually buy it but we can share it. We will gift it to you but this is shared ownership, and it comes with responsibilities. It comes with songs, it comes with hakas, it comes with ceremony. These must be honoured and you must come to New Zealand every year to come to the great waka ceremonies.' This is what the museum committed to; a new shared practice is the only way. It's hard and it's costly, but I think that these have to be worked into the budgets of these museums. They've made enough money off our collections, surely?
Divya:	That's a good ethical practice as well as being respectful, isn't it? So what period of time did that happen, the Leiden agreement? What year are you talking roughly?
Rosanna:	Well, now it's only a couple of years ago because I went to the opening. I mean that's such a good example. Also, I suppose that's an example of they've got – they're adding to a collection, whereas most of the museums it's still a very static collection, they're not adding to it. So, they've only just got the things from the past, so it's a constant presentation or re-presentation of the past in that they're very reluctant to bring on anything of the now. I suppose that's a lot of leftover kind of things with inauthenticity

	and change. They have the *real* artefacts – I've actually heard this, 'Oh those are great Rosanna, but we've got the real ones'!
Divya:	Oh my word. How did that make you feel?
Rosanna:	We'd just shown them these beautiful *taonga* with *tiki* that George had carved out of greenstone and whalebone. All my my *kākahu* (costume), my regalia. Then, with one little sweep of a hand, they totally dismissed us and what we'd done because 'they've got the real ones'. That whole sense of our value of taonga was taken away.
Divya:	Was that in the UK?
Rosanna:	No, it was actually in France. I find that the UK is much more politically correct than the Europeans. It's like, oh my goodness, there's still so much work to be done!
Divya:	The next question I have here is: what do you think is missing in these conventional representations? Does it overlap with the domain of emotion and affect? To me, it reminds me of something that Paul Gilroy talks about in terms of black African – British African American, which is that you're never allowed to be a black body and be part of a modern intellectual network. They're only comfortable representing people as representative of an era or of a culture, or of an ethnicity. It's very difficult to position somebody black as internationally mobile and modern. So, in terms of the conventional representations, what things have been obscured beyond this account, or including this account, of modernness?
Rosanna:	That's a big question! I suppose it's so hard to then replace with it factual or artefactual, so the emphasis is on looking at the cultural treasures as a whole by constantly reducing them to artefacts. They're not representing the diversity or embodying the overlaying of the Indigenous histories, of the Indigenous language, of the Indigenous people as a continuum. It is hard. With artefacts, you have to do a caption in 25 words. Institutional frameworks consistently limit possibilities of allowing engagement with a living culture. Sometimes it's purely about everyday design and formats. Usually these designers have very little understanding of the culture. You'll find that the curator gets pushed down. This is happening a lot in museums all over Europe. I saw that the position of the curator, the one who does actually have a connection to the culture, will be pushed aside at the expense of a very highly paid designer. It goes back to privileging aesthetics again, as opposed to content. So that's a worry. There are new ways of trying to disseminate things, this new information. But they're not able to catch up because of funding, because of space and because of the very static nature that the museum has for itself. I find that they themselves are frustrated.

Divya: Yeah, of course. So what's the thing about the designer and why is it that the designer has the primary weighting?

Rosanna: Who knows? Working with the likes of a really small museum – I was dealing with three people. You ask a question and it goes to three people. But in the British Museum, you ask a question, it goes through a whole process, through teams of people, and half the time the departments don't even speak to each other. So there's a lot of rupture in their own institutional priorities, I find. A lot [of it] is that because – and this will sound pretty tough – but because the institutions are full of non-Indigenous people representing Indigenous people. You get a whole lot of decisions based on somebody who may have studied these people, maybe for five years at university, and then they get a job and go into these institutions. Those connections are very hard for them to keep. So they're basing a lot of their knowledge on something they learned 20 years ago. There's disconnection, unless they're actually active in the field, or they keep a really good relationship with people that they did work with, it's clunky…

Divya: In terms of your own work, you're trying to counter or enrich the ways in which this dynamic culture is presented? Can you give us an example of how you're trying to challenge those accounts or processes that are not necessarily doing justice to the representation of Polynesia?

Rosanna: When I first started working with museums, I really felt that even just my actual presence inside the institution was like a political statement. Now when I do the research, I'm really looking at things that talk to me and stuff. So when I share it a bit more publicly it's quite interesting. The selfish part is because I look at body adornments, and anything to do with materials and fibres, I always try and share it with my own peers and then I can at least let people know that you can find that over there. Some works you do at the museums are very public, and some of them never see the light of day. Sometimes the audience itself is actually the *taonga*, the collections itself. Then at other times, I'll have an actual public that I will take on a journey with me hopefully – and again, that's the emotion. When I was in Berlin,[7] that was where I really saw how emotion can deeply affect and change people's attitudes towards the collections. That was interesting, because I did an activation. I actually managed to train them all not to say 'performance' because a 'performance' is different, and I do perform. But when I'm working in the museum space I'm creating a shared space through my body with the past. It's quite hard to get people to understand that because all they can see sometimes is a woman dressed up in regalia and grass skirts and tattoos, and they can't see past that. They can't see past what we've been

	trained to think. So I did this activation and I took the people on a procession and then with a prayer I talked to the taonga, I greeted the taonga. We walked up into the different levels of the museum acknowledging as much as I could. Then I ended up taking off a lot of my regalia and addressing one of the goddesses. Then I ended up totally undressing myself, and created an installation. So I basically then turned my living body into an artefact and walked off.
Divya:	You unravelled yourself so that you became an artefact in front of their eyes.
Rosanna:	Yes, It was quite interesting because I placed a lot of myself as object and one of them in a case and then some of them a wall. I found that they were unsettled although they are used to seeing things like that. They could understand that. They all of a sudden became unsettled; this fully naked body was hard for them to look at. Even though there were photos of naked bodies throughout the museum. So I was really challenging them in terms of what is an artefact, and – you know, by showing them how I am the museum too. There was a discussion the next day at this performance weaving centre and these very well-trained, very intellectual academics could not see past the exotic.
Divya:	Can you say a bit more about how they expressed that?
Rosanna:	We were going around the room and a lot of them kept using words like 'exotic', 'sexual', expressing … what is it … 'orientalism'? That's when I realized they were re-framing what I'd done through limited paradigms, albeit very well-established ones.
Divya:	So they framed you as something that they could then look at and talk about dispassionately?
Rosanna:	Yes, well, what's interesting is that I actually growled at them. Because I got to the point where I was like, 'is that all you can give me?' Now tell me what you saw! Then, what was interesting is they started to tell me how they felt. Then it got really interesting. There's still that real resistance to actually trust what they felt. They still had to put it through a very analytical, well-trodden framework. To me, that's when they missed a lot of the nuances that were going on. Then when we talked through what I'd done, they got a greater understanding. So that was – it really intrigued me because emotion is still a 'no-no' in the academic world. It's still not trusted as an analytical tool and it should be.
Divya:	Collectively people aren't really used to that path, are they? So how did they express their emotions? What sort of things were they saying? Or what did you feel? What were you feeling?
Rosanna:	Oh well that's when they started to talk about the powerful effect of the chants and that it moved them to tears. That's when they

started to talk about the fact that they felt another presence in the room. Within our culture, too, we've got terms for those things. You understand that *mura i roto i*, the spark in/of life – we understand *wairua*, the soul/spirit, the emotion, that *mana*[8] in the dignity and the gravity of the moment. We actually have terms that we can apply to these feelings. Actually, they can be used as powerful, analytical tools. I find this is my interest with working in the museums: how the academic world in general is. How do we acknowledge emotion in this academic framework? When does it get given some privilege, as opposed to just the pure factual? For me, that's when there is, and they had, a glass ceiling. When they only used the analytical tools that they had available to them, they had a very limited view of the experience. But when I made them open it up they were then able to bring in new ways of looking at what I'd done. Then something happened, something changed in their thinking. I felt that that then was successful. But if I hadn't had the talk, if I wasn't there to talk about it afterwards, they would have still been holding onto the activation as some highly exotic, erotic and oriental event…

Divya: It sounds like what you're saying is that through your activation it's not just you communicating the power or allowing folk to see and be with the taonga and the goddesses and the spirits, but you give people the tools to then … to firstly be able to see. It's like teaching someone the alphabet, then how to do words and then – so you had to do all of that in your activation. Is that right?

Rosanna: At one stage I was so concerned about the fact that people were only looking at it through a particular lens that I actually stopped performing and I stopped creating that work because I could see people couldn't get past the classic dusky maiden. Then I got to the point where I realized it had just stopped me being me. So I really had quite a fraught four years. But I was just like, 'oh, the message just isn't getting through'. They can't see past this exotic notion of what a Pacific body is. It really stopped me from being who I am. I had a real internal struggle for years. In the end I just got over it [laughs]. I had to, to move forward, just for my own sake. It was interesting that the museum space actually was quite helpful, because it's not a commodity-based space. So when you're making work, I don't have to worry if it's being sold. I really am making work that is response-based, that can be shared in different ways and not have commodity as the priority.

Divya: That's exactly the same way that all their original taonga are made, isn't it?

Rosanna: Absolutely. It wasn't pure commerce, maybe they were made for exchange, reciprocity, and they circulated through different communities. Yeah, so that's why I look at the museum space as really quite a positive living space of exploration and dynamicness. Then the flipside is that institutional framing is so heavy. It's so impenetrable that a lot of those messages get lost under this heavy academic didactic framework that is well established and it seems quite hard to move through the rigid western discourses. So coming back to New Zealand has been interesting because the focus is slightly different, but it's pretty much still the same. I was at Te Papa, our national museum. I made an appointment with Suzanne,[9] one of my Pacific sisters, to talk about seeing Kākahu the 'Twenty-first Century Cyber Sister', which the Pacific sisters made. We made her and we presented her in about, it would have been in the early 1990s, at Te Papa Museum.[10] We were so excited to be reconnected with our sister and we hadn't seen her for years. So we got her out of her box, and the first thing they did was ask us to put gloves on. I was just like, 'you're joking right?' They were talking about 'it's for conservation'. Then they got a lecture from me about how engaging physically with our bodies is part of the proper conservation of the object. Yeah, I was really shocked. We had made this. And to preserve her, she needed contact, physical care and nourishment. I sat there – I know, and I was actually really stunned because she had been disrespected anyway, by being locked away – and then I said, you want to talk about conservation? There were seeds which were very, very dry, I said, and without oil, without being oiled, these will crack and disintegrate. Of course, oil is not a certified conservation methodology. So I'm sitting there rubbing it all over my skin, getting the oils off my skin in there. Twenty-first Century Cyber Sister is her actual name. I was so shocked. This was only last year. I thought even with our own (there were two pacific island curators there) we'd failed her. But because they've been so well trained in the western museum methodologies, they were closed off. Afterwards, they let us touch her with our hands but I had to give them quite a firm lecture. If I wasn't so vociferous and so strong in why it was important to be actually physically touched by the maker and how this was vitally important to the actual conservation of these cultural treasures, if I'd been a lot shyer, we would have just had to sit there and touch it with plastic. That's just not good enough.

Divya: Good. When you oiled her, did you have another activation? How does it work?

Rosanna: Well it's – yes. Some of the seeds were very dry so they didn't have any oil and the best oil you have is on your skin. That's how you keep those things alive, is by constantly wearing them and they absorb all those beautiful oils from your skin and…

Divya: Were these seeds made into jewellery? Is that what you mean?

Rosanna:	Yeah, yeah. Then there were all sorts of other beautiful delights in there. I've got some images of that day. The other wonderful little activation I did was in Melbourne this year when I went over to see one of my G'nang G'near, which is the original patchwork type of jeans and jacket. Two white people spoke, and it was a beautiful mix of community voice and academic voice. But the focus and the privilege was firmly put on the Pacific Island voice. Boy, I enjoyed it. I loved it. Yeah, so I had two friends I realized would fit the garments. I didn't really tell the curators, but we got the box out and we came in and we chanted our way in. It's called G'near. *G'near*, so we opened up the box and then I'm slowly bringing things out and putting their oral histories back into them and updating the *kākahu* about what I've been doing. Then, I turned to two of my friends and said 'now we're going to wear them'. It was wonderful; it was amazing. Before one of the curators, before her very eyes, she started to understand what I had been saying that it is absolutely vital to bring the body into the act of conservation, and how this is a way that they can be kept alive and relevant, rather than just being in this consistent dormant state. So that was awesome. I mean that's where your emotion comes into it. There was magic; people feel things. All those words, emotions, experiences, atmospheres not really allowed in the analytical process.
Divya:	But it's also that these emotional and affective sorts of experiences are knowledge in themselves.
Divya:	That's the gap in the western understandings, isn't it? That embodied experience or things that we label emotions are part of the thinking, being and understanding process. They're not separate. It sounds like there are a lot of cultural knowledges that are missed in that museum space and academic space that are in that realm of the emotional.
Rosanna:	I think that without acknowledging the importance of the emotional, I think that there – it will be very hard to move forward in creating new ways of experiencing and allowing these new narratives into the academic world. It's – they are always talking about new ways of thinking and looking at things, and yet you have to do it with very old frameworks. So it's almost like oxymoron in itself. So unless you let these 'new' ways of thinking – and a lot of times it's not new thinking, a lot of times the knowledge bases are already there in the Indigenous indexes. But it is not privileged, and it hasn't been listened to, and it's been written about by others for others, the voice of the informant is left behind. So these are really important ways of allowing this new – well, this new – this reactivating a lot of knowledge that's already there, and giving it some *mana*.

Divya: It's also ironic, isn't it, that all cultural communities have this relationship with knowledge that's embodied and it's difficult to express it, isn't it? That's part of the reason why it's been negated. But also that that very way of being is something that's grasped for by new age spiritual stuff. Whether it's new age in the 80s or whether it's the equivalent in the twenty-first century. You know, like people are getting into mindfulness is that very process of taking your time, using your breath and feeling every bit of your body and the relationship to the broader earth and the physics of the atmosphere.

Rosanna: Yeah, understanding that every object is actually animate. There's no such thing as an inanimate object in a lot of Indigenous cultures. All these guys are going through these mind-blowing changes, and yet there are a lot of Indigenous cultures looking at them going, 'yeah'?

Divya: So it's cultural knowledge that's been around for centuries? But this bridge of seeing it as modern is a really key problematic. It's written off as an ancient knowledge that's still there. I think that's one of the blocks that you're talking about. You physically had to hold their hand and take them into a different way of being, didn't you?

Rosanna: Yeah, and I think that is why a lot of my focus has moved towards the – using the body as one of my primary sources because I realized that through the body work, through using the actual physical presence of myself, I was able to, I suppose, elicit a lot of this emotion. It's something that's a lot harder to do in a static work. I think that's why a lot of the work that I choose to do inside the museum is through my body.

Divya: It's interesting as well that you're saying that through the body, the body is a vector for challenging the very fact of them reducing you to your body.

Rosanna: Yes, I know. There's also a lot of invisibility in being a middle-aged woman and being a brown face, not that I'm particularly brown, but that sense of invisibility that you have inside these institutions. Sometimes, as I said, just physically being there was a statement in itself, let alone what I was actually doing in there.

Divya: We've talked about the missing knowledges because of this occlusion, in terms of thinking about Polynesia and Oceania. I remember when we first met you said that the Oceanic Gallery was the size of the dining room table at the British Museum. It doesn't allow a narrative that expresses the diversity or the vastness, let alone the life of all those sorts of cultures. So how do you conceive of Polynesian, Māori, Oceanic culture? How would you express it? What role does the unit of nation or national have in terms of you reactivating cultural heritage? You talk about the space of *taonga* or the space of the museum room, and your body.

	They are sort of units of space, yeah? So is Polynesia a nation? Can it be?
Rosanna:	That's a very contentious question, actually, especially because Polynesia, Micronesia and Melanesia are another construct that the West has given us. Our bodies were highly constructed into these sexual, free and easy beings. Then the actual vastness of the Pacific or the Moana–Nui–a–Kiwa was then cut into three. Those are really quite unhelpful in terms of a lot of the circulation of knowledge and the way that the Pacific nations have shared histories. So that's quite a problematic one, and a lot of times I'll get criticised for using Polynesia and I'm talking about Fijians and they're Melanesians. But for where I come from, our histories between Samoa and Fiji are so intertwined so that is a really unhelpful tag. A lot of debates are going on over here at the moment around 'how do we frame Oceania?' We need to break up this notion that we're just all these tiny little islands that are unconnected. The larger world needs to really see that Oceania is one third of the Earth's surface. We're incredibly connected through genealogies, through shared histories, through trade and exchange, through technology. So a lot of that made us even rethink our position of where we are in the world. That's why it is hard to see us reduced to these very small areas in the museums. So that's something we're still working through really. There are a lot of communities living outside the Pacific. So not just the genealogical connection, but the geographical space inside the body as well. So that when I was over in England I had that geography and genealogy inside me. Actually moving over there really made me think about that long and hard. What was I doing in England, and why was it relevant to enact and still think and feel as a *pacific* person? That's where the museum space was needed, that's why I ended up at the museum space to reconnect.
Divya:	Sure. So this volume is about heritage and emotion. Now, in terms of your manifesto, you can't reduce a culture to the way it was framed under colonisation and imperialism, and that removing it from a sense of modernity is the opposite in terms of finding its true meanings or perhaps an understanding of an authentic culture. But what that means, then, is that you dismiss, perhaps, heritage itself? I think, for me for example, when people think about Indian heritage or Asian heritage they have a completely different picture to the one that I have. They understand core values in relation to religious texts or religious architecture, or practices on the land. Whereas actually I don't really think about those things when I'm thinking about 'Indian', but how do you feel about that dismissing heritage altogether because it's rooted itself in a framework that's not – that's redundant? Or is that not true? You came to the

	museum because you were looking to connect with your heritage, but then because heritage means that you're in the past, perhaps heritage is the thing that needs to be removed?
Rosanna:	Yeah, well, that's where you need words like continuum. That's what I was thinking about. I think when you look at the way that I've been taught to look at that's – of space, in terms of the *Va*. It's not a linear space. We have these sayings that we walk into the past. We're actually – with the future behind us – so even the very way that we look at time and space is completely different. I think that's why the heritage aspect is really vital. But it's not so much looking at it – the past. It's enabling it too – because you – we know the past. You can see it. You can't see the future, that's why the future we have behind us. So I think this continuum is really important and that – and it's just, I suppose, having access to that as well. Because there's so many amazing historical treasures that I've seen that a lot of my peers and practitioners have not got access to over here. It's how we bridge those gaps now. Because there is a lot of good work happening and a lot of good practice out there. But it's how we make that a norm, not an exception, within the institutions.
Divya:	So it's returning slowly like returning heritage practice to where it should be. Connected, and giving life back to present generations?
Rosanna:	Yes, and I suppose for me, that should be one of the focuses of the museum. One of the main focuses. But their main focus is the preservation of the physical object.
Divya:	Conservation, don't forget conservation. [Laughter] Can you just describe *Va* again for the tape? Because you talked about different time and space, and I want people to understand that.
Rosanna:	Yeah, well the – so the – this framing of time and space. So we have *ta* and *va*. So *ta* is time, which is – which sort of punctuates and helps you interface space, and place yourself inside the space. But the space is not a linear space, it's a relational space that connects people and things together. When I apply this inside the museum space, it's where a lot of my practice made sense to me in terms of activating the *va*, the space, through the Pacific body, which is the vessel of the genealogical matter; this allows the past, and all that genealogical matter, to be present in the now through the sort of collapsing of time and space through the body; which is why I think the body has become so important to me. What you bring into the space, including people, has to be maintained and sustained. It helps you look at sustainable practice. It's not just an empty space but a very lively space that needs to be maintained and reciprocated and kept active, because it can go dormant.

Divya: Yeah. That's interesting actually because what happens if that genealogical matter isn't there, but someone identifies it as such? When I was talking to Kahutoi Tekanawa when she came in 2008 or 2007 she actually said that when she has her weaving classes – because her thing is that you must enable stewardship through practice. So you do the pedagogy one to one. I think she was suggesting that you could self-determine your genealogy, which then means that you can say, oh look, I am part of this stewardship of the land and stewardship of the culture.

Rosanna: Yeah, well I suppose I – I've done that myself in terms of – there's – a lot of the *taonga* I am not directly genealogically connected to. But I'm – geographically I feel a connection to it. Sometimes I literally adopt myself into these things. You're actually creating a relationship with the people, with the museum people, building a relationship with the actual cultural treasure. So sometimes the genealogical connection is very important, but so is any relationship; whether it's a curator, an academic, an Australian or a South African or an Asian engaging with other people's cultures. So I'd not sort of say that nobody else can be connected. I think it's important to acknowledge that enough of the genealogical connection holds much more emotion, I think. That affects that process and makes it mix that magic in, I suppose. I'm not saying that doesn't happen with other people. There's definitely other people's culture that I stand in awe of in those in museums. But that's when I realize I have become rather ethnocentric in my relationships.

Divya: But this is the flipside, isn't it? It's the problematic that we all have, right? We hate it when neo-fascists or the government are talking about who belongs, who doesn't belong and what the flag means, whether it's Britain or Australia or New Zealand. But at the same time, there's a real need for us to feel a sense of security through an account of culture, genealogy, language or whatever that is about a sense of connecting through place. Because we're not talking about bloodlines or phenotypes and brain–skull measures and stuff. But there's a doubleness. There's a constant flipping. So I don't think you necessarily have become ethnocentric. But I think you've expressed that doubleness very well. It's a real geographical thing, isn't it? Like so when you're in England sometimes it's narrated as you want to be in touch with your ancestors, yeah? So you walked into the museum space with that sense of time and space that's a western account. Actually it doesn't work. So when you're with the artefacts … the artefacts allow you an understanding of the present. That's what's given you your self-assuredness or your confidence or your sense of self, whatever it is you talk about. I think that's really beautifully articulated the way that you've done it.

Notes

1. Most recently Rosanna curated a SaVAge K'lub High Tea Ceremony on 15 November 2015: www.radionz.co.nz/national/programmes/nz-society/20151115
2. For further detail, please see https://muse.jhu.edu/journals/contemporary_Pacific/v020/20.1raymond.html
3. Founded in 2010 by Rosanna Raymond, the *SaVAge K'lub* presents twenty-first century South Sea SaVAgery, influencing art and culture through the interfacing of time and space, deploying weavers of words, rare anecdotalists, myth-makers, hip shakers, navigators, red faces, fabricators, activators, installators to institute the non-cannibalistic cognitive consumption of the other. More recently Rosanna Raymond set up a network of artists contributing to SaVAge K'lub events as part of a multi-art installation and performance event in 2015. For further information, see http://blog.qagoma.qld.gov.au/savage–klub/
4. London Maori community, Ngāti Rānana: www.ngatiranana.co.uk/
5. Exeter Museum Residency: www.rammuseum.org.uk/news/focusing–on–the–Pacific–collections
6. Museum Volkenkunde, in Leiden, Netherlands: http://volkenkunde.nl/en
7. Rosanna Raymond's Berlin Residency, http://indigeneity.net/residencies/, and Vimeo record: https://vimeo.com/110965423
8. *Mana* (as noun) translates to mean prestige, authority, control, power, influence, status, spiritual power, charisma – mana is a supernatural force in a person, place or object. *Mana* goes hand in hand with *tapu*, one affecting the other. The more prestigious the event, person or object, the more it is surrounded by *tapu* and *mana*. *Mana* is the enduring, indestructible power of the *atua* and is inherited at birth: the more senior the descent, the greater the *mana*. The authority of *mana* and *tapu* is inherited and delegated through the senior line from the *atua* as their human agent to act on revealed will. See: www.maoridictionary.co.nz/search?idiom=&phrase=&proverb=&loan=&keywords=mana&search=
9. Suzanne Tamaki: www.maoriart.org.nz/suzanne–tamaki–p–198.html
10. Te Papa Museum, Wellington, Aotearoa/ New Zealand: www.tepapa.govt.nz/pages/home.aspx

Index

9/11 attacks 55, 227, 228, 266
A Geography of Heritage 51
A Philosophical Enquiry into the Origin of our Ideas of the Sublime and the Beautiful 139
Abelson, Robert 49–50
Aboriginal culture 3, 37
absence/presence of affective heritage 239, 243, 244–5, 247, 249–51
acrophobia 144
Act of Abolition 120, 121, 122, 125
Agbetu, Toyin 127
Ahern, S. 124
Ahmed, S. 63, 118
Alderman, D. H. 51, 52
alienation 34, 37, 42, 129
Alpers, S. 38
al-Qaeda 266
alterity 2–3, 88
Altman, Kitia 80–1
Amazing Grace 121, 124–5
Ancient Monuments Protection Act 163
Ancient Monuments Society 262–3
Anderson, Ben 54, 57, 173, 202–3, 206, 215, 239
Anderson, Elijah 93, 98
anthropology 14–5
Anzac Hall (Australian War Memorial) 63–5, 70
AONBs (Areas of Outstanding Natural Beauty) 259, 261, 262–3, 264
Apartheid 20
Appadurai, Arjun 204, 211, 212, 215
Appelbaum, Barbara 81
Appiah, K. A. 94
Araeen, R. 33
architectural integrity 262–3, 264, 265
Armstrong, Louis 25
Ashe, Arthur 51

Atkinson, David 204
atmospheres 1–4: castles/ruins 154, 161, 163, 172–3, 175; museums 34–7, 39, 43–4, 54, 58–9, 62–3, 120, 121–6, 129; photography 201–3, 206–7, 208, 211–12, 214–15; social housing 237, 238–40, 242, 244–5, 248–51
AUF (Norwegian Labour Party youth league) 219, 222–3, 226, 230, 232, 233
Auschwitz-Birkenau concentration camp 20, 81
Australian War Memorial: audio-visual technologies 54, 55–7, 62–9, 70; and cultures of militarism 47–50, 57, 70–1; dioramas 57–62, 65, 70; haptic technologies 65–9, 70
authorised heritage discourse (AHD) 2, 239–41, 243–4, 247, 249–51, 257, 271
autoethnography 56, 165

Bagnall, Gaynor 96
Ballantyne, R. 20
Bamburgh Castle 162
Barthes, Roland 205, 207, 208
Baxandall, M. 36
Baxter, Anthony 186–7, 192, 193, 194
beaches 193–5
Bean, Charles Edwin Woodrow 57–8
Beck, Ulrich 93
Beirstadt, Albert 139
Benjamin, Walter 75, 77–8, 89, 203
Bennett, Tony 3, 33
Berenbaum, Michael 81
'beyond discourses' 13–4
bin Laden, Osama 266
Black Hawk Down 55
Black, Stephanie 185
Blair, Tony 266, 267
Böhme, Gernot 202

Bolton Castle 163, 167, 169
Bondi, Liz 95
Boyd, S. W. 182
Boym, Svetlana 158–9
Boynoff, S. 141
Braidotti, R. 145
Brandt, T. 136
Braun, Bruce 149
Breivik, Anders Bering 219, 222, 226, 230
British Museum 33–4, 38–43, 183, 278, 279, 283, 288
Brown, John 26
Browning, Christopher 85–6
Buck, Nathaniel 160
Buck, Samuel 160
Bunch, Lonnie 25, 26
Burke, Edmund 139, 161, 168
Burkitt, I. 117
Burra Charter (Australia ICOMOS) 258, 265
Bush, George W. 266–7
Bye, K. H. 229
Byker estate 237, 239, 240–9, 250–1
Byrne, Denis 24, 165

Camera Lucida 205, 207, 208
Carew Castle 155
Carter, S. 55, 56
'casing-in' culture 38
Castle Dangerous 162
Castle Otranto 161
castles/ruins: as cultural/heritage objects 154, 155, 156–7, 158, 160–1, 162, 164–5, 175; and embodied responses 160, 163, 165, 170–3, 175; and emotion 155–6, 159, 161, 163–5, 167–71, 173, 175; experience of 154–6, 159, 164–73; gothic 160–1, 163, 168, 171, 175; 'imagined' 173–5; and medievalism 157, 161–2, 163, 164; representation/ assemblage of 154, 155, 156–64; and tourism 155, 156–8, 160–5, 168, 173, 175
Charlie Hebdo attacks 220, 227, 233
Charter for the Preservation of Digital Heritage 201, 203
Chute Hill campground (Malakoff Diggins) 138–9, 142
Ciolfi, Luigina 210
Civil War (English) 158
Civilian Conservation Corps 141–2
Clark, Nigel 144, 147
Classen, C. 103
classical theory 159–60

coastal heritage landscapes: beaches 193–5; and 'destinations' 181–3, 184–5, 187–9, 192, 193, 194; 'development' of 179–80, 182, 184–5, 186, 187–98; and emotion 179–81, 182–5, 187–8, 189–90, 192–3, 195–7; theorising heritage 180–5; and tourism 179–80, 181–6, 187–90, 193, 194, 197; and Trump International Golf Links 179–80, 186, 190, 195; and violence 187–8, 189–91
Code, L. 39
collective memory 49, 51, 52–397, 120, 211, 239
colonialism/post-colonialism 2, 33–7, 41, 43–4, 51, 54, 70, 99, 185, 187–9, 192, 197, 279, 289
commemoration: and affective ecology 222–4, 225–6, 229, 233; and dissensus 221; grassroots memorials 221–2, 223, 224, 225, 227–8, 231–2, 233; and memory 220, 224, 230–1; narrative of 222, 224, 226; politicization of 224–5, 226, 228–33; and social media 221; and testimony 224–6; and tourism 230, 231, 233; and witnessing 221, 224–5, 226–7, 230, 233
conservation 192–3, 257–9, 260–1, 265, 267, 270–1, 273, 286–7, 290
Copenhagen attacks (2015) 233
Cosgrove, D. 140
'cosmohermeneutics' 98
cosmopolitanism: 'affective cosmopolitanism' 94–7; and contradiction 182; 'cosmopolitan canopies' 93–6, 98–100, 108; cosmopolitan engagements 100–1, 103, 108; and cross-cultural encounters/ landscapes 35, 93, 94–5, 96, 97–109; and embodiment 94–7, 98, 101–9; and emotion 94–5, 96–7, 98, 103, 104, 107–8; humanizing 107–8; and memory 94, 97, 102, 103–4; and narrative 95–6, 97, 105–6, 107; and objects 97, 101–4, 106, 108, 109
Cosmopolitanism: Ethics in a World of Strangers 94
Costello, Lisa 96–7
Council for British Archaeology 262–3
Cowper and Newton Museum 119–20, 121–5, 126, 129
Cowper, William 121, 123–5
Craggs, R. 245
Crang, M. 53, 114, 136, 183
Creel, L. 189, 194

Creets Bridge 259–68, 273, 274
critical discourse analysis 13, 21–2
Cromwell, Oliver 158
cross-cultural encounters/landscapes 35, 93, 94–5, 96, 97–109
Crouch, David 52, 175
Cubitt, G. 122
cultural memory 43, 52, 219
customs regulations 277
Cyclists' Touring Club 262–3

Dahlberg, Jonas 229, 230
Davidson, Joyce 95
Davidson, T. K. 220, 221
Davis, Fred 248–9
Dawney, Leila 95, 96, 98
de Certeau, M. 241
de Landa, M. 144–5, 148
Degan, M. M. 248
Delanty, Gerard 95–6
Deleuze, G. 55, 58, 61, 94, 95, 202
'democratizing memory' 204
DeSilvey, C. 136
'destinations' 181–3, 184–5, 187–9, 192, 193, 194
'development' of coastal heritage landscapes 179–80, 182, 184–5, 186, 187–98
'difficult' heritage 13, 19–24, 25–6, 220–1, 233
'difficult' knowledge 19, 21
'Digging up Utopia?' 52
digital heritage 201, 203–5, 215
dioramas 57–62, 65, 70, 83
Diprose, Rosalyn 101
'Discovery Zone' (Australian War Memorial) 68–9
disgust 169–70
dissensus 221
District Six Museum 20
Dittmer, J. 119, 124
'doing heritage' 2
double-consciousness 36, 41–2
Douglas, M. 14
Douglass, Frederick 26
Du Bois, W.E.B. 36, 41, 42
Duff, C. 239, 241
Dufrenne, Mikel 202
Dunstan, Don 79
Dunstanburgh Castle 156
'Dust Off' (Australian War Memorial) 66–7
Dwyer, O. J. 51, 52
Eastnor Castle 161

ecology, affective 220–1, 222–4, 225–6, 229, 233
Edensor, T. 156, 167, 173
Edkins, J. 228
Edwards, Elizabeth 205–6
Egenhoff, E. L. 140
Eichmann, Adolf 79
Ellington, H. 67
embodiment: and affective ecology 220, 221, 222, 225–6; and bodily performance 51–2; and castles/ruins 160, 163, 165, 170–3, 175; and complexity 114; and conservation 286–7; and cosmopolitanism 94–7, 98, 101–9; defining 12; and dioramas 58–62; embodied knowledge 12–3, 14, 16; embodied nationalism 49, 70; and empathy 47, 61–2, 64–5; geo-centric perspective 145, 147–9; haptic exhibits/technologies 65–9; and knowledge 287–8; and memory 47, 52–3, 67; and 'microbiopolitics' 54, 55–6; and militarism 47, 49, 70–1; and narrative construction 116–17; and objects/artefacts 64, 283–5, 286–7, 288; and polysense 12–3, 14, 15–6, 18, 19, 20–1, 25; and slavery museums/exhibits 117, 124–5, 126, 127, 128; and social housing 237–8, 242–5; and space 47, 64, 69; and vertigo 135, 136, 144; and visuality 53–7, 62–5, 70; and witnessing 224, 225–6
emotion: and affective ecology 220; and atmosphere 202–3, 215; and castles/ruins 155–6, 161, 163–5, 167–71, 173, 175; and coastal heritage landscapes 179–81, 182–5, 187–8, 189–90, 192–3, 195–7; and commemoration 227–8, 233–4; and cosmopolitanism 94–5, 96–7, 98, 103, 104, 107–8; and digital broadcasting of events 219; distinction from affect 35; and drivers of economies of heritage 4; and heritage management 258, 273–4; and narrative construction 114, 115–20; and photographic images 205–6, 215; and physiological responses 12–3; and polysense 12–3, 14, 15, 18–22, 25, 26; and representation of Māori and Polynesian culture 276, 278, 282–3, 284–5, 287; and slavery museums/exhibits 120–1, 122–5, 126–9; and social housing 238, 243–6; and *theatres of pain* 33
empathy 47, 61–2, 64–5

English Heritage (now Historic England) 157, 237, 262–3, 265, 273
Erskine, Ralph 237, 240, 246–7
Escobar, A. 188
ethnographic visuality 56–7
Evans, J. 239, 241
Exhibiting Māori 2

faciality 58, 61
Fairclough, G. 3
Featherstone, Mike 93
Federation of Polish Jews 79
Felman, Shoshana 86
'felt presence' 103–4
Feniger, Saba 80, 83
Figueroa, Esther 185, 194, 195
Fineberg, W. 87
Forbes, Michael 189, 191–2, 196
Forbes, Molly 189
'Forget Trauma? Responses to September 11' 228
Forsyth, Bill 186, 193
Fortepan 201–2, 203–5, 206, 207, 208–15, 216
Fortunoff Video Archive for Holocaust Testimonies 86
Freckleton, Marie 189
Fregonese, Sara 182

Gallipoli 49, 58–9
Gardner, Alexander 207
Gaus, Charlie 148
Gell, Alfred 101
genealogy 103, 289, 290–1
Geoghegan, H. 246
geographies of heritage 50–3, 135–8, 139–41, 144, 148–9
'geo-philosophy' 135, 137
geopolitics 33, 34, 35, 36–7, 47, 49–56, 184
Getting In gallery (Immigration Museum Melbourne) 107–8
Giaccardi, Elisa 203
Gilpin, W. 160, 165
Gilroy, Paul 183–4, 282
Ginsberg, R. 159–60
globalization 93
Goodman, N. 39–40
gothic castles 160–1, 163, 168, 171, 175
Graham, B. 51
grassroots memorials 221–2, 223, 224, 225, 227–8, 231–2, 233
Greater London Council (GLC) 269
Gregory, Kate 203

Grosz, E. A. 145
Ground Zero 226
Guattari, F. 58, 202

Hall, Stuart 1–2
Hansen-Glucklich, Jennifer 78, 80, 81
haptic technologies 65–9, 70
Hardy, D. 50
Harris, Andrew 81
Harrison, R. 2, 114–15, 119
Hartman, Geoffrey 86
haunting affect 250
Haywards Art Gallery (HAG) 268–9
'Heliborne Assault' (Australian War Memorial) 66
Henare, A. 4
Henry VIII, King 158
'hereness' 213
Heritage and Social Media 203
heteropathic memory 20
Hill, L. 137
Hirsch, Marianne 206
Hirst, Damien 269
historical memory 52
'History You Can't Erase' poster 81–2
Hodge, C. 249
Holocaust memorials/museums: and cosmopolitanism 96–7; and polysense 20–3; and testimony *see* testimony (Holocaust museums)
Holocaust Survivors Film Project Inc. 86
Holocaust Testimonies: The Ruins of Memory 86
'hot cognitions' 50
'hot interpretation' 50, 96
Howes, D. 14, 103
Humbug Creek Watershed 147
Hurricane Katrina 137
Hussein, Saddam 267
hypermedia 227

'ideal visitor' 80
Identity: Yours, Mine, Ours 100
IMM (Immigration Museum Melbourne) 98, 99–100, 101, 104–6, 107–8, 109
indexicality 205, 206, 207, 208, 212
industrial heritage 135–8, 139–41, 144, 148–9
industrialization 93, 135–6
Ingold, Tim 4
Insley, J. 60
Interpreting our Heritage 142
interpretive communities 245–6
'interruption' 96, 98

Index 297

Jackson, W. T. 147
Jacobs, J. M. 141
Jamaica Environment Trust 196
Jamaica for Sale 180, 184, 185–6, 187–8, 189, 190–1, 192–3, 194, 195–6, 197–8
Jenkins, Henry 203
Jenssen, A. T. 229
JHC (Jewish Holocaust Centre, Melbourne) 75, 78, 79–88
Johansen, Raymond 222–3
Johnson, Boris 269
Johnson, N. 51
Jones, P. 239, 241
Josem, Jayne 81, 84

Kadimah 79
kaitiakitanga (guardianship) 281
Kākahu 286, 287
Kallenberger, W. W. 140–1
Kapyong diorama (Australian War Memorial) 60–1
Karachi attacks (2014) 219–20
Karp, I. 39
Kearney, A. 52–3
Keating, Paul 49
Kerényi, Zoltán 212–15, 216
Kidd, J. 23
Kirkby Malzeard 259–66, 267–8
KKK (Ku Klux Klan) 26
Knudsen, B. T. 20
Kokoda campaign 49
KORO (Public Art Norway) 229
Kushner, T. 89
Kverndokk, K. 221

La Estrella Castle 172–3
Landsberg, A. 19, 26
Langer, Lawrence 86
Lanzmann, Claude 85
Latham, A. 241
Laub, Dori 86
Lavine, S. 39
Law, L. 196
Leaving Dublin exhibition (Immigration Museum Melbourne) 104–6
Lehrer, E. 21
Leib, J. I. 51
Leiden Museum Volkenkunde 281–2
Levinas, Emmanuel 78, 88, 89
Life and Debt 185
Lilford, Martin 84
Limerick, Patricia Nelson 143
Lincoln Presidential Library and Museum 54

Lindsay, Patrick 77
Lindstorm, S. 148
'living heritage' approach 203, 259, 265–6, 268, 273
'living museums' 75, 79–80
LLSB (Long Live South Bank) 268–73, 274
Local Heritage Initiative 261
Local Hero 186, 187, 193–4, 195
Lock, M. 14–5
London bombings (2005) 219, 227
London Docklands Museum 119–20, 125–8, 129
London, Sugar and Slavery Gallery (Docklands Museum) 125–8, 129
Lone Pine diorama (Australian War Memorial) 58–9
Longhurst, R. 56
Longstaff, Will 62
Lorde, Audre 118
Lorimer, Hayden 95
Lowenthal, David 51
Lowther Castle 161
Ludwig II, King 162
Lundø, Kristian Kragh 233–4

Macaulay, Rose 159
Macdonald, S. 15–6, 38, 94
MacLeod, S. 115
Madrid bombings (2004) 219, 227
Mah, A. 250
Maisel, Phillip 88
Malakoff Diggins State Historic Park 135, 138–44, 145–8
Mana Whenua exhibition (Te Papa) 102–4
Māori culture/heritage: and British Museum 33–4, 38–43, 278, 279, 283, 288; museum representations of 2, 33–4, 38–43, 276, 277–91; and 'otherness' 2; *taonga* 40, 42–3, 101, 103, 282, 283–5, 288; Te Papa Museum 98–9, 101–4, 107, 109
Martin Luther King, Jr Boulevards 51
Marx, K. 35
Massey, D. 145
Massumi, Brian 17, 18, 116, 202
materiality 201, 204, 206, 209–13, 215–16
Maxwell, John 192
McCarthy, C. 2
McCauley, Diana 185, 194, 195
McCormack, D. P. 35, 55, 56, 57, 147
McCue, David 192
McNulty, Tony 266
medievalism 157, 161–2, 163, 164

memory: collective 49, 51, 52–3, 97, 120, 211, 239; and commemoration 220, 224, 230–1; and cosmopolitanism 94, 97, 102, 103–4; cultural 43, 52, 219; 'democratizing memory' 204; and digital broadcasting of events 219, 224; and digital heritage 201, 204; and embodiment 47, 52–3, 67; heteropathic 20 ; historical 52; and photography 204, 205–6, 210–11; and polysense 12, 13, 16, 18–20, 22, 24, 25–6; popular 49, 52; prosthetic 19–20, 25; public 52; and slavery museums/exhibits 120; social 1–2, 52; and social housing 239, 243–51; 'surrogate memory' 206; and testimony 75–6, 78, 79, 80, 85–7; 'theatres of' 34
Memory Wound 220, 223–4
Memorylands 94
Menin Gate at Midnight (painting) 62–3
Menin Gate exhibition (Australian War Memorial) 62–3
'microbiopolitics' 54, 55–6
militarism 47, 49, 57, 70–1
Miller, Kei 197
Milton, C. E. 21
Mineral Information Service 140
Minore, Marzia 77
Modlin, E. A. 136
Monolithos Castle 172–3
Moran, Thomas 139
more-than-representational perspective 4, 95, 97, 109, 116, 129, 175, 222, 242, 248, 251
Morris, William 162, 163
Mowbray House 264
Muir, John 143
'mundane' objects/practices 23, 52, 168–9, 208, 213, 237–8
Munro, Susan 189
Museum of Science and Industry (Manchester) 96
museums: access to collections 279–80; affective
museum displays 34–6; and atmospheres 34–7, 39, 43–4, 54, 58–9, 62–3, 120, 121–6, 129; and colonialism/post-colonialism 33–7, 41, 43–4; connecting with communities 278, 279, 280; and cosmopolitanism *see* cosmopolitanism; customs regulations 277; and geopolitics 33, 34, 35, 36–7; negotiating space 278–9; and polysense 13, 15, 19–26; and race 33–44; representation of Māori and Polynesian culture 2, 276, 277–91; and

slavery *see* slavery museums/exhibits; and testimony *see* testimony; as *theatres of pain* 33–44; *see also under* specific museums

narrative: and audio-visual technologies 54–6, 64–5; and change in heritage landscape 1–4; and coastal heritage landscapes 179–80, 185, 186, 192, 193, 194; and commemoration 222, 224, 226; and cosmopolitanism 95–6, 97, 105–6, 107; and dioramas 58–9; and geographies of heritage 50, 52; geopolitical 36–7, 47, 49; as interpretative form 116–17; and polysense 12, 13, 16–7, 18–20, 21–3, 24; processes of narrative construction 114–29; and race 35–7, 39, 41–2; and 'relational' models of heritage 118–20, 129; and slavery museums/exhibits 115, 117, 119–29; and testimony 76, 77–81, 83, 86–9; and tourism 184–5
National Parks 259, 261
National Register of Historic Places 139
National Sports Museum (Melbourne) 54
National Theatre 268
Nava, Mica 98
'negative' heritage 13, 19–24, 25–6
Neuschwanstein Castle 162
Newton, John 121, 122, 123–5
Ngāti-Rānana 34, 278
NHLE (National Heritage List for England) 260
Nietzsche, Friedrich 144
Norham Castle 160
Northern Echo 266

objects/'artefacts' 3–4; and castles/ruins 154, 155, 156–7, 158, 160–1, 162, 164–5, 175; and circulation 183–4; and cosmopolitanism 97, 101–4, 106, 108, 109; and customs regulations 277 ; and embodiment 64, 283–5, 286–7, 288; and grassroots memorials 227–8; Māori *taonga* 40, 42–3, 101, 103, 282, 283–5, 288; materiality 201, 204, 206, 209–13, 215–16; and overlooks 142–4; and photography 202–3, 206–12, 213, 215–16; and polysense 19, 20–4, 25–6; and representation of Māori and Polynesian culture 277, 278, 279, 280, 282, 283–5, 286–7, 288; and slavery museums/exhibits 122, 123, 124, 125, 127–8, 129; and testimony 81, 88

Observations Relating Chiefly to Picturesque Beauty 160
Office of Homeland Security 266
Oklahoma City National Memorial 23–4
On the Genealogy of Morals 144
ordinary affects 237, 238–40, 242, 244, 249–51
'organic chauvinism' 144
Orientalism 43
orientalism 43, 284, 285
'otherness' 2–3, 34, 37–8, 41–2, 88
Over the Front: the Great War in the Air (Australian War Memorial) 63, 64–5
overlooks 135, 136, 138–44, 145–6, 148–9, 168–9
Oxford Castle 168

Pacific Sisters 40, 279, 286
Pain, R. 187
Pallasmaa, Juhani 202
Parker, G. 52
Parry, Corporal Ray 60, 62
Paterson, M. 65, 70
Patterson, M. E. 21
Paxman, Jeremy 269
Payne, Lewis 207–8
pedagogy 35, 54, 67, 75, 76–7, 80–1, 83–8, 126, 187, 291
'pedestrian speech acts' 241
Peirce, Charles Sanders 205
Pendlebury, J. 240
performative human body 12, 13, 14–5
Philosophical Enquiry into the Origin of Our Ideas of the Sublime and Beautiful 161
photography: and atmosphere 201–3, 206–7, 208, 211–12, 214–15; and cosmopolitanism 104–8; and digital heritage 201, 203–5, 215; ethnographic visuality 56; indexicality of photographs 205, 206, 207, 208, 212; materiality of 201, 204, 206, 209–13, 215–16; and memory 204, 205–6, 210–11; objects/object world of 202–3, 206–12, 213, 215–16; and polysense 20, 23; rephotography 212–15, 216; testimony 83–4, 87, 88
Pile, Steve 95
PIs (public inquiries) 267
'Pity for the Poor Africans' 124
Political Affect 137
'politics of listening' 76
Polynesian culture/heritage 276, 277–91

polysense: and affect 12–4, 15–20, 21–6; and anthropology 14–5; 'beyond discourses' 13–4; and 'difficult' heritage 13, 19–24, 25–6; and embodiment 12–3, 14, 15–6, 18, 19, 20–1, 25; and emotion 12–3, 14, 15, 18–22, 25, 26; and human sensorium 14, 16–7, 21; implications for critical heritage studies 24–6; and memory 12, 13, 16, 18–20, 22, 24, 25–6; and narrative 12, 13, 16–7, 18–20, 21–3, 24; and objects 19, 20–4, 25–6; and performative human body 12, 13, 14–5; and physiological responses 12–3; polysensory model 14–7
popular memory 49, 52
Poria, Y. 181
'positive heritage' objects 25
presence/absence of affective heritage 239, 243, 244–5, 247, 249–51
preserving voices 75
prosthetic memory 19–20, 25
Protevi, J. 137
public memory 52
punctum 207
Purity and Danger 14
Putin, Vladimir 267

Queen Elizabeth Hall (QEH) 268–9, 270, 273

race: and affective museum displays 35–6; and Apartheid 20; British Museum Māori exhibitions 33–4, 38–43, 278, 279, 283, 288; and exclusionary social relations 118; and geopolitics 33, 34, 35, 36–7; and narrative 35–7, 39, 41–2; and 'otherness' 34, 37–8, 41–2; and *theatres of pain* 33–4; *see also* colonialism/post-colonialism; cosmopolitanism; slavery museums/exhibits
Raymond, Rosanna 33, 34, 38, 40–1, 42–3, 276–91
Raz, Guy 26
Reddy, William 118
Redner, Harry 79
Reed, A. 245–6
reflective nostalgia 248–9
reflexive nostalgia 249
Reid-Henry, S. 188
'relational' models of heritage 118–20, 129
rephotography 212–15, 216
're-traumatization' 230
Ricketts, Simon 270
Ricoeur, Paul 224

Riley, W. 163
Ringerikes Blad 229
Rockman, P. 80–1
Rokeby 162
rose parades 221, 227
Rose, Gillian 206, 210, 248
'Royal Air Force Bomber Command' (Australian War Memorial) 66
Royal Albert Museum 280–1
Royal Festival Hall (RFH) 268
'ruin-gaze' 159
'ruin-porn' 251
Ruskin, John 162, 163

Said, E. W. 43
Saldana, A. 145
Samuel, Raphael 34
Sassen, Saskia 93
SaVAge K'lub 276, 277
Sawyer Decision (1884) 139
Schorch, P. 114, 119
Schubert, Franz 62
Scott, Walter 162
sensorium, human 14, 16–7, 21
Seremetakis, C. N. 15
Serres, Michel 144
Seward, W.H. 207
Sheller, Mimi 190
Sherman, D. J. 2
Shoah 85
Shoah Visual History Foundation 85–6, 88
Siana Castle 172–3
Simmel, Georg 158
Simon, R. I. 75–6, 77, 78, 89
slavery museums/exhibits: bicentenary of Act of Abolition 120–1, 122, 125; and embodiment 117, 124–5, 126, 127, 128; and emotion 120–1, 122–5, 126–9; narrative construction 115, 117, 119–29; and objects 122, 123, 124, 125, 127–8, 129; and polysense 25–6
slum clearance 237, 243, 246
Smirke, Robert 161
Smith, L. 2, 121, 201, 204, 240, 257
Smith, S. D. 76
Smithsonian National Museum of African American History and Culture 25–6
social housing: and atmosphere 237, 238–40, 242, 244–5, 248–51; and authorised heritage discourse 239–41, 243–4, 247, 249–51; capturing affect 241–9; claiming space 245–6, 249; and emotion 238, 243–6; listing of 240–1, 250, 251; and memory 239, 243–51; and ordinary affects 237, 238–40, 242, 244, 249–51; presence/absence of affective heritage 239, 243, 244–5, 247, 249–51; and reflective nostalgia 248–9; and slum clearance 237, 243, 246; vying for place 246–8
social justice 25
social media 203, 221, 269
social memory 1–2, 52
social traces 210
Sørenson, T. 63
Southbank Centre 268–72
sovereignty 193, 195
SPAB (Society for the Protection of Ancient Buildings) 262–3
Spinoza, Baruch de 94, 95, 221
Spivak, G. C. 37, 42
SSSI (Site of Special Scientific Interest) 186
Staiff, Russell 94, 97, 165
Stewart, K. 238, 244, 251
'sticky' heritage 24
Stoller, P. 15
Stoltenberg, Jens 219, 229
Stormer, N. 139–40
'Storypods' 87–8
Striking by Night (Australian War Memorial) 63–4
studium 207, 211
Sturken, M. 22
sublime aesthetics 139–40, 161, 168, 171, 175
'surrogate memory' 206
survivor guides (Holocaust museums) 75–6, 79–82, 83, 84, 87, 88
Sydney under Attack (Australian War Memorial) 63
Sylvester, C. 33
Szepessy, Ákos 204, 206
Sztajer, Chaim 83

ta 290
'Tail End Charlie' (Australian War Memorial) 47, 48
Tamási, Miklós 204, 206
taonga 40, 42–3, 101, 103, 282, 283–5, 288
Te Hau ki Turanga 102–3, 104
Te Papa (Museum of New Zealand Te Papa Tongarewa) 98–9, 101–4, 107, 109, 286
Tekanawa, Kahutoi 291
'terms of engagement' 98

terrorism 219–21, 222–3, 227, 228–9, 230, 233, 266
testimony (Holocaust museums): function/role of 76–8, 84–8; in JHC exhibitions 83–4; and 'living museums' 75, 80; and memory 75–6, 78, 79, 80, 85–7; and narrative 76, 77–81, 83, 86–9; and pedagogy 75, 76–7, 80–1, 83–8; preserving voices 75; 'Storypods' 87–8; and survivor guides 75–6, 79–82, 83, 84, 87, 88; video-testimony 75, 76, 78, 80, 81, 85–8
'The Bridge, HMAS Brisbane' (Australian War Memorial) 68
The Cartographer Tries to Map a Way to Zion 197
'The Geology of Morals' 144–5
The Māori Collections of the British Musuem 39
'The Negro's Complaint' 124
The Past is a Foreign Country 51
The Posthuman 145
The Silence of the Gods 40–1
theatres of memory 34
theatres of pain 33–4, 44
Thien, Deborah 95
Thinking Through Things 4
Thomas, Nicholas 279
Thoreau, Henry David 143
Thoughts on the African Slave Trade 121
Thrift, N. 18–9, 25, 35, 54–6, 118, 241
Tilden, Freeman 142–3
Timothy, D. J. 182
Tolia-Kelly, Divya P. 15, 18, 53, 114, 136, 183, 210, 276–91
tourism: and affective ecology 220; and castles/ruins 155, 156–8, 160–5, 168, 173, 175; and coastal heritage landscapes 179–80, 181–6, 187–90, 193, 194, 197; and commemoration 230, 231, 233; and heritage management 270, 271, 273; heritage tourism 51, 76; and slavery museums 122–3
'transhuman' framework 18–9
Treblinka concentration camp 83, 84
Treloar, John Linton 57–8
Trigg, Dylan 210
Tripping Up Trump 190, 196
Trump International Golf Links 179–80, 186, 190, 195
Trump, Donald 179, 186, 188–9, 190, 191–2, 193, 194, 195, 197
Truscott, J. 141

Turner, J. M. W. 160
'Twentyfirst Century Cyber Sister' 286
Tyson, Mike 269

Unfinished Symphony (Schubert) 62–3
United 93 55
United States Holocaust Memorial Museum 21–3
'universalist' approach 33
Uses of Heritage 201
Utøya 219, 220, 221–8, 229–34
Uzzell, David 49–50, 96

va 290
values-based approach 258, 262, 265, 272–3
VDS (Village Design Statement) 260–1, 262, 264, 266
vertigo 135–6, 139, 144, 149
video-testimony 75, 76, 78, 80, 81, 85–8
violence 187–8, 189–91; *see also* terrorism
visuality 53–7, 62–5, 70

walking interviews/tours 237–8, 241–51
Walpole, Horace 161
Warkworth Castle 174
Warne, Shane 54
Waterton, Emma 16, 53, 97, 119, 124, 203
Watkins, Carleton E. 140
Watson, Steve 16, 53, 97, 119
'ways of seeing' 34, 38–9
'Westnocentric' sensory repertoire 15
Westways 141
Wetherell, M. 117, 118, 129, 156
Whatmore, S. 135
White, R. 194
White, Walter 162
Wigan Pier 96
Wilberforce House Museum 122
Wilberforce, William 120, 121, 122
Williams, R. 238
Window to the Past 212–15, 216
Witcomb, Andrea 37, 76–7, 78, 87–8, 114, 119, 123–4, 203
witnessing 221, 224–5, 226–7, 230, 233

You've Been Trumped 180, 184, 185, 186–8, 189–90, 191–4, 195, 196, 197–8
Young, James E. 75, 83
youth heritage 259, 269–73
Yusoff, Katherine 144, 145

Zumthor, Peter 202, 203

For Product Safety Concerns and Information please contact our EU
representative GPSR@taylorandfrancis.com
Taylor & Francis Verlag GmbH, Kaufingerstraße 24, 80331 München, Germany

www.ingramcontent.com/pod-product-compliance
Ingram Content Group UK Ltd.
Pitfield, Milton Keynes, MK11 3LW, UK
UKHW021443080625
459435UK00011B/358